BRAIN

BRAIN
FETAL AND INFANT

Current Research on Normal and Abnormal Development

edited by

SAMUEL R. BERENBERG, M.D.

1977

MARTINUS NIJHOFF MEDICAL DIVISION - THE HAGUE

This volume is based on the proceedings of a conference held under the joint auspices of the Josiah Macy jr. Foundation and the International Children's Center in Paris. December 14-16, 1976.

The conference was dedicated to the memory of Dr. Nathalie Masse. for many years Director of Teaching at the International Children's Center.

ISBN 978-94-011-8221-8   ISBN 978-94-011-8884-5 (eBook)
DOI 10.1007/978-94-011-8884-5

HOMMAGE TO DR. NATHALIE P. MASSE

During the conference, the Macy Foundation presented a plaque as an enduring memorial to the contributions of Nathalie Masse, and the following comments were made at the ceremony:

"Nathalie Masse was Professor of Pediatrics to the world. More than a thousand pediatricians from scores of countries attended her classes in Europe, Latin America, Asia, and Africa.

"Nathalie's many travels were made easier by her assurance that she would return to a loving husband, and himself a distinguished physician, Paul Masse.

"Her professional life was intertwined with that of her patron and France's most distinguished medical statesman, Professor Robert Debré.

"She loved to recall leading the 'force de frappe', which, under the orders of Robert Debré, converted an abandoned chateau in the Bois de Boulogne into the splendid Centre International de L'Enfance that we meet in today.

"Nathalie Masse was lovely in her features; she was endearing and loved people; and she was a remarkably versatile scholar.

"She died in August, 1975.

"We mourn her loss, but today we wish to express our thanks and to rejoice for the children of the world who benefitted from the life of Nathalie Masse."

JOHN Z. BOWERS

CONTENTS

PARTICIPANTS

Jean Aicardi, M.D., Maître de Recherche à l'Institut National de la Santé et de la Recherche Médicale, Hôpital St Vincent de Paul, 74 Avenue Denfert Rochereau, 75674 Paris Cedex 14, France

Ellsworth C. Alvord, Jr., M.D., Professor of Pathology, Neuropathology Laboratory (RJ-05) University of Washington School of Medicine, Seattle, Washington 98195, USA [co-author for chapter 15: Cheng-Mei Shaw, M.D., Professor of Pathology, at above address]

Barry G.W. Arnason, M.D., Professor and Chairman, Department of Neurology, The Pritzker School of Medicine, University of Chicago, Chicago, Illinois, USA

Thomas E. Duffy, Ph.D., Associate Professor of Biochemistry in Neurology, Departments of Neurology and Biochemistry, Cornell University Medical College, 525 East 68th Street, New York, N.Y. 10021, USA [co-author for chapter 22: Robert C. Vannucci, M.D., Assistant Professor of Pediatric Neurology, Department of Pediatrics, Milton S. Hershey Medical Center, The Pennsylvania State University, Hershey, Pennsylvania, USA]

Edith Farkas, M.D., Service du Professeur Thieffry, Hôpital St Vincent de Paul, 74 Avenue Denfert Rochereau, 75674 Paris Cedex 14, France

Michel Hamon, M.D., Groupe de Neuropharmacologie Biochimique, INSERM U.114, Laboratoire de Neurophysiologie, Collège de France, 11 Place Marcelin Berthelot, 75231 Paris Cedex 05, France [co-author for chapter 16: S. Bourgoin, Ph.D., at above address]

Henry Hecaen, M.D., Directeur d'Etudes à l'E.H.E.S.S., Centre de Recherches de l'INSERM sur les Maladies du Système Nerveux Central, 2 ter rue de'Alésia, 75014 Paris, France

Michel Imbert, M.D., Laboratoire de Neurophysiologie, Collège de France, 11 Place Marcelin Berthelot, 75231 Paris Cedex 05, France

Richard T. Johnson, M.D., Eisenhower Professor of Neurology and Professor of Microbiology, School of Medicine, The Johns Hopkins University, 601 North Broadway, Baltimore, Maryland 21205, USA

A. Jost, M.D., Professeur, Laboratoire de Physiologie du Développement du Collège de France et de l'Université Pierre et Marie Curie, 9 Quai St Bernard, 75230 Paris Cedex 05, France

Jacques Legrand, M.D., Professeur, Laboratoire de Physiologie Comparée, Université des Sciences et Techniques du Languedoc, Place Eugène Bataillon, 34060 Montpellier Cedex, France

Paul Mandel, M.D., Professor of Biochemistry at the Medical School, Centre de Neurochimie, 11 rue Humann, 67085 Strasbourg Cedex, France [co-authors du CNRS, Faculté de Medicine for chapter 1: Henry Dreyfus, Ph.D., Attaché de Recherche à l'INSERM; Suzanne Harth,

Ph.D., Chargé de Recherche au CNRS; Louis Freysz, Ph.D., Chargé de Recherche au CNRS; Paul-Francis Urban, Ph.D., Chargé de Recherche au CNRS - all at above address]

Robert Y. Moore, M.D., Professor, Department of Neurosciences, School of Medicine, University of California at San Diego, La Jolla, Calif. 92093, USA

Fred Plum, M.D., Ann Parrish Titzell Professor of Neurology, Department of Neurology, The New York Hospital-Cornell Medical Center, 525 East 68th Street, New York, N.Y. 10021, USA

Heinz F.R. Prechtl, M.D., Director, Department of Developmental Neurology, University Hospital, Oostersingel 59, Groningen, The Netherlands

Arthur L. Prensky, M.D., The Allen P. and Josephine B. Green Professor of Pediatric Neurology, Mallinckrodt Department of Pediatrics and Department of Neurology, Washington University School of Medicine, 500 S. Kingshighway Blvd, St. Louis, Missouri 63110, USA; Director of the Division of Pediatric Neurology, St. Louis Children's Hospital, St. Louis, Missouri, USA [co-author for chapter 9: Laura Hillman, M.D., Assistant Professor of Pediatrics, Edward Mallinckrodt Department of Pediatrics, Washington University School of Medicine, St. Louis, Missouri]

Dominick P. Purpura, M.D., Professor and Chairman, Department of Neuroscience, Albert Einstein College of Medicine, Bronx, N.Y.,10461, USA

Th. Rabinowicz, M.D., Assoc. Professor, Head, Division of Neuropathology, 1 Ch. des Falaises, Lausanne, Switzerland [co-authors for chapter 3: G. Leuba; D. Heumann]

Geoffrey Raisman, M.D., National Institute for Medical Research, The Ridgeway, Mill Hill, London NW7 1AA, United Kingdom

Roger N. Rosenberg, M.D., Professor of Neurology and Physiology, Chairman, Department of Neurology, Southwestern Medical School, Health Science Center at Dallas, The University of Texas, 5323 Harry Hines Boulevard, Dallas, Texas 75235, USA

Arnold Scheibel, M.D. and Madge Scheibel, Departments of Anatomy and Psychiatry and Brain Research Institute, UCLA Center for the Health Sciences School of Medicine, Los Angeles, California 90024, USA

Louis Sokoloff, M.D., Chief, Laboratory of Cerebral Metabolism; National Institute of Mental Health; U.S. Department of Health, Education, and Welfare; Public Health Service; Alcohol, Drug Abuse, and Mental Health Adm. Bldg 36, Room 1A-27, 900 Rockville Pike, Bethesda, Maryland 20014, USA

Constantino Sotelo M.D., Laboratoire de Neuromorphologie, Groupe U.106, INSERM, 123 Boulevard de Port-Royal, 75014 Paris, France

Jean-Louis Valatx, M.D., Chargé de Recherches, INSERM U. Département de Médicine Expérimentale, Université Claude Bernard, 8 Avenue Rockefeller, 69373 Lyon, France

Claude G. Wasterlain, M.D., Associate Professor of Neurology, University of California, Los Angeles, Calif.; Chief, Neurology Division, School of Medicine, San Fernando Valley Medical Program, Veterans Administration Hospital, Sepulveda, California 91343, USA

Bruno E. Will, M.D., Laboratoire de Psychophysiologie, Université Louis Pasteur, 7 rue de l'Université, 67000 Strasbourg, France [co-author for chapter 19: Thomas E. Duffy, Ph.D., Associate Professor of Neurology and Biochemistry, Cornell University Medical College, New York, New York]

ABBREVIATIONS

ATP	=	Adenosine Triphosphate
B_2cAMP	=	Dibutyryl Cyclic Adenosine Monophosphate
BudR	=	Bromodeoxyuridine
C.A.	=	Chronologic age
Cer	=	Cerebroside
$CerSO_4$	=	Sulphatide, cerebroside sulphate
CGluT	=	UDP-glucose: ceramide glucosyltransferase
CGluGalT	=	UDP-galactose: glucosylceramide galactosyltransferase
CNS	=	Central nervous system
CPT	=	CDP-choline 1-2 diglyceride choline phosphotransferase
CST	=	Cerebroside sulphotransferase
DNA	=	Deoxyribonucleic acid
DNase I	=	Deoxyribonuclease I
EEG	=	Electroencephalogram
EPT	=	CDP-ethanolamine 1-2 diglyceride ethanolamine phosphotransferase
ERG	=	Electroretinogram

Ganglioside nomenclature of Svennerholm (1963)

G_{M3}	=	Monosialosyl-lactosylceramide
G_{M2}	=	Monosialosyl-N-triglycosylceramide
G_{M1}	=	Monosialosyl-N-tetraglycosylceramide
G_{D3}	=	Disialosyl-lactosylceramide
G_{D1a}	=	Disialosyl-N-tetraglycosylceramide
G_{D1b}	=	Disialosyl-N-tetraglycosylceramide
G_{T1}	=	Trisialosyl-N-tetraglycosylceramide

G_{Q1}	=	Tetrasialosyl-N-tetraglycosylceramide
Gg	=	Ganglioside
GH	=	Growth hormone
Gly	=	Glycoprotein
5-HIAA	=	5-Hydroxyindole acetic acid
MAO	=	Monoamine oxidase
MGD-SO$_4$	=	Monogalactosyl diacylglycerol sulphate
NaCl	=	Sodium Chloride
NeuNAc	=	N-Acetylneuraminic acid
M	=	Molar
PAPS	=	3′-Phosphoadenosine-5′-phosphosulphate
PC	=	Phosphatidylcholine [as used in chapter 1]
PC	=	Purkinje cell [as used in chapter 16]
PE	=	Phosphatidylethanolamine
PI	=	Phosphatidylinositol
PS	=	Phosphatidylserine
PSTh	=	Post stimulus time histogram
PTU	=	Propylthiouracil
RNA	=	Ribonucleic acid
RNase	=	Ribonuclease
SA	=	Specific activity
SM	=	Sphingomyelin
TCA	=	Trichloroacetic Acid
T3	=	Triiodothyronine
T4	=	Thyroxine
mM	=	mille Molar

CHAPTER ONE

LIPIDS OF CHICK RETINA DURING ONTOGENESIS*

PAUL MANDEL, M.D., PH.D., HENRY DREYFUS, PH.D.,
SUZANNE HARTH, PH.D., LOUIS FREYSZ, PH.D.
AND PAUL-FRANCIS URBAN, PH.D.

INTRODUCTION

Biochemical studies of the ontogenesis of the brain encounter many difficulties due to morphologic and functional heterogeneity in the central nervous system (CNS) and due to differences in the developmental timing of the various brain regions when studying the whole brain. It is also difficult to establish which is the initial response to the factor under analysis and which are the secondary responses. The retina offers a rather simple developmental system which nevertheless is applicable to the comprehension of CNS. In the retina, the number of parameters is smaller, and it is possible to investigate specific structures with relatively little interference from neighbouring structures. The retina has a relatively simple morphology consisting roughly of alternating layers of cell bodies and synaptic regions, ordered from the scleral to the vitreal side as follows: photoreceptor region (outer and inner segment of rods and cones), outer synaptic layer (outer plexiform layer), region of horizontal, bipolar and amacrine cell bodies, zone of inner synapses (inner plexiform layer), ganglion cells. Non-neuronal cells, the so-called Müller cells, span the full thickness of the retina, and constitute the chief glial component of the retina. Microdissection of fresh or frozen tissue allows one to pool enriched fractions of a given cell type. Due to the dimensions of the whole intact retina, it has been termed an "instant" tissue slice which may be immersed in an appropriate medium, and maintained in physiological state by perfusion. The versatility of this isolated system allows one to study various parameters of metabolism, to analyse the bioelectrical response (electroretinogram: ERG) to its natural physiologic stimulus: light, under various

* This investigation was supported in part by the Centre National de la Recherche Scientifique and by the Institut National de la Santé et de la Recherche Médicale (Contrat no. 75-1-215-1).

The authors thank Drs. A. Farooqui, N. Neskovic, R. Pieringer, A. Preti and L. L. Sarlieve for assistance with certain aspects of this work. They are grateful to Dr. J. Hesketh and Mr. B. Curtis who provided helpful assistance in the preparation of this manuscript.

adaptation conditions, and to investigate the influence of important biochemical compounds (putative neurotransmitters, their antagonists and agonists, as well as pharmacologic drugs) on the functional behavior of the retina. Such a perfusion system of an isolated retina allows one to correlate the effects due to compounds introduced in the perfusion medium, the electroretinogram modifications, and the biochemical changes at various levels in the layers of the tissue.

Taking into account these advantages offered by the retina compared to brain, we have studied the biochemical correlates during retinal development. This paper deals chiefly with lipid metabolism during retinal development. Chick retina was chosen for this purpose, since chick development is well documented, and the time scale of appearance of the ERG and of the morphologic changes are well established.

MATERIAL AND METHODS

Biochemical analyses and enzyme assays were performed as described elsewhere (Burton, 1956; San Lin and Schjeide, 1969; Edel-Harth et al., 1973; Sarlième et al., 1974; Dreyfus et al., 1974, 1975a, 1975b, 1976, 1977a, 1977b; Freysz et al., 1977; Harth et al., 1977).

RESULTS: CHANGES OF BIOCHEMICAL COMPOUNDS DURING CHICKEN RETINA ONTOGENESIS

Four fundamental stages in the development of the chicken retina can be distinguished by morphologic and histologic studies (Romanoff, 1960). The first stage up to the 8th day of embryonic life is characterized by the active proliferation of undifferentiated cells. During the second stage, 8th to the 10th day, the retina undergoes a cellular readjustment and the inner and outer plexiform layers begin to appear. The third stage starts at about the 12th day of embryonic life and corresponds to the final differentiation of the various retinal layers. This can be subdivided into two stages (Hughes and Lavelle, 1974). Between the 10th to the 15th day the plexiform layers are well differentiated and inner segments can be observed. The period from the 15th day to hatching is chiefly characterized by the formation and elongation of the outer segments of the photoreceptor cells. This period also corresponds to the functional activity of the retina (Witkovski, 1963). The fourth step concerns the post hatching period and is characterized by the end of the maturation of the retina. The cell density continues to increase but at a slower rate during this latter stage.

1. Deoxyribonucleic acid (DNA)

The increase in tissue DNA offers a measure of cell proliferation, and the DNA content in adult tissue gives an evaluation of final cell content. In the retina, DNA increased by a factor of 3 between the 8th and 12th day of embryonic life (Fig. 1). This indicated that the number of retinal cells increased 3-fold in 4 days. Between the 14th embryo day and hatching the DNA content decreased by about 60%. This suggested that more than half of the retinal cells are eliminated during the differentiation of the various retina layers when functional activity is acquired. After hatching, the content of DNA increased but reached only about 80% of the cell number present at the 12th day of embryonic life.

The changes in DNA content during retinal ontogenesis are quite different from those in the brain. Mandel et al. (1964) and Freysz (1969) have reported that the increase of DNA in brain is regular till the 8th day of post-hatching life and then increases at a lower rate up to the 60 days, when adult values are

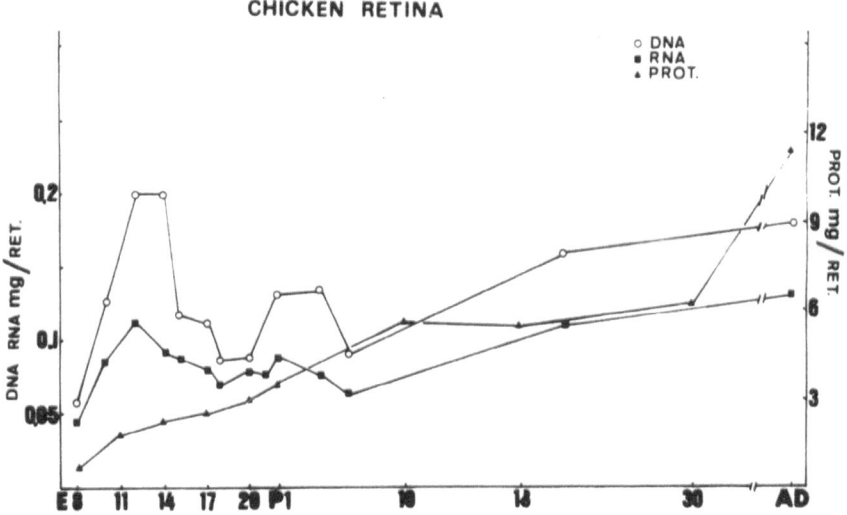

Fig. 1. Protein (▲), DNA (○) and RNA (■) content in chick retina during development. E, embryonic age; P post-hatching age; Ad, adult.

reached. It would be of interest to know which cells are eliminated from the retina before hatching, by which cells they are replaced thereafter and what kind of regulatory mechanisms are involved. The chemical data reported here did not answer this question. However, the cell proliferation after hatching might be ascribed to glial like cells.

2. *Ribonucleic acid (RNA)*

During chicken retina ontogenesis the variations of RNA content were similar to those of DNA (fig. 1). The amounts of RNA increased 3-fold between the 8th and the 12th day of embryonic life, and decreased thereafter up to hatching. However, the decrease of RNA observed before hatching (about 30%) was less than that of DNA. After hatching, the rate of increase of RNA was similar to that of DNA. Like DNA the neosynthesis of RNA during chick development was different in the retina from that in the brain. In brain, RNA increased regularly up to hatching and at a lower rate thereafter (Mandel et al., 1964).

3. *Proteins*

Fig. 1 also shows that proteins increased 6-fold between the 8th day of embryonic life and the 10th day of post-hatching life. The content of proteins did not change between the 10th and the 30th day after hatching. However, after the 30th day, proteins increased 2-fold. Wet weight increased similarly during retinal development. This high increase in protein and wet weight is due essentially to cell maturation, since the proliferation is low.

4. *Phospholipids and cholesterol*

Since phospholipids are localized mainly in membranes, the study of their evolution during development provides a good estimate of cell membrane formation. Total phospholipids increased 16-fold from the 8th day of embryonic life to adulthood (table 1). Four periods can be distinguished for phospholipid increase. The first one, up to the 10th day of embryonic life, was characterized by a large increase in phospholipids which doubled from the 8th to the 10th day of embryonic life. During the second period, from the 10th embryonic day up to hatching the content of phospholipids continued to increase but at a lower rate. The third period up to the 24th day after hatching, showed a large increase (of 100%) of sphingomyelin whereas the content of

Table 1. Phospholipids, cholesterol and ganglioside-NeuNAc during chicken retina development.

Age (days)	Total lipid* phosphorus	Individual phospholipids*					Cholesterol**	(Mol) Phospholipids / (Mol) Cholesterol	Ganglioside-NeuNAc***
		PC	PE	PS	PI	SM			
Embryo									
8	6.2	3.5	1.2	0.7	0.4	0.2	40	1.94	0.9
10	13.4	5.8	3.8	1.5	0.9	0.7	105	1.58	2.3
14	16.5	8.7	4.0	1.6	1.0	0.8	110	1.86	3.2
18	19.5	9.4	5.1	1.9	1.2	1.1	115	2.11	3.5
20	20.5	10.1	5.9	2.0	1.4	1.2	126	2.02	4.5
Hatched									
1	25.0	11.4	6.5	2.7	1.3	2.2	188	1.66	5.7
4	26.0	11.8	7.0	2.8	1.4	2.2	190	1.71	6.0
18	35.0	18.2	8.3	3.2	1.8	2.4	216	2.02	6.5
24	41.0	20.3	9.8	3.9	2.1	3.1	265	1.92	7.5
35	57.0	27.2	14.1	6.0	3.2	5.0	345	2.06	9.9
Adult	97.0	45.6	27.7	9.1	5.5	8.2	510	2.37	13.8

* lipid phosphorus/retina
** cholesterol/retina
*** ganglioside-NeuNAc/retina
Results are expressed in µg

other phospholipids increased at a lower rate. The last period after the 24th day was characterized by a large increase of all phospholipids (from 122% for phosphatidylcholine to 185% for phosphatidylethanolamine).

The distribution of the various phospholipids did not change significantly during retinal development, except for sphingomyelins (table 1). Their rate of increase changed from 3.2% on the 8th embryonic day to 8.5% at adulthood. The highest rate of increase for these compounds is observed during hatching.

The profile of accumulation of cholesterol (table 1) during retina development is similar to that of total phospholipids. Thus the molar ratio of phospholipids to cholesterol did not change significantly. The increase of phospholipids and cholesterol during chick development was slower in the retina than in brain (Freysz et al., 1971). More than 70% of phospholipids were present at the 30th day of post-hatching life for brain; it is less than 50% in the retina. Moreover, in the brain the increase of cholesterol is different from that of phospholipids, the molar ratio phospholipids cholesterol decreased from 1.42 at the 8th day of embryonic life to 0.46 for the adult animal (Freysz et al., 1971).

5. Enzymes involved in the biosynthesis of phosphatidylethanolamines and phosphatidylcholines

Phosphatidylethanolamine (PE) and phosphatidylcholine (PC) represent 80-85% of total membrane phospholipids. The study of the enzymes involved in their biosynthesis during retinal ontogenesis could therefore enlarge our knowledge concerning the regulatory mechanisms which control the biosynthesis of membranes.

5.1. Activity of CDP-ethanolamine 1-2 diglyceride ethanolamine phosphotransferase (EC 2.7.8.1)

CDP-ethanolamine 1-2 diglyceride ethanolamine phosphotransferase (EPT) catalyses the last step of the synthesis of phosphatidylethanolamine according to the reaction:

$$\text{1-2 diglyceride} + \text{CDP-ethanolamine} \xrightarrow{\text{EPT}} \text{phosphoethanolamine} + \text{CMP}$$

The evolution of the specific activity (SA) of this enzyme during retina ontogenesis is reported in fig. 2. The SA increased from the 8th to the 18th day of embryonic life, when the maximum was reached. Its value was 2-fold higher that of the 8th day. The SA decreased thereafter up to adulthood. However, this decrease was due to the large increase in retinal proteins since the total enzymatic activity per retina increased up to the 10th day of post-hatching life. During the development of the retina, the apparent Km for CDP-ethanolamine, calculated with retina homogenates, varied (Table 2). It is about twice higher for th retinas of 8-day-old embryos and adults than embryo 18-day-old. The enzyme had the highest apparent affinity for CDP-ethanolamine at a period when its specific activity was maximum.

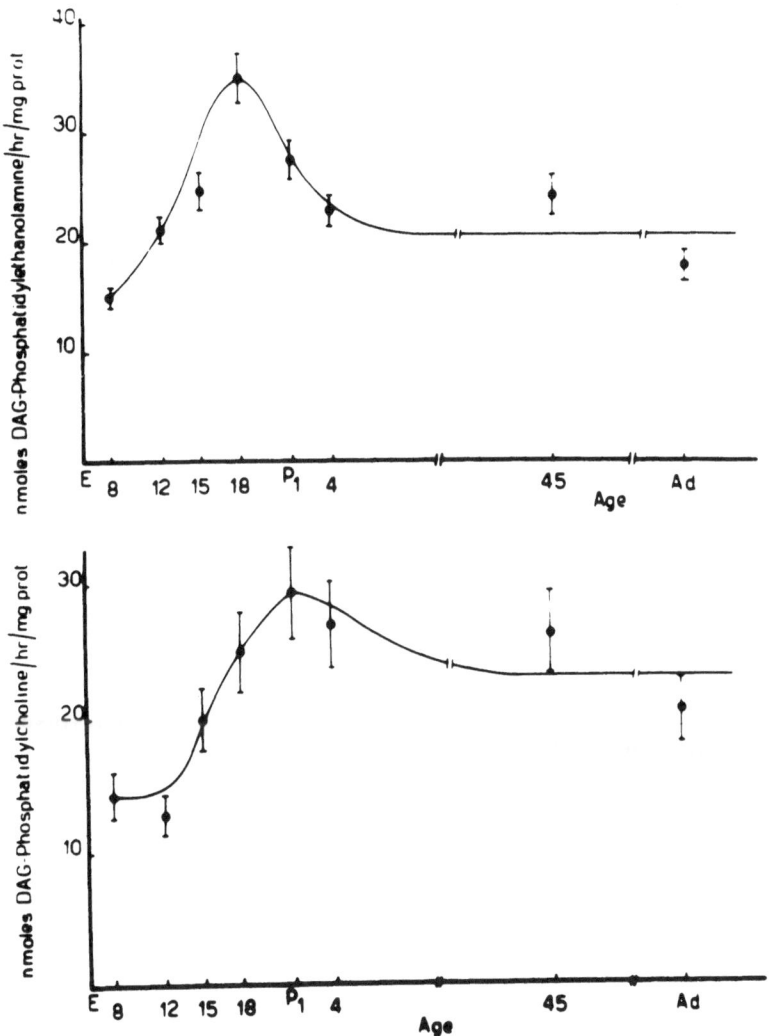

Fig. 2. Ethanolamine- and choline phosphotransferase activities in chick retina during development. Each value is the mean ±S.D. of 5 experiments. E, P, Ad, see legend to Fig. 1.

5.2. *Activity of the CDP-choline 1-2 diglyceride choline phosphotransferase (EC 2.7.8.2)*

The synthesis of phosphatidylcholine is catalysed by the CDP-choline 1-2 diglyceride choline phosphotransferase according to the reaction:

$$1\text{-}2 \text{ diglyceride} + \text{CDP-choline} \xrightarrow{\text{CPT}} \text{phosphatidylcholine} + \text{CMP}$$

The profile of the evolution of the specific activities of this enzyme during retinal ontogenesis is different from that obtained for ethanolamine phosphotransferase (fig. 2). The maximal SA was reached at hatching with a value twice that observed at the 8th day of embryonic life. SA decreased thereafter. As for ethanolamine phosphotransferase, the decrease of SA of choline phosphotransferase during post-hatching life corresponds to a large increase in the protein content. The apparent Km of choline phosphotransferase for CDP-choline determined on retina homogenates did not change during ontogenesis (table 2).

6. *Quantitative changes of retinal gangliosides in chick during development*

The amount of ganglioside *N*-acetyl neuraminic acid (NeuNAc) per retina was low compared to the levels of phospholipids and of cholesterol (table 1). During development, five phases of total ganglioside-NeuNAc accumulation occurred in the chick retina; the gangliosides appeared between 8 and 11 days of embryonic life, the amount plateaued between 11 and 16 days, there was a second increase from 16 days up to hatching, another plateau until 18 days and a final period of increase up to adulthood. Similarly, the levels of gangliosides per chick brain during development have been examined (Dreyfus et al., 1975a). In brain, ganglioside-NeuNAc accumulation increased slowly in the embryonic period and there was a brief rapid burst noted at hatching followed by a regular continual increase until adulthood. As in the retina, this latter accumulation paralleled brain growth and only minor changes in the patterns of several gangliosides occurred during the post-hatching life, when NeuNAc levels are expressed per g wet weight (Dreyfus et al., 1975a).

Chick retinal ganglioside patterns were studied during ontogenesis and remarkable variations in the amount of individual gangliosides were found as shown in fig. 3. In the earliest embryonic period studied (8-10 days), the amount of the different gangliosides was small. Expressed as % ganglioside-NeuNAc, the disialogangliosides, G_{D3} and G_{D1b} were present in the highest

Table 2. Variation of the apparent Km of phosphoethanolamine and phosphocholine diglyceride transferase during the development of chicken retina.

Age	Km (μM)*	
	Phosphoethanolamine diglyceride transferase	Phosphocholine diglyceride transferase
Embryo 8-day-old	36.6 ± 4.5	22.8 ± 3.0
Embryo 18-day-old	18.8 ± 2.8	24.3 ± 4.0
Chicken 1 day (hatching)	21.0 ± 3.1	19.4 ± 4.1
Adult	31.6 ± 4.6	22.0 ± 4.8

* Means ± S.D. of 3 experiments.

amounts (fig. 3 left side). The most important changes in the ganglioside patterns were noted during the embryonic period and at hatching time. $G_{D_{1a}}$ showed a complementary evolution to that of G_{D_3}. G_{D_3} was the most important ganglioside in the 8-day-old embryo and accounted for 50% of the total

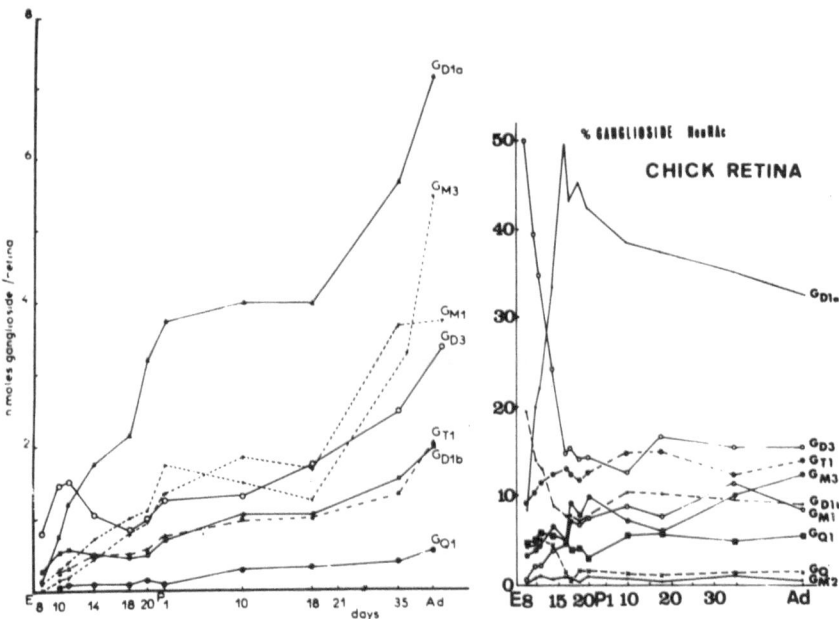

Fig. 3. Ganglioside patterns in chick retina during development. Left side: nmol gangliosides/retina; right side: % ganglioside-NeuNAc. E, P, Ad, see legend to Fig. 1.

Table 3. Action of chick retinal endogenous neuraminidase on the sialic acid from gangliosides (Gg) and glycoproteins (Gly).

Incubation time	Embryo 12 days		Embryo 15 days		Embryo 18 days		Chick 4 days		Adult	
	Gg	Gly	Gg	Gly	Gg	Gly	Gg	Gly	Gg	Gly
hours										
0	100 (570)	100	100 (647)	100	100 (851)	100	100 (874)	100	100 (670)	100
0.5	84.1	98	80	98	77.9	93	82.2	99	81.2	96
1.0	71	98	65	96	60.8	92	65.9	98	70	95
2.0	52.3	95	52.5	95	49.4	90	51.1	93	57	92
4.0	42.6	91	45.0	93	44.1	89	42.6	90	48.3	91
8.0	42.0	90	44.5	90	43.5	88	40.7	89	46.4	91

Values represent the percentage of remaining ganglioside-NeuNAc and glycoprotein-NeuNAc after various incubation times.
Values in parentheses are corrected by the factor 1.2 and are expressed in nmol g wt. wt. They represent the ganglioside-NeuNAc in retinas of embryos and chickens of various ages.
NeuNAc. N-Acetylneuraminic acid.

ganglioside-NeuNAc. Its level decreased dramatically to only 15% in the 17-day-old embryo. This decrease slowed down until the 10th day after hatching, followed by a small increase until the 18th day and finally remained at an almost constant level. $G_{D_{1b}}$ followed a similar change, falling from 20% to 7% of total ganglioside-NeuNAc between the 8th and 20th day of embryonic life. $G_{D_{1a}}$ showed a comparable evolution. In the 8-day-old embryo, it accounted for 8% and increased to 50% of total gangliosides in the 17-day-old embryo. Ten days after hatching the $G_{D_{1a}}$ decreased to 38% and finally remained constant. After hatching there are less variations of ganglioside distributions.

Fig. 3 (right side) shows the ganglioside pattern evolution in chick retina without taking into account retinal growth. This allows the drastic changes in gangliosides observed during the embryonic period to be seen more clearly.

7. *Ontogenetic studies on retinal ganglioside metabolism*

Evolution of the activities of enzymes involved in the primary steps of retinal ganglioside synthesis during chicken ontogenesis

UDP-glucose: ceramide glucosyltransferase (CGluT) and UDP-galactose: glucosylceramide galactosyltransferase (CGluGalT) catalyse the first two steps of the synthesis of gangliosides according to the reactions:

$$\text{Ceramide} + \text{UDP-glucose} \xrightarrow{\text{CGluT}} \text{glucosylceramide} + \text{UDP}$$

$$\text{and Glucosylceramide} + \text{UDP-galactose} \xrightarrow{\text{CGluGalT}} \text{lactosylceramide (dihexosylceramide)} + \text{UDP}.$$

The evolution profile of CGluT (fig. 4) showed a rapid increase in activity from the 8-day-old embryo to a maximum for 10-day-old embryos. Then the activity decreased rapidly with a minimum 4 days after hatching followed by a second maximum at 10 days. Thereafter it remained at a relatively low activity accounting for 30% of the maximum activity. The pattern of the evolution of CGluGalT (fig. 4) showed a similar picture to that of CGluT with a maximal activity at day 11 of embryonic life. No second maximum was detected and the enzymatic activity was lower in adulthood, representing around 20% of the maximal activity.

This study indicated a high enzymatic activity mainly in the embryonic period during which important morphologic modifications occur resulting in differentiation of the neuronal tissue and in the period of appearance of functional activity.

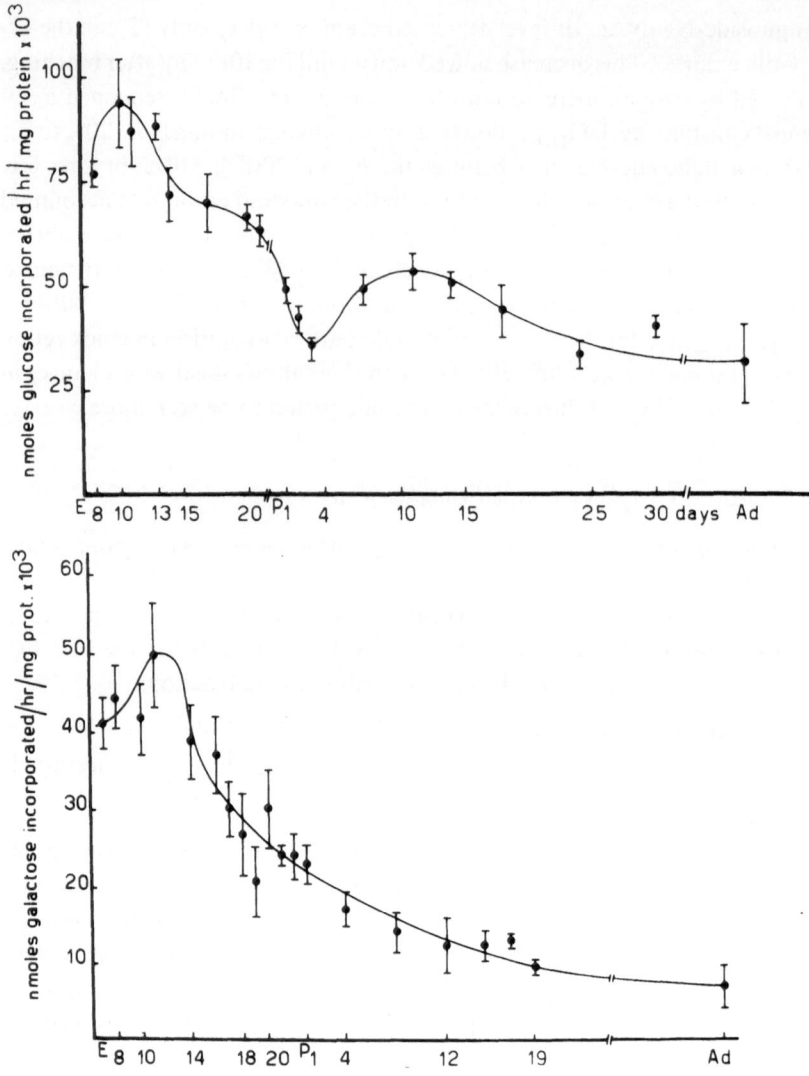

Fig. 4. UDP-glucose: ceramide glucosyltransferase and UDP-galactose: glucosylceramide galac-tosyltransferase activities in chick retina during development. Each value is the mean ±S.D. of 2-4 determinations. E, P, Ad, see legend to Fig. 1.

8. *Implication of catabolic enzymes in the regulation of the ganglioside pool in chick retina during ontogenesis*

We examined in vitro the degradation of gangliosides by the endogenous neuraminidases.

In chick retina, only "particle-bound" neuraminidase was measurable at all stages of retinal development. Total activity, in the presence of endogenous and exogenous substrates ($G_{D_{1a}}$ = 0.09 mM final concentration) was measured (fig. 5). "Endogenous" and "total" neuraminidase activities were low at the 8th day of embryonic life and increased in parallel reaching the maximum levels at the 18th day of embryonic life. Thereafter the activities decreased slightly but remained high during adulthood.

As shown in table 3, on the basis of nmol/g wet weight, only about 10% of glycoprotein-NeuNAc (Gly-NeuNAc) was released whereas 60-70% of ganglioside-NeuNAc (Gg-NeuNAc) was liberated at all stages studied. It seems therefore that in retinal homogenates NeuNAc bound to gangliosides is a more accessible substrate for the endogenous retinal neuraminidase than is the glycoprotein bound one.

We have also determined the effect of endogenous neuraminidase on the distribution of the gangliosides for different ages of animals at different times of incubation (fig. 5). During the first hour of incubation G_{M_1} and $G_{D_{1b}}$ increased whereas those of other gangliosides decreased quickly. After 1 hour only G_{M_1} accumulated.

9. *Ontogenetic studies on retinal sulpholipids*

Two sulphoglycolipids, sulphatide (Cer-SO_4) and monogalactosyl diacylglycerol sulphate (MGD-SO_4) are present in the retina. Detectable amounts of Cer-SO_4 were present in 11-day-old embryo. MGD-SO_4 appeared only after 15 days of embryonic life (fig. 6). The rate of increase was higher for the sulphatide than for monogalactosyl diacylglycerol sulphate particularly in post-hatching period.

Thin-layer chromatography of crude chicken retinal glycolipids also showed the presence of cerebrosides. Compared to brain, the retina contained 15 times less cerebrosides. The amounts of sulphatide were also about 8 times higher in the brain than in the retina (Dreyfus et al., 1977b). An attempt was made to study the activities of enzymes which synthesize and desulphate these sulpholipids. The general profile of retinal 3'-phosphoadenosine-5'-phosphosulphate: cerebroside sulphotransferase (PAPS-CST) activity during ontogenesis showed highest activity during the pre-hatched stage (fig. 6)

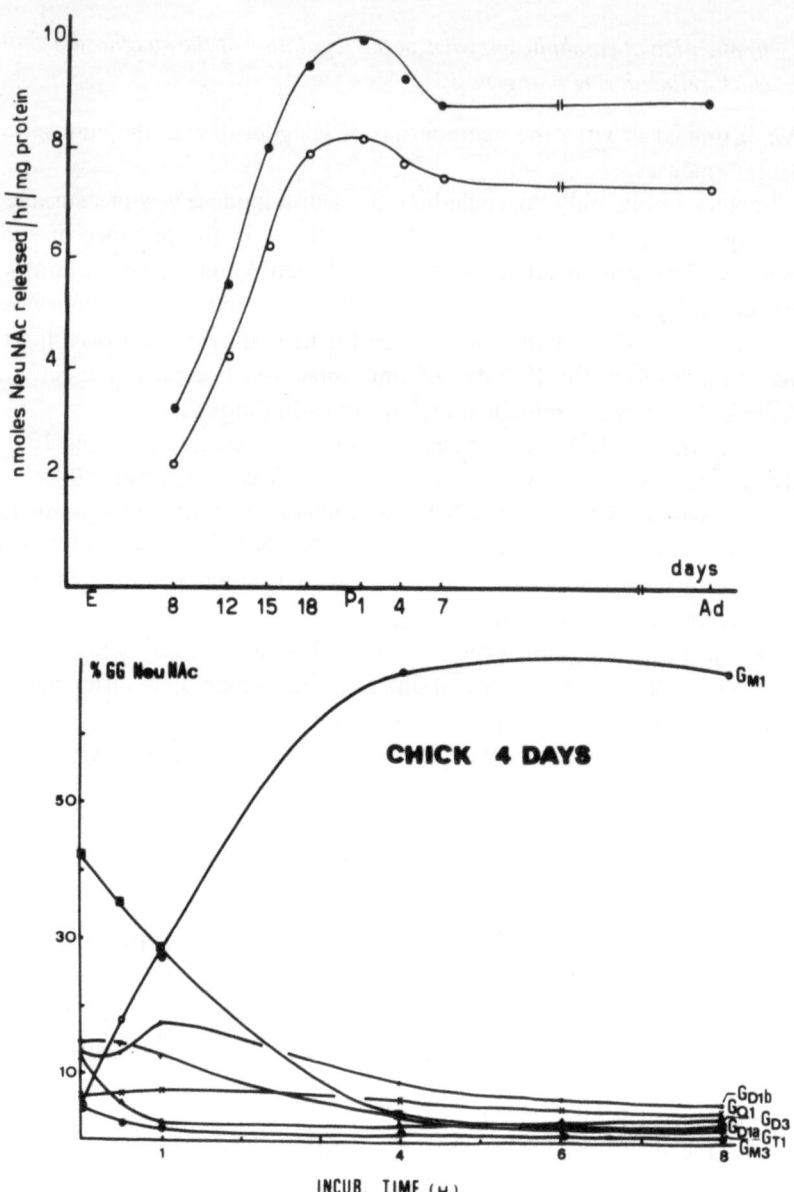

Fig. 5. "Endogenous" (○) and "total" (●) (presence of added 0.055 μmol G_{D1a}) particlebound neuraminidase activities in chick retina during development. Each value is the mean of 3 experiments.
Ganglioside pattern (% of ganglioside-NeuNAc) of chick retina (4-day-old) after various periods of incubation. The same picture was observed for other ages. E, P, Ad, see legend to Fig. 1.

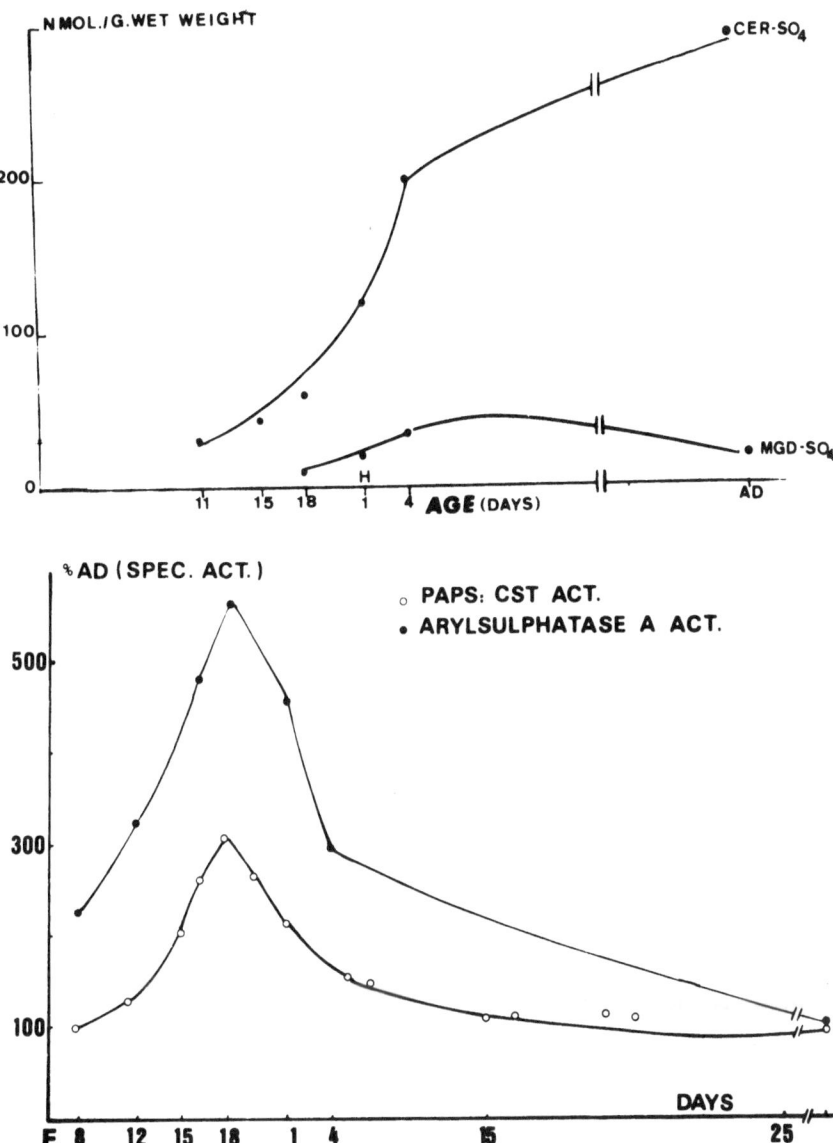

Fig. 6. Sulphatide (Cer-SO$_4$) and monogalactosysl diacylglycerol sulphate (MGD-SO$_4$) content in chick retina during development.
3'-Phosphoadenosine-5'-phosphosulphate:galactocerebroside sulphotransferase (O) and aryl-sulphatase A (●) activities in chick retina during development. Results are expressed as % of adult specific activity. E, P, Ad, see legend to Fig. 1.

which covers the period of functional activity. Arylsulphatase A activity rose sharply during embryonic life, reached a peak at the 18th day and dropped slowly after hatching. The developmental profile of this enzyme is similar to that of PAPS-CST. Both the enzymes showed highest activity around the hatching period. The development of PAPS-CST and arylsulphatase A activities during chick brain ontogenesis was similar to that in retina (Dreyfus et al., 1977b).

DISCUSSION

DISCUSSION

The studies on the retinal phospholipids during ontogenesis have allowed us to classify the different phospholipids into two groups which differ in their rate of synthesis during ontogenesis. The first group includes PC, PI and PS which increase in content 16-fold between the 8th embryonic day and adulthood, and PE which increases 23-fold during the same period. Sphingomyelin which increases 41-fold during the developmental period studied, forms the second group. A similar pattern has been observed by Freysz et al. (1971) for chicken and Dittmer (1967) for rat brain phospholipids. The increase in the various phospholipids at different rates could be related to the synthesis of new cellular membranes.

The high increases of PC, PS, PE and PI before the 10th embryonic day correspond to the formation of membranes of the external and internal plexiform layers. The important increase in sphingomyelins during the period of hatching when the rod outer segments are formed (Hughes and LaVelle, 1974) is consistent with the high content of these phospholipids in these structures (Dreyfus et al., 1974).

Retina synthesizes phosphatidylethanolamine and phosphatidylcholine by the cytidine pathways as has been reported for the brain (McCaman and Cook, 1966; Freysz et al., 1972; Radominska et al., 1972). Moreover, results support the hypothesis that the activities of both enzymes, choline and ethanolamine phosphotransferase, reside in two different proteins (Van den Bosch, 1974). Both enzymes seem to be located in the rod outer segments.

Although extensive work has been devoted to the study of gangliosides since their discovery by Klenk (1939) very little is known about the biologic and physiologic roles of these compounds in retinal and nervous tissue.

In previous work we have shown that there are many differences in the distribution of gangliosides in retinas of different species (Edel-Harth et al.,

1973; Urban et al., 1975; Dreyfus et al., 1975a; Dreyfus, 1976). The question arises, are all these differences in ganglioside patterns in the retina correlated with structural and functional differences? In order to obtain a better insight into retinal ganglioside pattern, their changes during chick retina development have been investigated.

Pronounced changes in both brain and retinal gangliosides were observed during development. The accumulation of gangliosides has been associated with arborization of dendrites and formation of synapses (Vanier et al., 1971). Net accumulation of gangliosides in the two nervous tissues is quite different (Dreyfus et al., 1975a) and yet the modifications in ganglioside patterns during development are similar.

It is difficult to correlate the evolution of ganglioside patterns with the morphologic changes occurring in the retina during development, due to multiple morphologic changes at different levels in the retina and in view of overlapping periods of successive events. Nevertheless, the patterns of increase of the absolute quantities of each ganglioside reveal the highest rate of increase for $G_{D_{1a}}$. This increase seems to parallel that of plasma and synaptic membranes.

The two first steps of retinal ganglioside biosynthesis were studied in vitro during chick retina ontogenesis.

Maximum activity of CGluGalT occurred at the 11th day of embryonic life, 1 day after the corresponding CGluT. The enzymes levels are high during the earliest stage of synaptogenesis e.g. formation and development of inner and outer plexiform layers.

The attachment and removal of ganglioside-sialic acid may be regulated by two different enzymes: sialyltransferases and neuraminidases, involved in the synthesis and catabolism of gangliosides, respectively.

We demonstrated a "particle-bound" neuraminidase activity in the chick retina. We found that the enzyme was not saturated by endogenous substrates. As we found in the retina, the neuraminidase activity in chick brain was low during the early embryonic period and the activity in adult animals remained high (Schengrund and Rosenberg, 1974). In contrast to the two transferases previously studied, the maximal activity of neuraminidases was not attained until the retina was morphologically mature.

We have also shown that in the retina the NeuNAc of gangliosides rather than that of glycoproteins is the primary substrate for neuraminidase, indicating that the ganglioside-NeuNAc is more accessible to the enzyme.

The membrane bound particulate neuraminidases, which are closely associated with the sialyltransferases of membranes, may be involved in the regulation of membrane structure and function. The fact that neuraminidases are present and active at the beginning of embryonic life suggests that they participate very early in the catabolism of sialocompounds (chiefly ganglio- sides) and that they contribute to the stability of the membrane by maintain- ing a surface charge. They could allow the liberation of NeuNAc placed in terminal position of the glycosidic chain of gangliosides and so render them accessible to other synthetic and catabolic enzymes (Gatt, 1967). One may hypothetize that the physiologic role of gangliosides is related to both their localisation in membranes and their negative charge due to the sialic acid.

Though the retina is a poorer source of sulpholipids and cerebrosides than the brain, it contains high amounts of these glycolipids compared to non nervous tissue (Green and Robinson, 1960) with the exception of kidney. Actually, Wisniewski and Bloom (1975) and Reale et al. (1975) reported the presence of myelin in rabbit retina, so that chick retina might also contain myelinic structures which could explain our results. These myelinic structures are formed very early during retinal development and reach their highest concentration near hatching. Our results clearly show that both retina and brain enzymes involved in sulphatide metabolism exhibit the highest activity during the active period of myelination. However, the ratio of cerebrosides to sulphatides, was very different in the retina from that in brain (Dreyfus et al., 1977b).

The other assumption which would explain the occurrence of sulpholipids in the retina is the involvement of sulphatide in the transport of sodium ions. In fact, Bentley et al. (1976) suggested that the presence of sulphatide in basal plasma membrane of ciliarly unpigmented epithelium of rat eyes, is related to sodium transport and aqueous humor secretion by this tissue.

GENERAL CONCLUSIONS

Interest in phospholipids and more recently in gangliosides has increased due to the suggestions of their implications in neurotransmission. Their subcel- lular localization and their implication in specialized membrane structures have been investigated by various authors (Lapetina et al., 1967; Hawthorne and Kai, 1970; Lunt et al., 1971; Morgan et al., 1971). Furthermore gan- gliosides exhibit duality in their physico-chemical properties leading to binary and tertiary complexes, and also complexes with compounds like serotonin or acetylcholine (Burton and Howard, 1967; Hayashi and Katagiri, 1974; Ochoa and Bangham, 1976). These properties reinforce the probability of an important functional role during synaptic transmission where theoretical and experimental data have been reported (Rahmann et al., 1976).

Developmental studies have shown interesting patterns during certain periods of growth and important correlation can be eventually established with the appearance of functional activity.

SUMMARY

Ontogenic studies of chicken retina lipid, DNA, RNA and protein content have been undertaken. Cell proliferation as judged by retinal DNA showed the major increase during the first 12 days of embryonic life. This is quite different from brain. RNA content changes were similar to those of DNA. Retinal protein content increased two-fold between the 8th day of embryo and 10th day of post-hatching life.

In contrast to brain, the molar ratio phospholipids cholesterol is similar throughout ontogenesis. The phospholipid patterns did not change noticeably during retinal development, whereas drastic changes of ganglioside patterns occurred chiefly during embryonic life; G_{D_3} was progressively replaced by another disialoganglioside, $G_{D_{1a}}$.

Study of the activity of choline- and ethanolamine phosphotransferases during development as well as the determination of Km values for CDP-bases, indicated that the synthesis of phosphatidylcholine and phosphatidylethanolamine occurs through two distinct enzymes. Maximal activities were noted during photoreceptor elongation indicating a location of these enzymes in the outer structures of the retina.

Both glycosyltransferases, catalyzing the formation of precursors of gangliosides showed maximal activity during the first part of embryonic life, which is different from the profile determined for endogenous particulate neuraminidase. This catabolic enzyme showed maximal activity at hatching and remained high until adulthood. Under the conditions used, gangliosides were preferentially degraded by endogenous neuraminidase, compared to the glycoproteins; G_{M_1} was the most resistant ganglioside and is an important end product of this reaction.

As in brain, cerebrosides and sulpholipids were found in the retina but in lower amounts. PAPS-CST and arylsulphatase A showed maximum activity at hatching, a period of intense myelination in brain. The presence of these "myelin markers" may indicate the presence of myelin in the retina.

REFERENCES

Bentley, J.P., Feeney, L., Hanson, A.W. and Mixon, R.N.: Sulfated glycolipids in ciliary body epithelium. Invest. Ophthal. 15: 575-578 (1976).

Burton, K.: A study of the conditions and mechanism of the diphenylamine reaction for the colorimetric estimation of deoxyribonucleic acid. Biochem. J. 62: 315-323 (1956).

Burton, R.M. and Howard, R.E.: Gangliosides and acetylcholine of the central nervous system. VIII. Role of lipids in the binding and release of neurohormones by synaptic vesicles. Ann. N.Y. Acad. Sci. 44: 411-432 (1967).

Dreyfus, H.: Recherches sur le rôle des lipides membranaires de la rétine envisagée comme modèle de structure nerveuse intégrée ("Doctorat ès-Sciences" thesis, University Louis Pasteur of Strasbourg, 1976).

Dreyfus, H., Edel-Harth, S., Urban, P.F., Neskovic, N. and Mandel, P.: Enzymatic synthesis of lactosylceramide by a galactosyltransferase from developing chicken retina. Exp. Eye Res. 25: 1-7 (1977a).

Dreyfus, H., Harth, S., Urban, P.F. and Mandel, P.; Preti, A. and Lombardo, A.: On the presence of a "particle-bound" neuraminidase in retina. A developmental study. Life Sci. 1: 1057-1064 (1976).

Dreyfus, H., Pieringer, J.A., Farooqui, A.A., Harth, S., Rebel, G. and Sarliève, L.L.: Sulpholipid metabolism in developing chicken retina. J. Neurochem. in press 1976b.

Dreyfus, H., Urban, P.F., Edel-Harth, S. and Mandel, P.: Developmental patterns of ganglio-sides and of phospholipids in chick retina and brain. J. Neurochem. 25: 245-250 (1975a).

Dreyfus, H., Urban, P.F., Edel-Harth, S., Neskovic, N.M. and Mandel, P.: Enzymatic synthesis of glucocerebrosides by UDP-glucose: ceramide glucosyltransferase during ontogenesis of chicken retina. Lipids 10: 542-544 (1975b).

Dreyfus, H., Urban, P.F., Edel-Harth, S., Bosch, P., Rebel, G. and Mandel, P.: Effect of light on gangliosides from calf retina and photoreceptors. J. Neurochem. 22: 1073-1078 (1974).

Edel-Harth, S., Dreyfus, H., Bosch, P., Rebel, G., Urban, P.F. and Mandel, P.: Gangliosides of whole retina and rod outer segments. FEBS Lett. 35: 284-288 (1973).

Freysz, L.: Distribution et renouvellement des phosphatides du système nerveux central: Évolution au cours de l'ontogenèse ("Doctorat ès-Sciences" thesis, University Louis Pasteur of Strasbourg, 1969).

Freysz, L., Bieth, R. and Mandel, P.: Cinétique de la biosynthèse des phosphatides du cerveau de poulet durant la période embryonnaire et post-natale. Biochimie 53: 399-405 (1971).

Freysz, L., Horrocks, L.A. and Mandel, P.: Effects of deoxycholate and phospholipase A_2 on choline and ethanolamine phosphotransferases in chicken brain microsomes. Biochim. Biophys. Acta in press (1977).

Freysz, L., Lastennet, A. and Mandel, P.: Phosphocholine diglyceride transferase activity during development of the chicken brain. J. Neurochem. 19: 2599-2605 (1972).

Gatt, S.: Enzymatic hydrolysis of sphingolipids. V. Hydrolysis of monosialoganglioside and hexosylceramide by rat brain β-galactosidase. Biochim. Biophys. Acta 137: 192-195 (1967).

Green, J.P. and Robinson, J.D.: Cerebroside sulfate (sulfatide A) in some organs of the rat and in a mast cell tumor. J. Biol. Chem. 235: 1621-1624 (1960).

Harth, S., Dreyfus, H., Urban, P.F. and Mandel, P.: Direct thin layer chromatography of a total lipid extract. Analyt. Biochem. submitted (1977).

Hawthorne, J.N. and Kai, M.: Metabolism of phosphoinositides, in LAJTHA, Handbook of Neurochemistry, Vol. 3, pp. 491-508 (Plenum Press, New York, 1970).

Hayashi, K. and Katagiri, A.: Studies on the interactions between gangliosides, protein and divalent cations. Biochim. Biophys. Acta 337: 107-117 (1974).

Hughes, W.F. and LaVelle, A.: On the synaptogenic sequence in the chick retina. Anat. Rec. 179: 297-302 (1974).

Klenk, E.: Beiträge zur Chemie der Lipoidosen. I. Niemann-Pick'sche Krankheit und amaurotische Idiotie. Z. Physiol. Chem. 262: 128-143 (1939).

Lapetina, E.G., Soto, E.F. and De Robertis, E.: Gangliosides and acetylcholinesterase in isolated membranes of the rat brain cortex. Biochim. Biophys. Acta 135: 33-43 (1967).

Lunt, G.G., Canessa, O.M. and De Robertis, E.: Association of the acetylcholine-phosphatidyl inositol effect with a "receptor" proteolipid from cerebral cortex. Nature New Biol. 230: 187-189 (1971).

Mandel, P., Rein, H., Harth-Edel, S. and Mardell, R.: Distribution and metabolism of ribonucleic acid in the vertebrate central nervous system, in RICHTER, Comparative Neurochemistry, pp. 149-163 (Pergamon Press, Oxford, 1964).

McCaman, R.E. and Cook, K.: Intermediary metabolism of phospholipids in brain tissue. III. Phosphocholine glyceride transferase. J. Biol. Chem. 241: 3390-3394 (1966).

Morgan, I.G., Wolfe, L.S., Mandel, P. and Gombos, G.: Isolation of plasma membranes from rat brain. Biochim. Biophys. Acta 241: 737-751 (1971).

Ochoa, E.L.M. and Bangham, A.D.: N-Acetylneuraminic acid molecules as possible serotonin binding sites. J. Neurochem. 26: 1193-1198 (1976).

Radominska-Pyrek, A. and Horrocks, L.A.: Enzymatic synthesis of 1-alkyl-2-acyl-sn-glycero-3-phosphorylethanolamine by the CDP-ethanolamine: 1-radyl-2-acyl-sn-glycerol ethanolaminephosphotransferase from microsomal fraction of rat brain. J. Lipid Res. 13: 580-587 (1972).

Rahmann, H., Rösner, H. and Breer, H.: A functional model of sialo-glycomacromolecules in synaptic transmission and memory formation. J. Theor. Biol. 57: 231-237 (1976).

Reale, E., Luciano, L. and Spitznas, H.: Zonulae occludentes of the myelin lamellae in the nerve fibre layer of the retina and in the optic nerve of the rabbit: a demonstration by the freeze-fracture method. J. Neurocytol. 4: 131-140 (1975).

Romanoff, A.L.: The avian embryo. Structural and functional development (The Macmillan Company, New York 1960).

San Lin, R.I. and Schjeide, O.A.: Micro estimation of RNA by the cupric ion catalyzed orcinol reaction. Analyt. Biochem. 27: 473-483 (1969).

Sarlième, L.L., Neskovic, N.M., Rebel, G. and Mandel, P.: PAPS-cerebroside sulphotransferase activity in developing brain of a neurological mutant of mouse (MSD). Exp. Brain. Res. 19: 158-165 (1974).

Schengrund, C.L. and Rosenberg, A.: Gangliosides, glycosidases, and sialidase in the brain and eyes of developing chickens. Biochemistry 10: 2424-2428 (1974).

Urban, P.F., Harth, S. and Dreyfus, H.L.: Gangliosides and phospholipids from frog and duck retina. Exp. Eye Res. 20: 397-405 (1975).

Svennerholm, L.: Chromatographic separation of human brain gangliosides. J. Neurochem. 10: 613-623 (1963).

Van den Bosch, H.: Phosphoglyceride metabolism. Ann. Rev. Biochem. 43: 243-277 (1974).

VanierM.T., Holm, M., Öhman, R. and Svennerholm, L.: Developmental profiles of gangliosides in human and rat brain. J. Neurochem. 18: 581-592 (1971).

Wells, M.A. and Dittmer, J.C.: A comprehensive study of the postnatal changes in the concentrations of the lipids of developing rat brain. Biochemistry 6: 3169-3175 (1967).

Wisniewski, H.M. and Bloom, B.R.: Experimental allergic optic neuritis (EAON) in the rabbit. J. Neurol. Sci. 24: 257-263 (1975).

Witkovski, P.: An ontogenetic study of retinal function in the chick. Vision Res. 3: 341-355 (1963).

PROTEIN SYNTHESIS BY DIFFERENTIATING NEUROBLASTS AND GLIOBLASTS IN CELL CULTURE: A MODEL SYSTEM FOR ANALYSIS OF GENETIC NEUROLOGIC DISEASE

ROGER N. ROSENBERG, M.D.

Quantitation of developmental events in the mammalian nervous system has achieved a high degree of sophistication in recent neuroanatomic and neurophysiologic studies (Jacobson, 1976). Neuronal recognition, neuronal and glial migration, cell sorting-out, cell agglutination, cellular morphologic differentiation and developmental as well as reactive synaptogenesis are well-documented events occurring during embryonal and fetal life. Quantitative studies of the regenerative capacity and plasticity of adrenergic neurons after discrete lesions in the rat septal nuclei have also been reported (Moore, et al., 1971; and Raisman, 1969).

Defects in this orderly developmental process do occur with some frequency. Anencephaly, arrhinencephaly, porencephaly, lissencephaly, heterotopias, microcephaly, polymicrogyria, and hypoplasia are well-recognized neuroanatomical architectonic irregularities which functionally are translated into syndromes of mental retardation, perceptual disorders, focal motor or sensory neurological deficits and epilepsy. Genetic neurologic disorders similarly may result in these types of neuroanatomic defects as well as in biochemical abnormalities. Single gene mutations inherited as an autosomal recessive disorder have been associated with single enzyme defects resulting in loss of biologic activity of the particular enzyme. The molecular defect responsible for many such autosomal recessive genetic neurologic disorders have been described in recent years and each syndrome is due to a different single enzyme defect. Thus for the amino acidopathies, leukodystrophies, sphingolipidoses, disorders of heavy metal metabolism, and the mucopolysaccharidoses, a single enzyme alteration has been described to explain the biochemical and clinical manifestations of each syndrome (Brady, 1975, 1976; Kolodny, 1976; Stanbury, et al., 1972).

The problem is more complex for the autosomal dominant genetic neurologic disorders. They are more common as pediatric neurologic disorders than are the recessively inherited diseases and thus far they do not appear to be clearly related to primary enzyme defects as are the recessively inherited diseases. Examples of several common and severe dominantly inherited genetic neurologic diseases which may be present in infancy or childhood include neurofibromatosis, tuberous sclerosis, dystonia musculorum deformans, Huntington's disease, striatonigral degeneration (Joseph's disease), Friedreich's ataxia, olivopontocerebellar degeneration, microcephaly, and idiopathic major motor generalized seizures. It is promising to note that biochemical changes have been documented in neurofibromatosis with increases in activity of nerve growth factor in patient serum (Schenkein, et al., 1974) and in Huntington's disease with decreases in gamma-aminobutyric acid and glutamic acid decarboxylase activity in the corpus striatum of patient brain (Perry, et al., 1973; Bird, et al., 1973). However, these biochemical changes are probably not the basic and primary molecular defects and thus, further research is warranted.

It may be that, in these common dominantly inherited disorders, the defects may reside not in enzyme protein but rather in non-enzymatic structural membrane or specialized receptor membrane protein. Data have recently become available, obtained from brain tissue of Huntington's disease patients, indicating defects in receptor binding for serotonin and acetylcholine in the corpus striatum (Enna, 1976). Thus a clear and important precedent is cited to investigate receptor protein species in dominantly inherited neurologic disease. Thus separation and identification of as many species of cellular protein as possible would be important to seek possible molecular defects. To begin such potential investigations with brain tissue or skin fibroblasts from patients with dominantly inherited neurologic disorders it became necessary to develop techniques which would resolve neuronal and glial proteins with a high degree of precision and sensitivity. A high degree of resolution of individual proteins from neural tissue was achieved using neuroblastoma, glioma, and hybrid neuroblastoma-glioma clones grown in cell culture and extracted proteins separated in the second dimension by acrylamide slab gels (Prashad and Rosenberg: Prashad et al., 1977).

The Ajax strain of mouse neuroblastoma C1300 tumor cells has been cloned into individual cell lines which possess high specific activities of choline acetylase or tyrosine hydroxylase and thus are cholinergic or adrenergic, respectively (Rosenberg, 1973). Additional clones have been described which are long or short neurite formers or have neurophysiologically active membranes and generate mature action potentials. Thus the neuroblastoma cell culture system offers a unique opportunity to study the molecular properties of a modified neuroblast, devoid of other contaminating brain cell types. Similarly, the C-6 rat glioma cell line has been useful for studying molecular glial properties, as well as neuroblast-glioma trophic interaction (Newburgh and Rosenberg, 1973). In addition, a hybrid cell line of Sendai virus fused glioma-neuroblastoma cells has been developed to investigate the regulatory properties of the genome of one cell type on the other one. The cell culture approach is also most useful to control the state of cellular morphologic and molecular differentiation and thus to carefully examine aspects of the program of genetic differentiation.

In these studies we have used high resolution, two-dimensional polyacrylamide gel electrophoresis to determine the pattern of protein synthesis of neuroblastoma cells, glioma cells, and hybrid cells in cell culture. The data indicate that with this new approach it is possible to resolve total cell protein into at least one-hundred discrete spots, identify specific functional proteins as tubulin, and determine as different and separate the protein patterns of neuroblasts from glioblasts (Prashad and Rosenberg, 1977).

Neuroblastoma, glioma, or hybrid cells were grown in monolayer cultures as previously described in growth media enriched with 10% fetal calf serum for six days (Rosenberg, 1973). The medium was then aspirated and the cells were rinsed twice with cold 0.9% NaCl. The cells were scraped from the plates, treated with RNase A, sonicated and treated with DNase I. Then solid urea was added to 9M and an equal volume of lysis buffer was added. Isoelectric focusing gels (pH 5 to 7) were then run at 400 volts for 15 hours at room temperature to separate proteins on the basis of electric charge. Each sample contained 200 μg protein. Whole cell proteins were then separated by second-dimensional electrophoresis on 10% polyacrylamide, 0.1% sodium dodecylsulfate-slab gels. Gels were run at 80 volts until the dye front reached the bottom of the gel. The gels were fixed in 50% TCA for 1 hour and stained in 0.1% comassie blue in methanol: H_2O: acetic acid (5:5:1) for 30 minutes and destained in 7.5% acetic acid. Proteins were separated in the second dimension on the basis of molecular weight and molecular weights of protein spots range within 2,000 daltons from different gels.

In the 70,000 to 200,000 dalton range, 40 to 50 protein spots are detected. The predominant proteins are found in the 40,000 to 70,000 dalton range. There are approximately 120 protein spots, 35 of which are very heavily stained. The protein patterns from neuroblastoma (clone N-18), glioma (clone C-6), and hybrid cells are shown in figure 1A, B, and C, respectively. It is important to note the totally different pattern between neuroblastoma and glioma cells indicating the technique has the resolvable ability to identify the protein pattern of each of the two major brain cell types. Although neuro-blasts and glioblasts express many common proteins, the gel patterns indicate that several proteins are synthesized by only one cell type. Interestingly the neuroblastoma-glioma hybrid cell expresses mainly a neuronal pattern al-though some glioma proteins are synthesized. Thus the neuronal genome under the specialized conditions of a hybrid cell apparently exerts a dominant and inhibitory regulatory role over the glioma cell genome and the two dim-ensional gel system is able to measure these differential synthetic characteristics.

In comparing protein patterns from cultures of cells which are undifferen-tiated from those which are induced to differentiate by the addition to the culture medium of 0.1 mM dibutyryl cyclic adenosine monophosphate (B_2 cAMP), it is significant that only 2-3% of the proteins are quantitatively changed while 97-98% are unchanged (Prashad and Rosenberg, 1977). Thus in neuroblastoma or glioma cells during differentiation, synthesis of large amounts of new proteins may not be required but rather the modification of existing proteins may be essential for the reorganization of proteins, as for example, the conversion of tubulin to microtubule in neurite formation.

Further, in other investigations, Prashad and Rosenberg (1977) found that B_2 cAMP added to mouse neuroblastoma cultures induced cAMP-binding protein levels by 5-fold and this increase in cAMP-binding protein occurred 20 hours after B_2 cAMP treatment. The phosphorylation of endogenous proteins by using (γ-^{32}P) ATP was studied using the same neuroblastoma culture system and these phosphoproteins were analyzed by two-dimensional polyacrylamide gel electrophoresis. These data indicate that there is more phosphorylation in B_2 cAMP treated cells compared to control cells. These results indicate that B_2 cAMP induced cellular differentiation increases cAMP-binding and protein kinases and these protein kinases are involved in the phosphorylation of proteins. Phosphorylation modifies these proteins and this modification may play a biologic role, such as increasing neurotrans-mitter enzyme activity and the conversion of tubulin to microtubules for neurite formation.

These studies have a clear bearing on the biochemical approach to the investigation of brain tissue from children with anatomic developmental ab-normalities and dominantly inherited genetic neurologic disorders. The two dimensional acrylamide slab gel separation of proteins from brain tissue or skin fibroblasts obtained from children with Huntington's disease or tu-berous sclerosis or dominantly inherited idiopathic epilepsy may be of great value in seeking molecular markers of disease. From a more general view of brain development this biochemical approach will be useful for quantifying the genetic program of neuronal and glial differentiation.

REFERENCES

Bird, E.D., McKay, A., Rayner, C. and Iversen, L.: Reduced glutamic-acid decarboxylase activity of post-mortem brain in Huntington's chorea. The Lancet. May 19, 1973: 1090-1092.
Brady, R.O.: The lipid storage diseases: new concepts and control. Annals of Internal Medicine. 82: 257-261 (1975).
Brady, R.O.: Inherited metabolic diseases of the nervous system. Science 193: 733-739 (1976).
Enna, S.J., Bird, E.D., Bennett, J., Bylund, D., Yamamura, H., Iversen, L. and Snyder, S.H..: Huntington's chorea. Changes in neurotransmitter receptors in the brain. New Eng. J. Med. 294: 1305-1309 (1976).
Jacobson, M.: Neuronal recognition in the retinotectal system in S.H. Barondes, editor, Neuronal Recognition, pp. 3-23 (Plenum Press, New York 1976).
Kolodny, E.H.: Current concepts in genetics: lysosomal storage diseases. New Eng. J. Med. 294: 1217-1220 (1976).
Moore, R.Y., Bjorklund, H. and Stenevi, V.: Plastic changes in the adrenergic innervation of the rat septal area in response to denervation. Brain Res. 33: 13-35 (1971).
Newburgh, R.W. and Rosenberg, R.N.: Glucose metabolism in mixed glioblastoma and neuroblastoma cultures. Biochem. Biophys. Res. Comm. 52: 614 (1973).
Nyhan, W.L.: Patterns of clinical expression and genetic variation in the inborn errors of metabolism in W.L. Nyhan Heritable Disorders of Amino Acid Metabolism: Patterns of Clinical Expression and Genetic Variation, pp. 3-14 (Wiley & Sons, New York 1974).
Perry, T.L., Hansen, S. and Kloster, M.: Huntington's chorea – deficiency of gamma-aminobutyric acid in brain. New Engl. J. Med. 288: 337-342 (1973).
01Prashad, N. N. and Rosenberg, R.N.: Dibutyryl cAMP-induced protein changes in differentiating mouse neuroblastoma cells (manuscript in preparation).
Prashad, N., Wischmeyer, B., Evetts, C., Baskin, F., and Rosenberg, R.N.: Cell differentiation 6: 147-157, 1977.
Society, 1977. In press.
Raisman, G.: Neufonal plasticity in the septal nuclei of the adult rat. Brain Res. 14: 25-48 (1969).
Rosenberg, R.N.: Regulation of neuronal enzymes in cell culture in Gordon Sato, Tissue Culture of the Nervous System, pp. 107-134 (Plenum Press, New York 1973).
Schenkein, I., Bueker, E.D., Heison, L., Axelrod, F. and Dancis, J.: Increased nerve-growth stimulating activity in disseminated neurofibromatosis. New Engl. J. Med. 290: 613-614 (1974).
Stanbury, J.B., Wyngaarden, J.B. Fredrickson, Frederickson, H.S.: Inherited variation and metabolic abnormality in Metabolic Basis of Inherited Disease, pp. 3-28 (McGraw-Hill Co., New York 1972).

Fig. 1. Two-dimensional acrylamide 10%, SDS 0.1% slab gel separating whole cell protein on the basis of molecular weight. See text for details. The top of each gel shows proteins of 200,000 daltons and the bottom of the gel, proteins of 20,000 daltons. (A) Neuroblastoma clone N18TG2, deficient in HGPRT (hypoxanthineguaninephosphoribosyl the enzyme defect associated with the Lesch-Nyhan syndrome; (B) glioma clone C6, BudR resistant; (C) neuroblastoma-glioma hybrid cell comprised of parental lines N18TG2 and C6-BudR and fused by inactive Sendai virus. It is significant to note the clear difference in the protein patterns of the neuroblastoma and glioma cell lines and the similarity of the patterns of the hybrid and neuroblastoma cells. Thus the technique is able to resolve protein differences from different cell types derived from the nervous system and to portray the dominance and inhibitory regulation of the neuronal genome on the glial genome as expressed through the hybrid cell system. ⟶

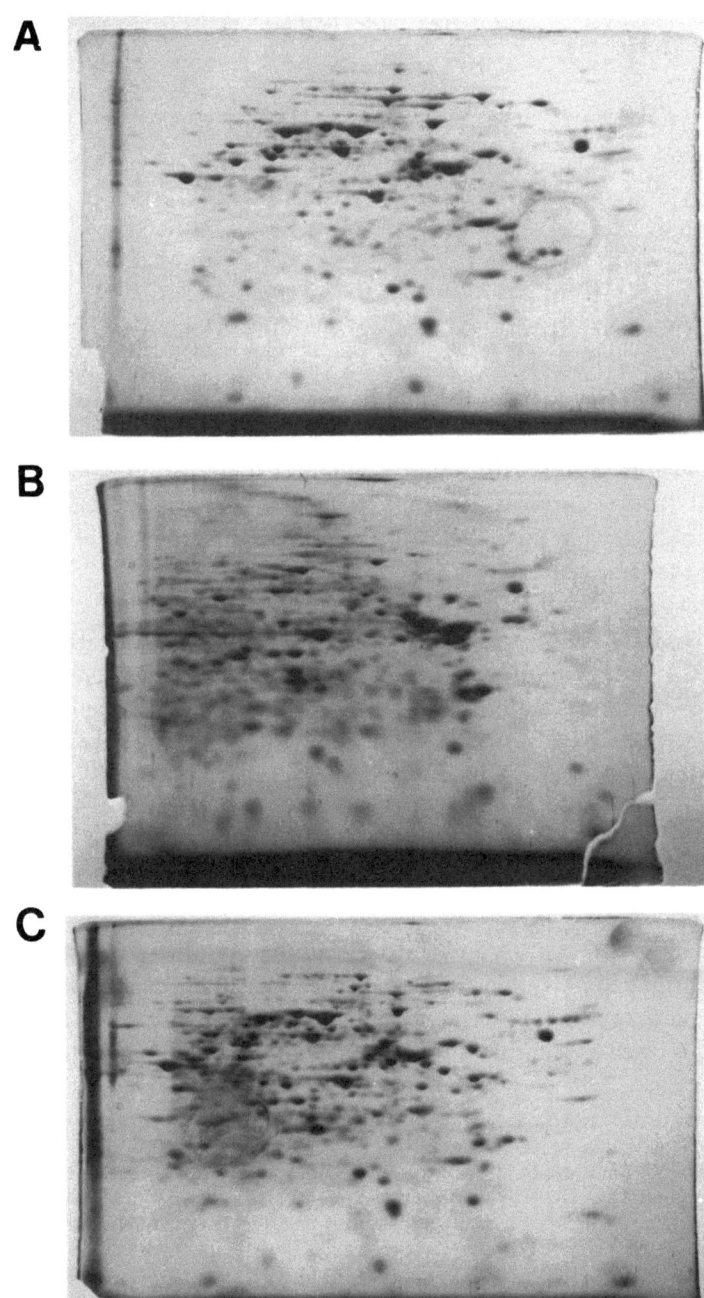

CHAPTER THREE

MORPHOLOGIC MATURATION OF THE BRAIN:

A QUANTITATIVE STUDY*

TH. RABINOWICZ, G. LEUBA, AND D. HEUMANN**

The first part of this study deals with the postnatal evolution of four parameters of cortical maturation: the evolution of the depth of the cerebral cortex both in the human and in the mouse brain, the evolution of the size of pyramidal cells from the precentral gyrus in human, the dendritic development of the pyramidal cells in mice visual cortex and the development of dendritic spines of the same cells.

The second part of this work deals with the differences in cortical growth between the right and the left hemisphere. The evolution of cortical depth will be studied in four areas of the human brain and in three areas of mice brain. Differences in the evolution of the left and right hemisphere will also be evaluated by the study of the neuronal densities in homologous areas of the human brain.

TECHNIQUE

The human cerebral cortex is studied by cutting out blocks taken from 44 areas according to Economo and Conel. These blocks are cut out perpendicularly to the long axis of the cerebral gyrus. Age and absence of any lesions of the brain, as well as clinical and pathologic data, are ascertained to allow a

* This work is dedicated to R.D. Adams, Bullard Professor of Neuropathology, Harvard Medical School, Boston, Mass.
** The authors are indebted to Mrs. J. McD.-C. Petetot and Mrs. M.L. Meier for histoquantitative and technical assistance; to Mrs. M.D. Mello (Boston) for the drawings from the 8-year-old child; to Mr. B. Maurer for photographic work and to Miss M. Laufer for typing the manuscript.

This work was supported by NIH Grants M-156-C4-C6, HD 00326-07-08 (Th. Rabinowicz) and M151 (J.L. Conel) and by the Swiss National Science Foundation Grants No. 3,641,071 and No. 3,434,074.

normative study. The cases used in our study as well as in Conel's work also showed normal school records.

Our standardized histologic and quantitative techniques are the same as those used by Conel to establish his atlases of the cerebral cortex of full-term newborn, children of 1, 3, 6, 15 months, of 2, 4 and 6 years. We established the same quantitative data of the premature of the 8th month and, in a preliminary way, from the premature of the 7th and 6th month. We also established together with Conel the data of the child of 8 years and on Conel's material, of a child of 10 years and of one adult of 22 and another of 35 years. Our data on the prematures of the 6th and 7th month as well as of children of 8 and 10 years and adults of 22 and 35 years should be considered as preliminary since they are based mostly on single cases.

The cellular density is histologically established in human cases on cubes of 100 micra from a section of 25 micra thickness. Results are multiplied by 4 to obtain a theoretical cube of 100 micra sides. Cortical thickness of the whole cortex as well as of each cortical layer, length and width of neurons as well as the numbers of each type of neurons are established on cresyl violet stained standardized paraffin embedded sections. Cajal and Golgi-Cox impregnations have been done as well as myelin staining according to Weigert, (Rabinowicz 1964, 1967).

In mice the brains were cut frontally in serial sections after standardized paraffin embeddings. Sections are cut at 25 micra thickness and stained with cresyl violet. Measurements of cortical thicknesses as well as of cellular densities were done in the same way as for humans. The cell densities are established by counting the number of nuclei and are corrected with Abercrombie's correction. All the data were obtained at 10 different places for each area and each cellular type on 10 animals of the same age (male, Swiss Albino mice).

Quantitative data have been obtained on mice by G. Leuba and D. Heumann. The following areas according to Krieg were studied: 2, 3, 4, 7, 20, 41, 17, 18 and 18a. Colgi-Cox Van der Loos impregnations were made in order to obtain quantitative data on the dendritic development of the pyramidal cells of layers III and V, as well as of the spines of the apical dendrites of these neurons.

D. Heumann established the total number of neurons and of glial cells in an almost total neocortical volume. This volume is limited anteriorly by the fornix, posteriorly by the subiculum, laterally by the entorhinal fissure and medially by the corpus callosum. The evolution of this cortical volume as well as its cytoarchitectonic data were studied at 5, 10, 30, 60 and 180 days in mice. This total neocortical volume is mentioned here under the name of Volume H.

THE EVOLUTION OF THE DEPTH OF THE HUMAN CEREBRAL CORTEX

This parameter was evaluated in 44 neocortical areas. Considering as an example the evolution of the cortical thickness of the precentral gyrus at the level commanding the movements of the trunk (fig. 1), it appears that a rapid

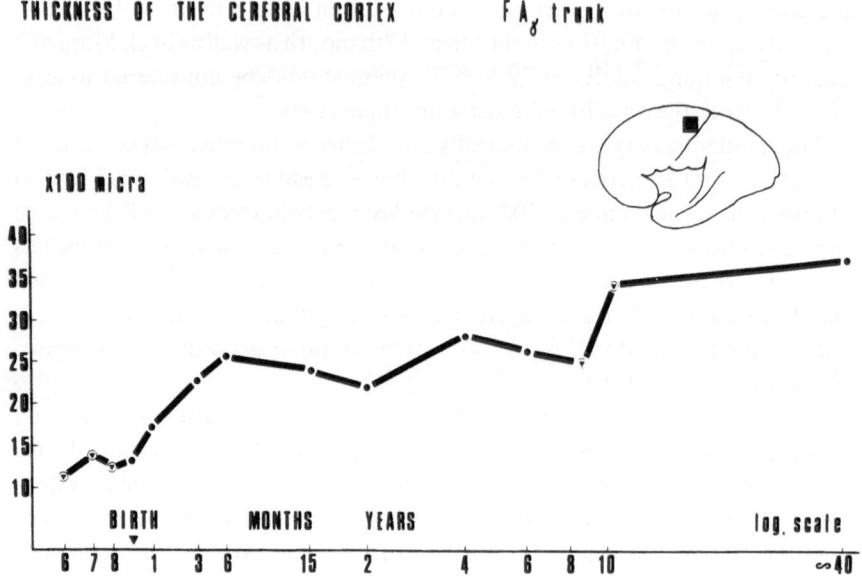

Fig. 1. Evolution of the cortical depth in the precentral gyrus FA gamma at the level for the trunk. Around 15-24 months and 6 to 8 years cortical thickness decreases. Data not yet available between 10 and 40 years. Human cases (from Conel, Economo and ourselves).

increase of the cortical depth lasts till about 6 months postnatal except between the premature of the 8th month to the full-term newborn. Between 6 months and 2 years the thickness of the cortex decreases, being the thinnest at 24 months. A second period of increase appears between 2 and 4 years. The cortical depth then decreases once more between 6 and 8 years and from that point on increases till adulthood. Thus at least two periods of thinner cortex exist between 15 months and 2 years and between 6 and 8 years. This phenomenon can be seen in the great majority of the areas we studied except in the visual areas (fig. 2).

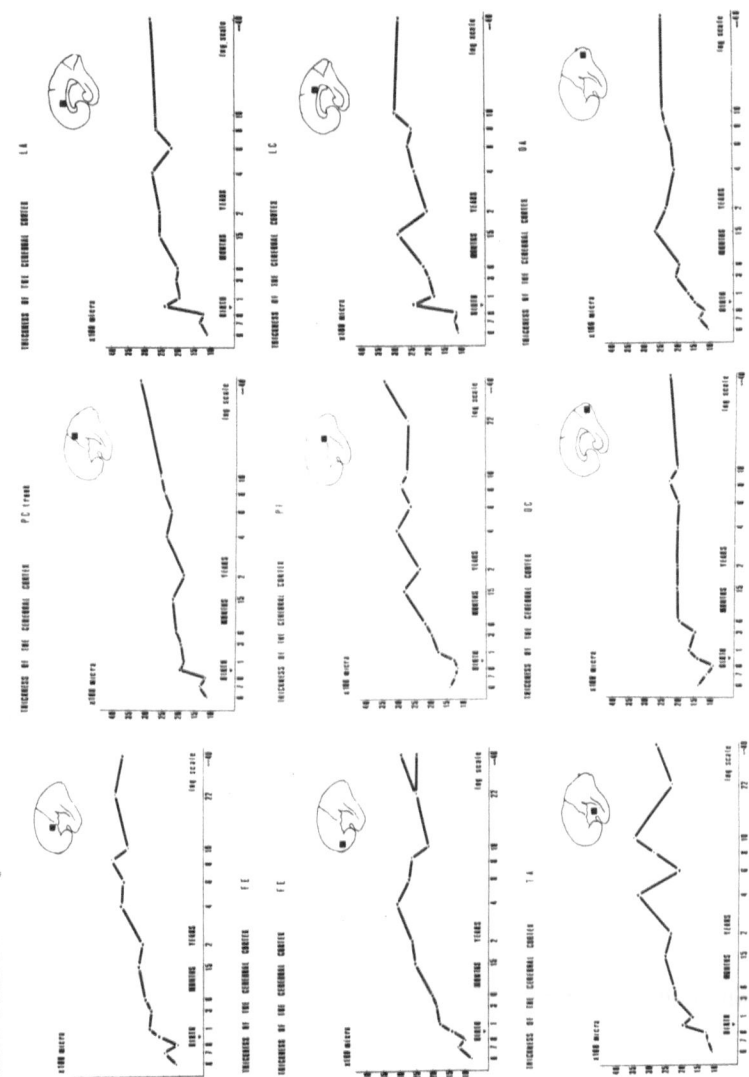

Fig. 2. Evolution of the cortical depth in 9 areas taken as examples. Note that in almost all the areas (except visual areas OC and OA) the periods around 15-24 months and 6-8 years show a thinner cortex. Human cases (from Conel, Economo and ourselves).

The visual area is the earliest neocortical area to attain an almost adult thickness around 6 to 15 months while the peristriate area OA reaches that point between 1 and 2 years.

We have already mentioned the fact (Rabinowicz, 1974, 1976) that, if one compares the rather homogeneous evolution through time of the motor area for speech (FCBm) with the posterior speech areas PF or TA, it is striking how these posterior centers for speech show an irregular evolution. We have assumed that this variability may represent a considerable non-

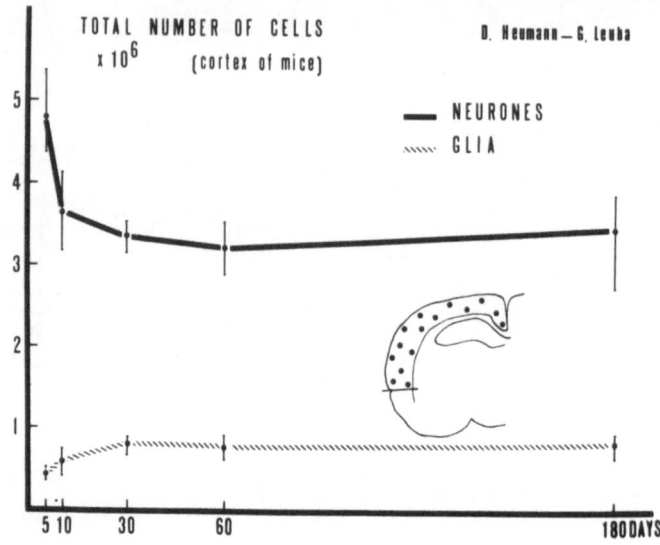

Fig. 3a. Evolution of the total number of neurons in mice total cortical Volume H (see text). Note the rapid decrease in the number of neurons in the first 10 postnatal days followed by rather stable values from 30 days on. The total number of glial cells increases until about 30 days. See also: Rabinowicz 1976.

homogeneity of the cases in the posterior speech center. As for each age the number of children studied is relatively low (between 4 and 9), individual variability may produce non-homogeneous curves and may well represent a non-homogeneous growth of some areas as compared with others (for example the posterior speech area vs. the motor speech area).

At adult age a personal variability was also present, for instance in the prefrontal area FE or in the hippocampal areas not shown here. These facts have been previously noticed by Von Economo.

The series of graphs showing the temporal evolution of cortical thickness shows that there are at least two periods in childhood when the cortex is thinner: between 15 and 24 months and between 6 and 8 years. In some graphs these periods may be difficult to recognize because of the variability of some areas (fig. 2).

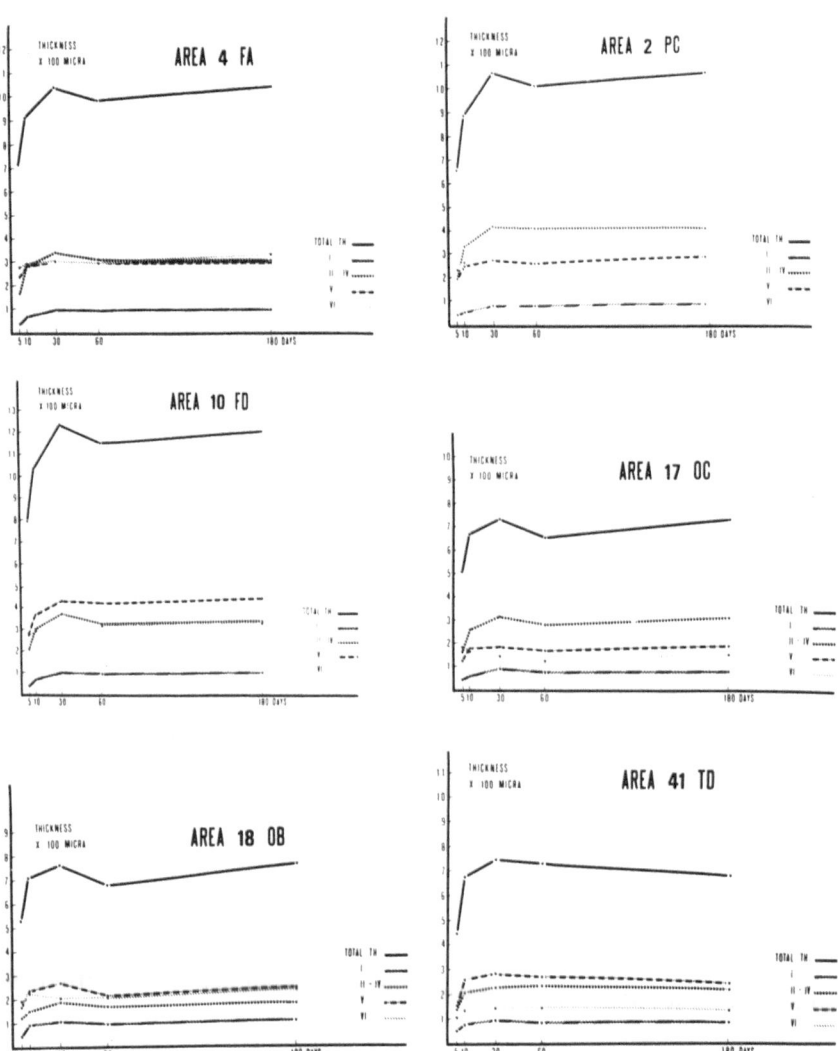

Fig. 3b. Evolution of cortical depth in 6 areas in mice. The total cortical thickness shows, in all the areas except area 41, that at 60 days the cortex is thinner (upper line on all areas). In all areas 30 days corresponds to the deepest cortex.

The following facts drew our attention to this phenomenon: one was the temporal evolution of pyramidal cell size in the precentral gyrus of the human cerebral cortex and the other was the decrease of the cortical thickness in our mice at 60 days and in rats at 90 days (Diamond and coll.) (fig. 3a).

Later the evolution of the size of the pyramidal cells will be seen. The transient decrease of cortical depth in humans as well as in mice is not an isolated fact. At the same time, there is a slight increase in neuronal density. It is probable that this temporary increase in neuronal density is not a real increase in the number of neurons. As is known there is no secondary neuronal proliferation after 6 months in utero in the human and after birth in mice. Hence this must be only a concentration into a smaller volume of a number of neurons which did not change very much, thus indicating that this phenomenon represents only a relative increase in neuronal density.

The same phenomenon also appears in the total cortical Volume H of the mice, which represents more than 80% of the total neocortical volume. Figure 3b shows that the real number of neurons is modified very little between 30 and 180 days.

We do not yet fully understand the reasons for this transient cortical thinning. It seems to be a general phenomenon appearing in most cortical areas in the human brain as well as in mice and in rats. It probably is not a technical artifact since, at the same periods, there is an increase in cellular densities of the same areas, showing that there is a higher concentration of neurons in a decreasing volume.

In mice the transient cortical thinning is found throughout the cortex but is not as marked in all the different areas and layers. It seems to be slightly greater in layers II, III and IV and in the pyramidal cell layers. This may indicate that some important growth phenomena are occurring in the pyramidal cell layers and/or in their neuropile.

EVOLUTION OF THE SIZE OF THE PYRAMIDAL CELLS IN THE HUMAN CEREBRAL CORTEX

The evolution of the shape and size of the pyramidal cells during development (fig. 4) shows interesting features. Once more these features correlate with the periods between 15 and 24 months and between 6 and 8 years. At these two periods the pyramidal cells, mainly those of Betz and the great pyramidal cells in the precentral gyrus, show an almost equilateral triangular shape the width being almost the same as the height (fig. 5). Also worth noticing is the fact shown in fig. 5 that in the premature of the 6th month the shape of Betz cells is already quite equilateral. We do not yet understand these phenomena. Perhaps they are related to the dynamics of the dendritic growth.

LENGTH AND WIDTH OF PYRAMIDAL CELLS (FA $_\gamma$ trunk)

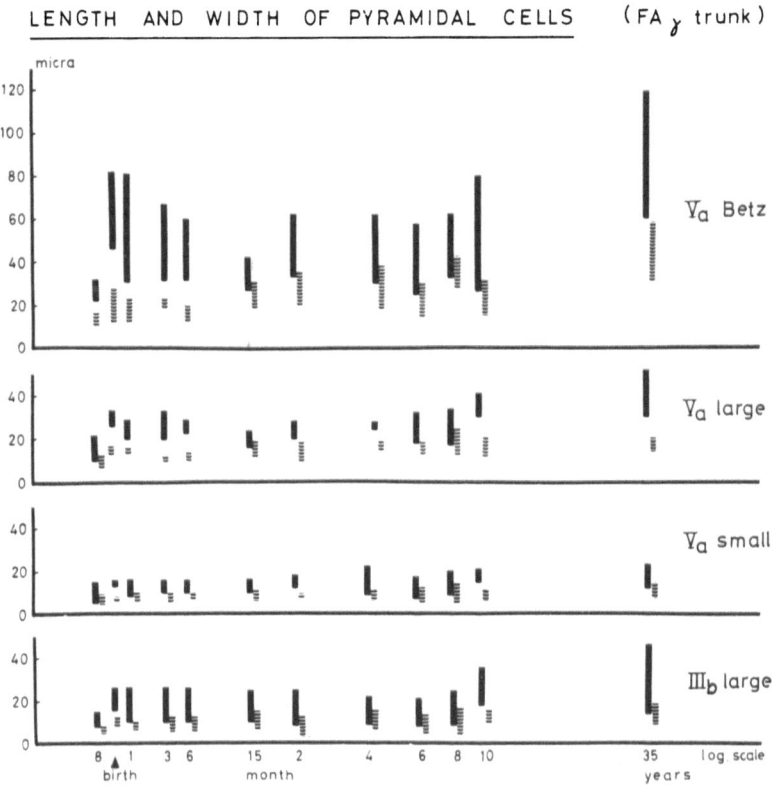

Fig. 4. Evolution of length and width of pyramidal cells of layers V and III in FA gamma trunk. Maximal and minimal values given only. Width in dotted lines, length in solid lines. Measurements on standardized paraffin embedded 25 micra thick cresyl violet stained sections. Values from 30 cells by Conel (from newborn up to 8 years old child) and by ourselves at the other ages.

The evolution of Betz cells, of the great and small pyramidal cells of layer Va as well as of the great pyramidal cells of layer IIIb has been studied quantitatively in the precentral gyrus (at the level commanding the movements of the trunk) from the premature of 8 months on to the adult of 35 years. At 15 months Betz cells are almost as wide as high but smaller than at 6 years where the same phenomenon reappears. Still at 4 years these cells are quite wide. The same phenomenon is seen also in the big pyramidal cells of the Vth layer but is less pronounced in the small pyramidal cells of the same layer. Moreover the latter cells are broader between 6 and 8 years and remain more so at 35 years. In the IIIb layer the increase in length appears very early and is

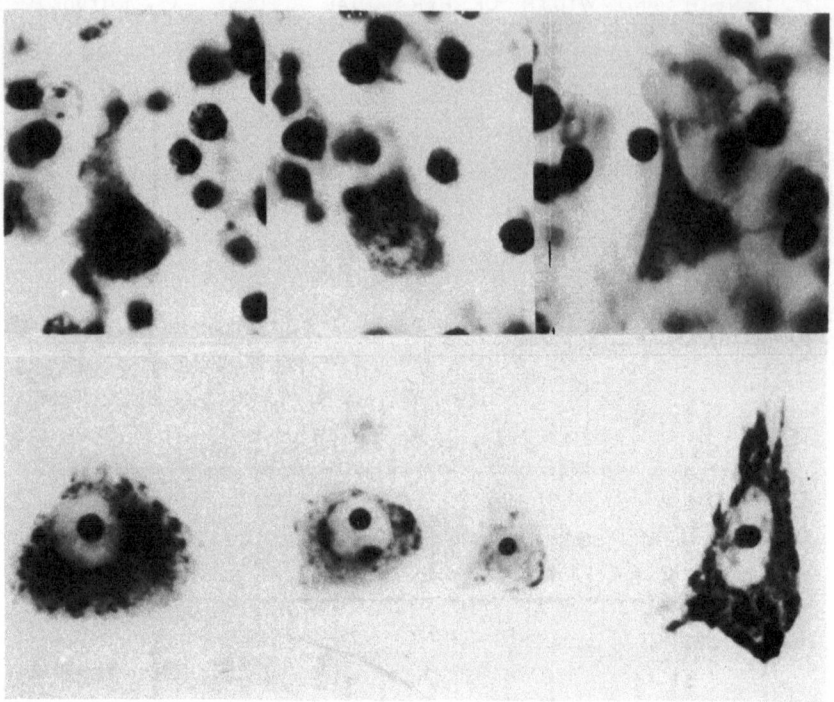

Fig. 5. Examples of Betz cells at 3 different ages. *Upper row* shows Betz cells from FA gamma in a premature of the 6th month: broad and short pyramidal cells without Nissl substance. *Lower row* shows a rather big but equilateral triangularly shaped Betz cell at 6 years in the lower left corner, with big dark Nissl bodies at the periphery. In the middle lower row a smaller Betz cell also at 6 years. In the right lower corner a Betz cell from an adult of 41 years. All magnifications: 1000 ×, oil immersion objective. See also fig. 4.

found around birth, while the increase in width continues progressively, with a slightly smaller cell around 6 years. At 8 years pyramidal cells are rather broad and begin to increase much more in length than in width, as shown in figure 5.

Thus it appears that, mostly in the Vth layer, the pyramidal cells show a growth pattern with a different shape at two times, while in the IIIrd layer the decrease in length is only slight and is present only once around 6 years. It still remains to be shown that in all the other areas the pyramidal cells behave in the same way as in the precentral gyrus. As it would be highly unprobable that only the precentral gyrus and its pyramidal cells show this particular pattern of growth and as we have other data showing that during these two periods some remodelling in the cerebral cortex is going on, we could also assume that there might be some relationship with the development of the children at these periods: between the end of the first postnatal year and the second year many important new functions appear as well as between 6 and 8 years which is also an important moment in the life of children.

THE DENDRITIC DEVELOPMENT IN MICE

The visual area 17 in mice was studied by G. Leuba at 10, 15, 21, 30, 60 and 180 days using Golgi-Cox Van der Loos impregnations. The dendritic pattern of the pyramidal cells in both layer V and III was studied using an ocular with 8 concentric circles. The number of crossings between these circles and the dendrites were recorded for each circle (fig. 6). Curves were established by plotting the number of dendritic intersections for each circle beginning from the central perikaryon. At each age 10 cells were examined from each of 10 animals i.e. a total of 100 cells at each age and on each layer.

In the Vth layer at 10 days the maximum of intersections is found near the cell body at 28 micra. At 30 days the maximum is found at around 47 micra. The same happens at 60 days but at that time there are also many intersections at 56 and at almost 66 micra from the cell body. It appears thus that at 60 days the densest part of the dendritic pattern is broader, having about 20 micra and goes further away than at 30 days and 180 days. At this latter time the pattern reverts to what it was at 30 days, showing that there was a transient extension of the dendritic pattern of the pyramidal cells of layer V.

The maximum of dendritic intersections was found to be at about 50 micra in adult animals as described by Eayrs (1959) and by Cordero and coll. (1976).

Things are slightly different in the IIIrd layer. The region of maximum branching appears slightly later (at 10 and 15 days) than in the Vth layer. The most remote point is attained at 30 days but the number of branchings increases until 60 days. There is no transient extension of the dendritic pattern as in the Vth layer. Moreover there is no apparent difference at 60 and 180 days of the density of the dendritic pattern, as was found in the Vth layer.

The fact that the neuropile of the IIIrd layer grows slower has previously been shown (Leuba and coll.) using the criteria of the cell densities, whose decrease is slower in the IIIrd than in the Vth layer. Two interesting phenomena have thus been observed. The first is a transient extension of the den-

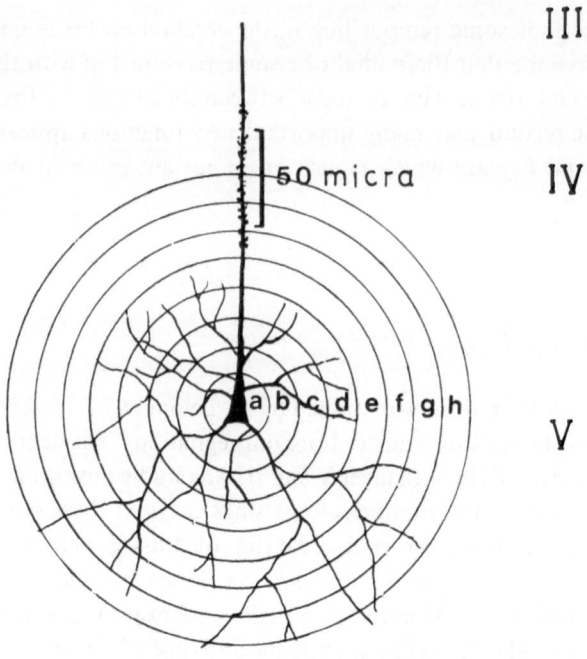

Fig. 6a. Evaluation of the number of dendrites around the perikaryon of pyramidal cells of layer V or of layer III. Golgi-Cox Van der Loos impregnation. Concentric circles in the ocular are roughly equidistant: values given in fig. 7. The number of crossings between dendrites and circles are proportional to the development of the dendritic network in both width and depth (see fig. 7). *b.*Estimation of the number of spines on the apical dendrite of the pyramidal cells of layer V and on those of layer III. Spines are counted on dendrites of 1 to 1.5 micra diameter and on dendrites of 1.6 to 3 micra diameter. Countings done at 100 micra from the perikaryon, on a segment of 50 micra. Values obtained on 10 neurons on each of 10 animals. One curve represents 100 measurements per layer per age (see fig. 8).

dritic pattern in the Vth layer at 60 days, followed by a partial regression (plasticity of the dendritic pattern). The second phenomenon is the later dendritic growth of the IIIrd layer (compared to the Vth) without a dendritic regression. The question must be raised, if this absence of regression is not a consequence of a lack of data between 60 and 180 days or of an earlier growth of the IIIrd layer (see below).

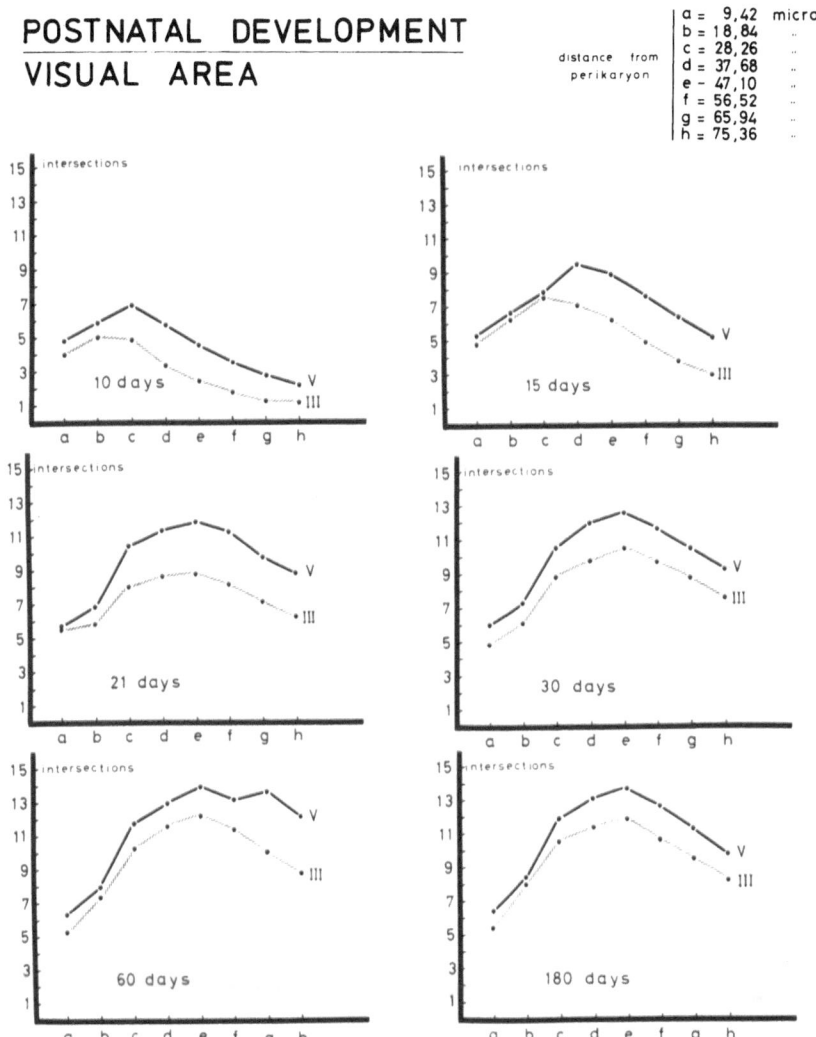

POSTNATAL DEVELOPMENT

VISUAL AREA

distance from
perikaryon

a = 9,42 micra
b = 18,84   "
c = 28,26   "
d = 37,68   "
e ‑ 47,10   "
f = 56,52   "
g = 65,94   "
h = 75,36   "

Fig. 7. See also fig. 6. At 10 days in mice area 17 the maximum number of intersections is at 28 micra from the perikaryon in layer V and at 19 micra for layer III. Slower speed of dendritic development in layer III compared to layer V. Layer III shows also less dendrites even at 180 days. Note a transient extension between 47 micra and 66 micra of the dendrites of the Vth layer at 60 days.

Sequences of Development of the Dendritic Spines in Mice

Spines were counted on the apical dendrite of pyramidal cells of the Vth layer and of the IIIrd layer on a segment of 50 micra length at 100 micra from the perikaryon. Countings were done by G. Leuba on Golgi-Cox Van der Loos impregnated sections of 100 micra thickness using an oil immersion objective 100 × and an ocular 20 × (magnification 2000 ×). 10 cells were counted from each of 10 animals. Countings were made for two groups, one in which the apical dendrite had diameters of 1 to 1.5 micra and a second in which the apical dendrites showed diameters between 1.6 and 3 micra. Fig. 8 shows the evolution of the number of spines depending on whether they are on thin or thicker dendrites and in the IIIrd or the Vth layer.

The IIIrd layer develops earlier. At 10 days the number of spines is less in the IIIrd layer for both thin and thicker dendrites than in the Vth layer. At 15 days this is no longer true. From 30 days on the number of spines is higher in the IIIrd than in the Vth layer.

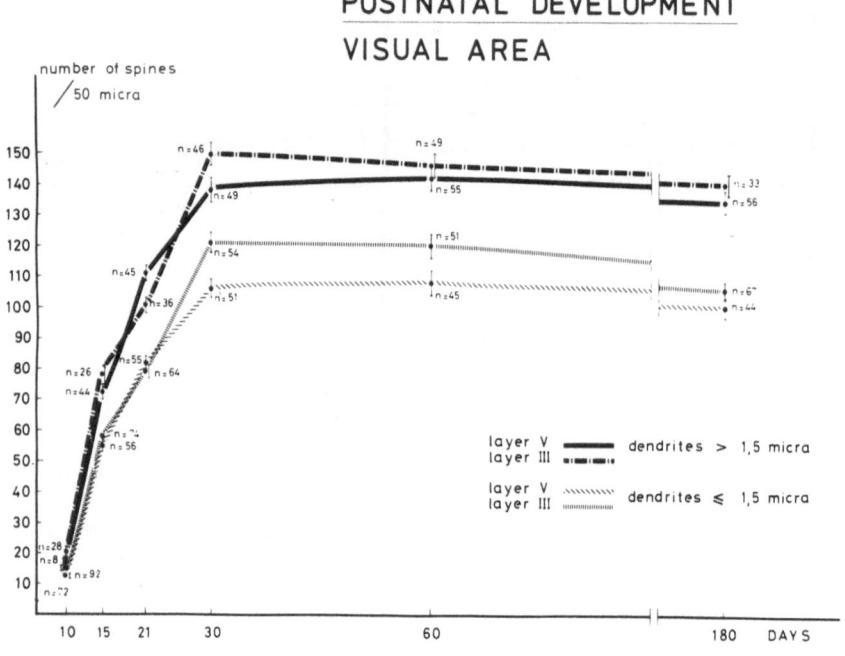

Fig. 8. Evolution of the number of spines in mouse area 17. Small numbers refer to the number of dendrites used for each value. Note a maximum of spines in layer III at 30 days and in layer V at 60 days. Note the overall decrease in the spine-density beginning at 30 days for the IIIrd layer and at 60 days for the Vth layer.

In the IIIrd layer the maximum number of spines is found at 30 days-more in the dendrites of more than 1.6 micra than in those of less than 1.5 micra. There is a slight decrease in the number of spines between 30 and 60 days, which decrease lasts until 180 days. The decrease in the number of spines is faster for the dendrites of the IIIrd layer than for those of the Vth. The number of spines is also greater on the wider dendrites for both the IIIrd and the Vth layers.

The pattern of growth of the pyramidal cells of the Vth layer is slightly different. The increase in the number of spines is slower until 30 days; after that the spine density is almost the same as at 180 days. Still from 60 days on a general decrease in the number of spines is noticeable and the curve is almost the same for dendrites of more than 1.6 micra as for those of less. Statistically this decrease is significant only for the thinner dendrites of the IIIrd layer. Thus a maximum of spines as well as of thicker dendrites was present around 30 to 60 days and then decreased. Once more, as for the extension of the dendritic network of the Vth layer, transient modifications appeared on the apical dendrite-spines system with a maximum at 60 days. This corresponds to what was shown in the rat after 60 days by Feldman and Dowd (1974) up to 600 days.

It is also interesting to note that, if the number of spines is higher in the IIIrd layer than in the Vth, this may be due to the fact that the countings of the apical dendrites of the pyramidal cells of the IIIrd layer were done at the level of the IInd layer which shows a higher cell density. Things are similar with the pyramidal cells of the Vth layer whose spines were counted mostly in the IVth layer. As a consequence the evolution of the number of spines reflects also the relationships between the granular layers and their underlying pyramidal cells.

Summing up our observations on the development of the mouse cerebral cortex one may say that there exists a period of remodelling between 30 and 60 days. At 60 days the cerebral cortex is thinner than at 30, the cell density is higher, the dendritic network is more developed and the dendritic spines, after a maximum at 30 days, are beginning to decrease. The slower development of the IIIrd layer compared to the Vth is now well established and is seen through all the criteria we have studied. This is well in accordance with what we know from the early development of the layers (Angevine and Sidman 1961, Berry and Rogers 1965, Bruckner and coll. 1976, etc.). Thus postnatal maturation follows the embryonal sequence of settlement.

On the other hand the mouse cerebral cortex does not show a real stabilisation at the end of its period of more rapid maturation, that is at 30 days.

There is a rather extensive process of remodelling around 60 days indicating great plasticity at an age generally considered as adult. As a matter of fact we should admit that the period between 30 and somewhere after 60 days corresponds most probably to the adolescence of mice.

Finally the sequence of cortical events in mouse brain seems to be as follows: first, a period of cell multiplication with cell migration and the beginning of a settlement. Then some of the nerve cells disappear while the others grow their dendrites and spines. The neuronal densities are then stabilizing slowly while dendrites and spines continue their development. This is followed by a remodelling of the dendritic network part of which disappears with spines still multiplying. Then in turn part of the spines disappear. All these phenomena overlap more or less.

DIFFERENCES IN GROWTH PATTERN BETWEEN THE RIGHT AND THE LEFT HEMISPHERE

Differences between the right and the left hemisphere have already been noticed by Beck in adult humans in homologous areas. We tried to gather some information on differences in growth velocity between the right and the left hemisphere in children compared to one adult. We used our own data from the premature of the 8th month, those of Conel and ourselves for a child of 8 years and our own data from a child of 10 years and an adult of 35 years. Data have been obtained from all the 44 neocortical areas in both the premature of the 8th month and the child of 8 years while for the child of 10 years and the adult we studied only about 10 areas. For the other ages of Conel's atlases we do not have any data from areas of both hemispheres.

The same type of research was also done on mice serially cut brains on cresyl violet stained sections after paraffin embedding.

BILATERAL EVOLUTION OF CORTICAL DEPTH IN THE HUMAN BRAIN

Quite important differences may exist between homologous areas (fig. 9) as early as in the premature of the 8th month. We show here 7 areas: the precentral gyrus at the level commanding the muscles of the trunk (FA gamma), the motor speech center (FCBm), the frontal pole (FE), the primary visual area (OC), the inferior part of the posterior speech area (TE) and the auditory speech area (TA). They all present a deeper cortex on the right side than on the

PREMATURE of 8th month (case A 831/60)

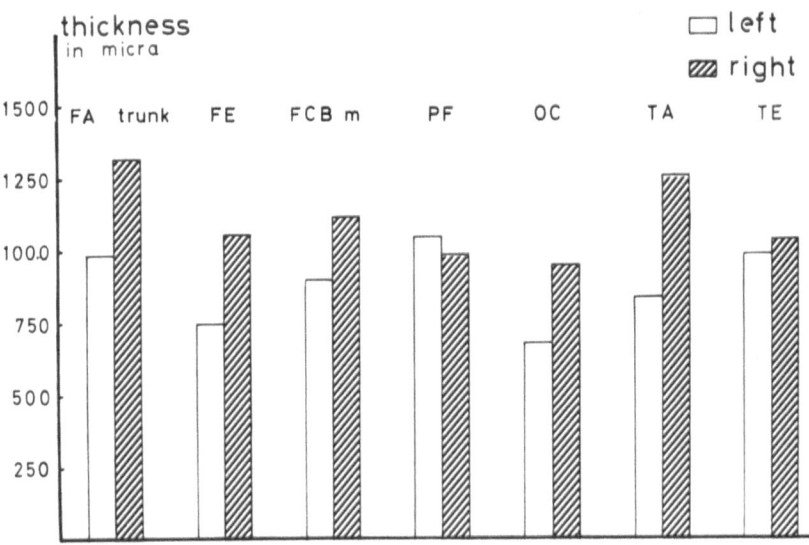

Fig. 9. Differences in depth of left and right areas of a premature of the 8th month. In 6 out of 7 areas the right homologous area is thicker than the left. Only the posterior speech center PF shows a thicker cortex on the left side.

left in the premature. There is one exception among these 7 areas, namely PF the central part of the posterior speech center which is slightly deeper on the left side than on the right. Differences are considerable between the left and the right in area TA, an auditory speech center, and in the precentral motor area FA. The latter fact raises the question as to whether the child would have been left-handed because of its clearly thicker (better developed?) right motor area. At this time we do not know if there is a relationship between, for example, left-handedness and the better development of the right precentral gyrus. Moreover there are fibers of other frontal origins in the pyramidal tract and probably from other lobes. Also Yakovlev and Rakic showed that there are several patterns of decussation of the pyramidal tract and that in adults 80% of the cases they studied anatomically showed a greater volume of the right pyramidal tract in the spinal cord. The anatomical basis of left- or right-handedness is still not yet settled but a number of cortical parameters and their combinations, with perhaps other features, for example the number of fibers of the pyramidal tract, makes it a difficult problem to solve.

Evolution of Cortical Depth in Four Homologous Areas
We tried to obtain information on the temporal evolution of homologous
areas. This was only possible for four moments during life: in the premature
of the 8th month, in children of 8 and 10 years and in one adult of 35 years.
Thus our data can only be indicative and preliminary. Still it appears that the
behaviour of homologous areas through time is not uniform (fig. 10).

In the precentral gyrus at the level of the trunk there was a clear difference
in thicknesses between right and left sides in the premature. This disappeared
in the child of 8 years and was back in the child of 10 years and the adult of 35
years. In this area the difference in depth was maintained in three out of the
four cases studied here and remarkably enough, the right precentral gyrus is
thicker than the left at most ages.

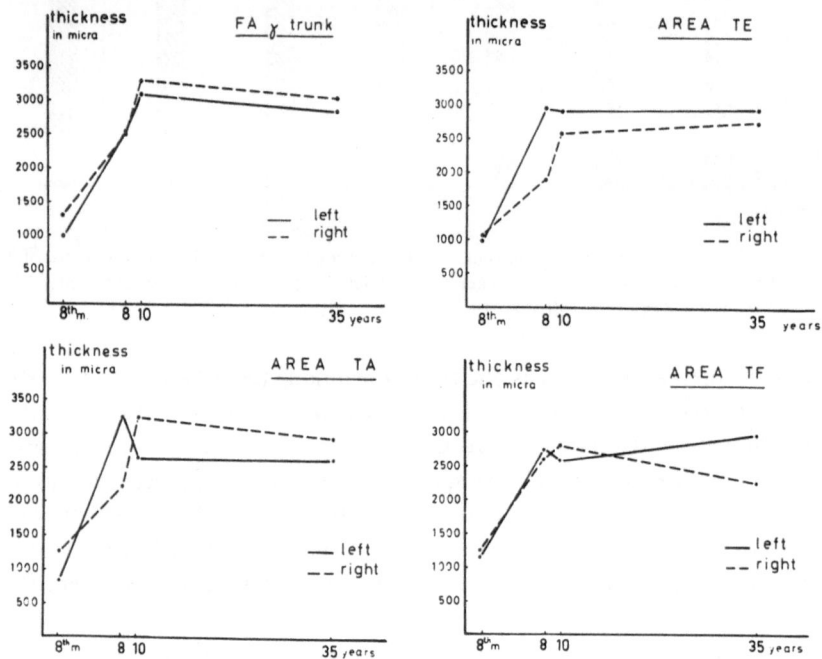

Fig. 10. Bilateral evolution of 4 areas of a premature of 8 months, children of 8 and 10 years and
an adult of 35 years.
FA gamma shows at 8 years the same depth in both the right and the left side. Right side thicker
than the left at 10 and 35 years.
TE is thicker at 8, 10 and 35 years on the left side except in the premature.
TA shifts from left to right side at 10 years, the left side getting thinner.
TF shows the same phenomenon as TA the left side becoming thinner than the right at 10 years.
At 35 years this is reversed again.

Area TE which is the anterior part of the posterior speech center shows, in the premature of the 8th month, a slightly thicker cortex on the right than on the left side. From 8 years on, in all three cases the left cortex was thicker than the right. The difference is noticeable at 8 years and seems to be less at 10 years and at 35 years. This curve seems to correspond to the fact that it is generally considered that the posterior speech center is more important in the left hemisphere.

In the two other temporal areas we show here, *area TA* auditory speech area and *area TF* (internal superior temporal area) there is an interesting phenomenon which appears between 8 and 10 years. At this time the left side presents a decrease in depth which makes it thinner at 10 years than at 8 years whereas the right side was thicker at 8 years. This inversion of the relative thicknesses remains stable in the auditory speech center TA while it reverses once more in the internal superior temporal area TF.

The evolution of these two areas shows that there may be a time during development when the differences in cortical depth may shift from one side to the other in homologous areas.

It is obvious that there is no possibility at present to draw any conclusions from so few cases. Still one may say at least that there does not seem to be homogeneity at the hemispheric level, that is that the cerebral cortex would be thicker in all the areas on one side. As we have seen it, in the premature of 8 months and in the child of 8 years there is no hemisphere whose cortex is thicker on one side than on the other. Once more there are differences at the level of areas but no general differences even at the level of lobes. Thus neither hemisphere is better developed in terms of cortical depth than the other and the evolution in the imbalance between homologous areas may shift from one side to another during development.

Evolution of Cortical Depth in Mice
D. Heumann and G. Leuba studied the evolution of the cortical depth in mice in areas 17, 18a and 41 at 10, 30, 60 and 180 days in both right and left hemispheres. Figure 11 shows, in each of the 10 studied animals, studied, the differences in thickness between the left and right hemispheres, the left being taken as reference. At a first look differences between animals may sometimes be important: up to 10% more on the right than on the left side. Few animals show the same thickness on both sides of homologous areas. Generally speaking the mouse neocortex is thicker on the right than on the left. Nevertheless, during development some differences do appear depending on the areas. Both visual areas 17 and 18a show greater differences between right and left sides,

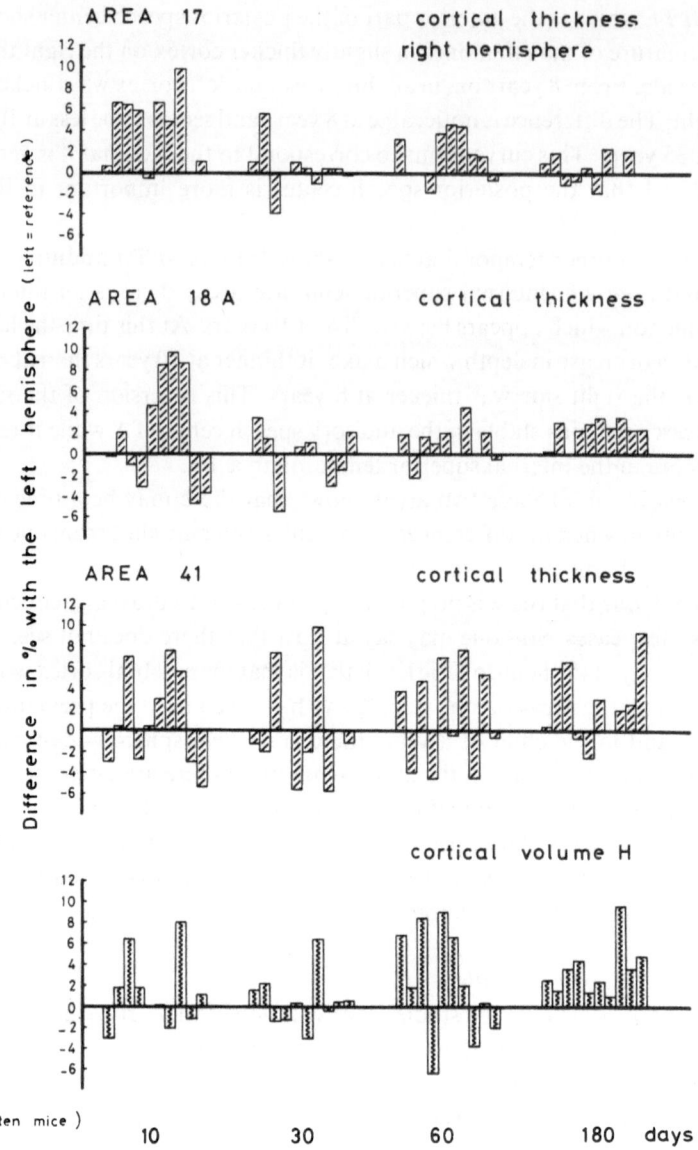

Fig. 11. Differences in cortical depth of 3 areas in mice. Differences are given in % for each of ten animals, the left side being taken as reference. Thus the right hemisphere is represented only. In all 3 areas great "personal" variability in younger animals. At 180 days a majority of animals show a thicker right side for homologous areas. The same is true for Volume H.

while auditory area 41 shows better differences between the sides. In each of the three areas studied it is interesting to notice that at 180 days there is a general dominance of the right hemisphere whose areas are thicker.

The total cortical Volume H which corresponds to better than 80% of all neocortical areas in mice shows quite clearly that the right neocortex is thicker than the left. This fact has already been observed by Diamond in rats. It is noteworthy that variability is considerable at 10 and 30 days while at 60 days a slight dominance appears. The latter is clear at 180 days mostly in favor of the right hemisphere.

NEURONAL DENSITIES

In the *premature of the 8th month* the following areas are shown here both on the left and on the right side: the precentral gyrus at the level of the trunk (FA gamma), the motor area for speech (FCBm), the frontal pole (FE), the post-central gyrus at the level of the trunk (PC), the primary visual area (OC) and the auditory speech area (TA). (Fig. 12.)

The neuronal densities have been compared in the precentral gyrus (FA gamma) in layers II to VI. Neuronal densities are slightly less in layers II, III and V on the left side while layers IV and VI show a reversed situation. It is interesting to see that in the premature of the 8th month even one of the most advanced areas shows only slight differences between the right and the left at the level of the layers.

In the motor area for speech FCBm the differences in cell densities are more important between the right and the left sides. Here cell densities are slightly less in the right hemisphere than in the left. It is an open question whether the motor speech area in our premature would be "dominant" on the right side while motricity at FA gamma is nearly balanced on each side.

The prefrontal polar area FE presents slightly greater maturation on the left side for layers II, III and V while it is greater on the right side for layers IV and VI. This corresponds more or less to what has already been shown in the precentral gyrus.

In the postcentral gyrus PC at the level of the trunk layers II and IV (granular layers) shows slightly better maturation on the right side in terms of neuronal density while layers III, V and VI are mostly balanced. In this layer there seems to be some sort of balance between left and right, at least for the pyramidal cell layers, while for the granular cell layers there remain some noticeable differences.

The primary visual area OC shows slightly better maturation on the right than on the left except for the pyramidal layer V, where the maturation is almost balanced.

The auditory speech area TA also shows slightly better maturation on the right than on the left, with a balance for the IIIrd layer only.

Generally speaking in the 8th month old premature for the six areas we have shown, differences in cell densities between left and right sides are not

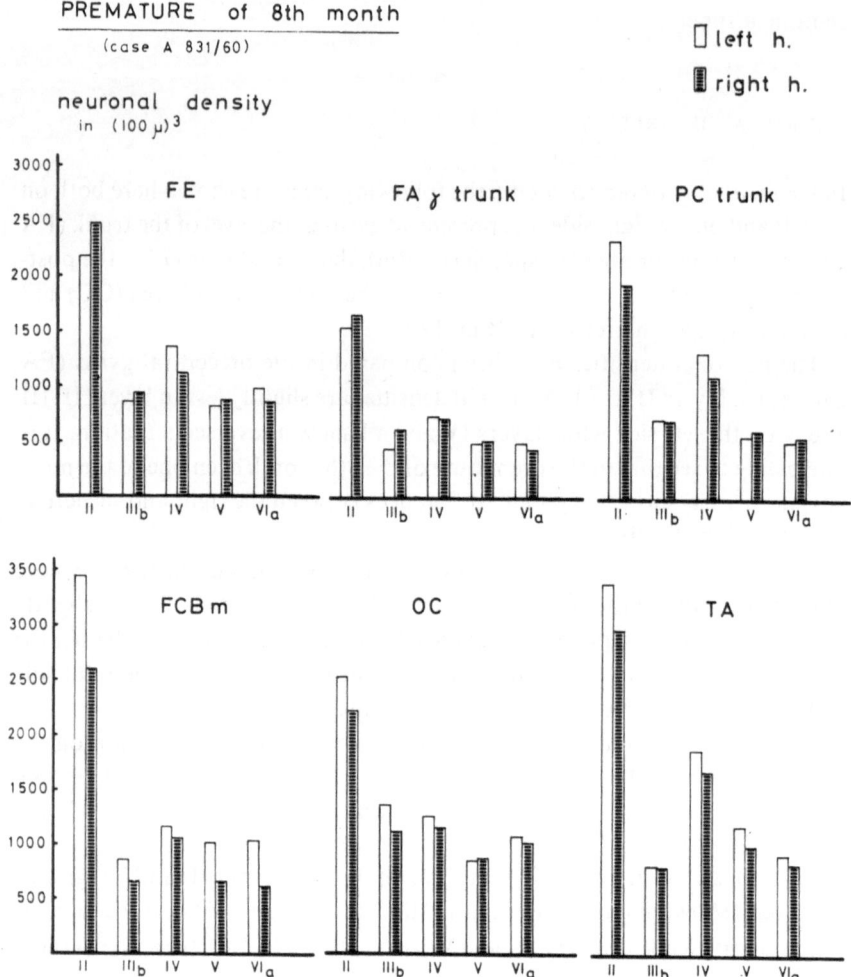

Fig. 12. Differences between right and left sides of the neuronal densities in 6 areas in a premature of the 8th month. See text.

very important except in the motor speech area FCBm. Moreover, differences exist not only between right and left at the level of a whole area but also inside each area there are differences between the layers, if one compares for instance the IInd right layer with the IInd left layer of a given area.

We tried to follow the evolution of cell densities in three temporal layers: TA, TC a temporal polar area and TE a temporal internal superior area. (Fig. 13a, b and c.)

The chronologic evolution of area TA (from the premature of the 8th month to the children of 6 and 10 years and to the adult of 35 years) appears as

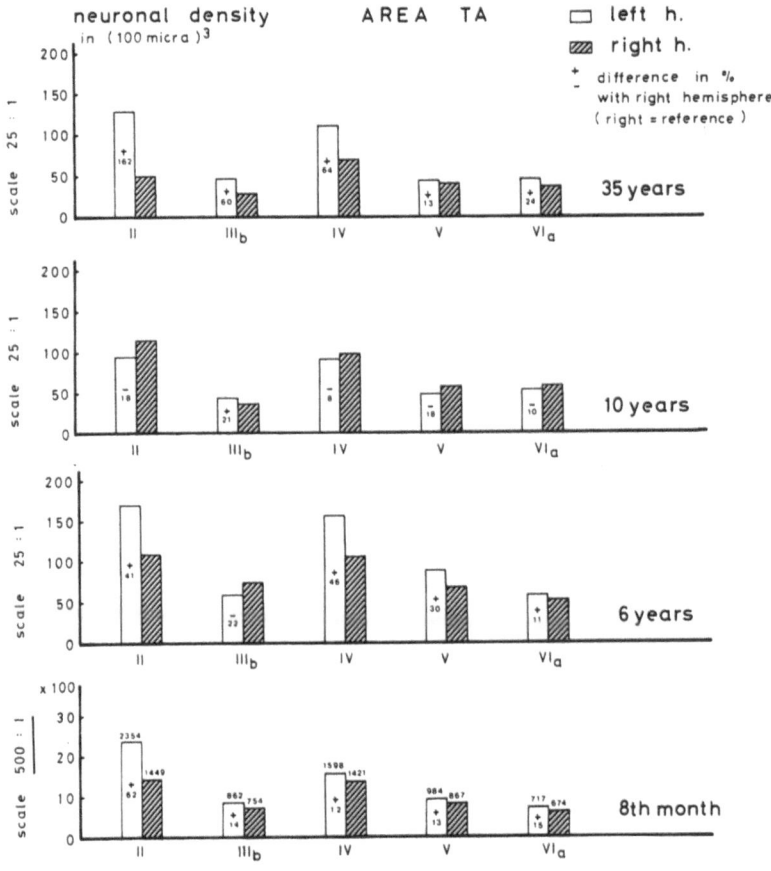

Fig. 13a. Area TA at 8th months prematurity, 6 years, 10 and 35 years. Neuronal densities for right and left sides, layers II through VIa. Cell densities mostly higher in the left side (=better maturation of the right side) except for the child of 10 years.

50   TH. RABINOWICZ, G. LEUBA, D. HEUMANN

follows: there is a higher cell density which means a lesser maturation on the
left side in layer II for the premature, the child of 6 years and the adult while at
10 years it is slightly inverted. For the IIIrd layer, differences are less between
left and right side, the right showing slightly better maturation than the left
except at 6 years. In the IVth layer the same is true except at 10 years. In the
Vth and in the VIth layer the same is also true except at 10 years but both for
the Vth and the VIth layer differences are sometimes slight between right and
left. One child of 10 years, shows an almost completely inverted pattern as
compared with the other 3 cases. (Fig. 13a.)

The polar temporal area TG shows almost the same evolution as area TA.
(Fig. 13b.)

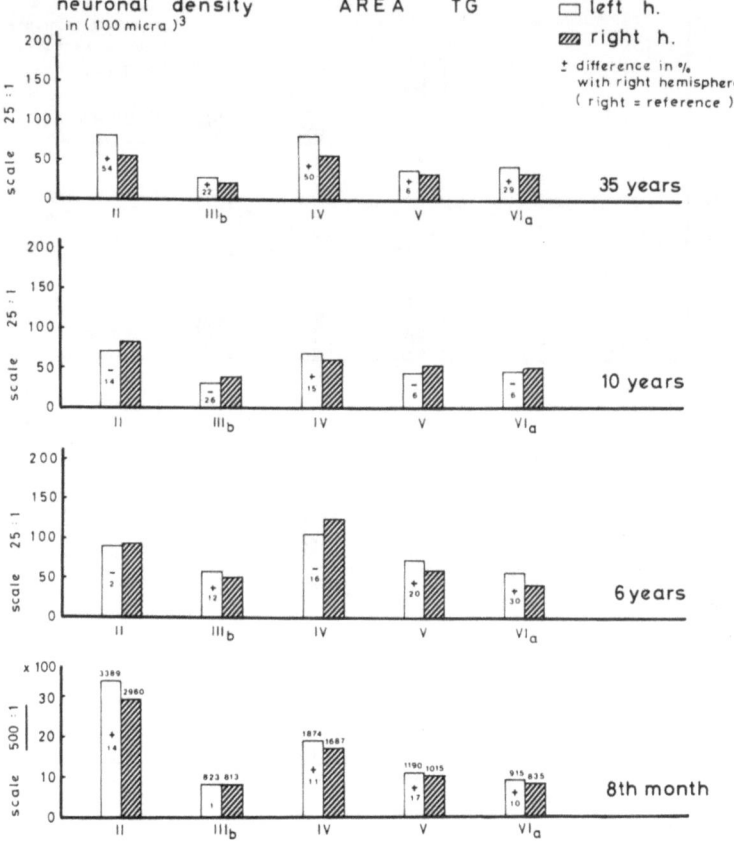

Fig. 13b. Area TG. See fig. 13a. Here it is the child of 6 years whose cell densities are higher on the
right side. The 3 other children show a better maturation (lower cell density) on the left side.

Area TE seems to be rather well balanced in the premature as well as in the children of 6 and 10 years while differences between right and left are more pronounced in the IInd and IVth layers in the adult. (Fig. 13c.)

Generally speaking for these three temporal areas a dominance seems to exist for a better matured right side except for the child of 10 years. Here also, as we have seen for the thickness of the cortex, there is no general trend and differences in maturation are shown not only at the level of an entire area but also at the level of the layers within each area studied.

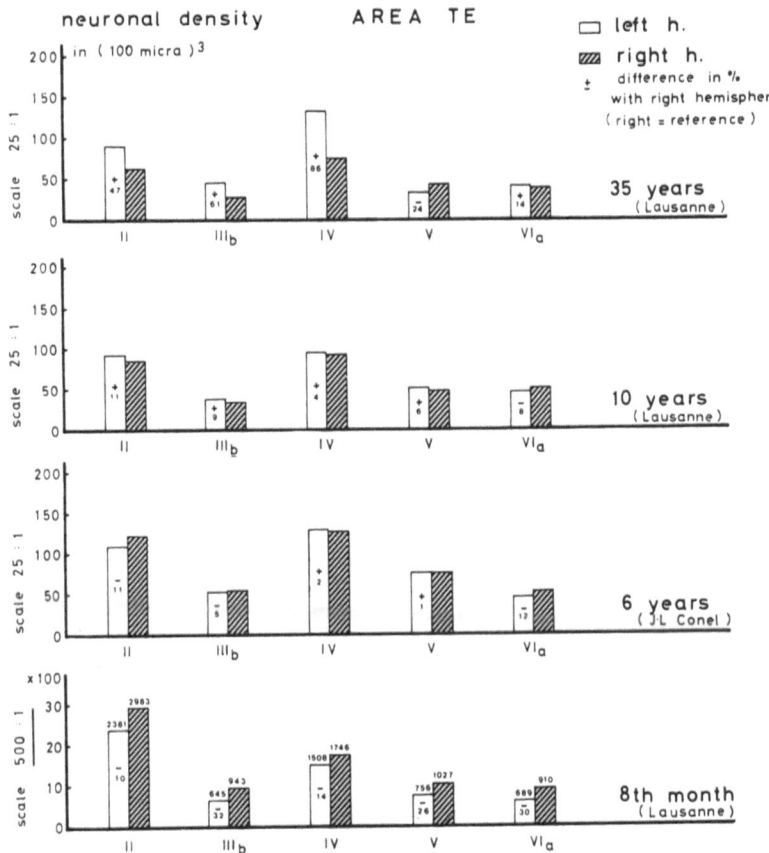

Fig. 13c. Area TE. See fig. 13a. Neuronal densities higher in the right side of the premature of the 8th month. The other 2 children are rather well balanced at 6 and 10 years. Right side less dense for layers II, III, IV and VI in the adult case.

Our *preliminary conclusions* would be that in humans there is no overall dominance of one hemisphere nor of one lobe over the other in terms of quantitative morphology. Still, differences are clear between the right and the left side only at the level of the areas. As we could see during development and as far as we know actually, dominance may change sides between 6 and 10 years showing that some areas may revert their degree of maturation from one side to another. This introduces a supplementary element which would be that dominance between hemispheres is not only unevenly distributed but moreover is evolutionary.

REFERENCES

Angevine, J.B. and Sidman, R.L.: Autoradiographic study of cell migration during histogenesis of cerebral cortex in the mouse. Nature 25: 766-768 (1961).

Beck, E.: Unterschied zwischen links und rechts im cytoarchitektonischen Bau der vorderen Zentralwindung und die Frage der Linkshirnigkeit. Dtsch. Zschr. Nervenhk. 163: 183, 214 (1950).

Berry, M. and Rogers, A.W.: The migration of neuroblasts in the developing brain cortex. J. Anat. 99: 691-709 (1965).

Bruckner, G., Mases, V. and Biesold, D.: Neurogenesis in the visual system of the rat. An autoradiographic investigation. J. comp. Neurol. 166: 145-162 (1976).

Conel, J.: The postnatal development of the human cerebral cortex, 8 vol. (Harvard University Press. Cambridge 1939-1967).

Cordero, M.E., Diaz, G. and Araya, J.: Neocortex development during severe malnutrition in the rat. The American Journ. of clin. nutr. 29, no. 4: 358-365 (1976).

Diamond, M.C., Johnson, R.E. and Ingham, C.A.: Morphological changes in the young, adult and aging rat cerebral cortex, hippocampus and diencephalon. Behav. Biol. 14: 163-174 (1975).

Eayrs, J.P. and Goodhead, B.: Postnatal development of the cerebral cortex in the rat. J. Anat. 93: 385-402 (1959).

Economo, C. von and Koskinas, G.N.: Die Cytoarchitektonik der Hirnrinde des erwachsenen Menschen (Springer, Wien 1925).

Feldman, M.L. and Dowd, C.: Aging in rat visual cortex: light microscopic observations in layer V pyramidal apical dendrites. Anat. Rec. 178: 355 (1974).

Heumann, D., Leuba, G. and Rabinowicz, T.: Postnatal development of the mouse cerebral neocortex. II. Quantitative cytoarchitectonics of visual and auditory areas. (To be published.)

Krieg, W.J.S.: Connections of the cerebral cortex. I. The Albino rat. B. Structures of the cortical areas. J. comp. Neurol. 84: 221-276 (1946).

Leuba, G., Heumann, D. and Rabinowicz, T.: Postnatal development of the mouse cerebral neocortex. I. Quantitative cytoarchitectonics of some motor and sensory areas. (To be published.)

Rabinowicz, T.: The cerebral cortex of the premature infant of the 8th month. In: Progress in Brain Research vol. 4: Growth and maturation of the brain, 39-92 (1964), P. Purpura and J.P. Schadé Eds. Amsterdam, Elsevier.

Rabinowicz, T.: Quantitative appraisal of the cerebral cortex of the premature of 8 months; in Minkowski, Regional development of the brain in early life, p. 91-124 (Blackwell, Oxford 1967).

Rabinowicz, T.: Techniques for the establishment of an atlas of the cerebral cortex of the premature; in Minkowski, Regional development of the brain in early life, p. 71-89 (Blackwell, Oxford 1967).

Rabinowicz, T.: Some Aspects of the Maturation of the Human Cerebral Cortex. In: Pre- and Postnatal Development of the Human Brain. Mod. Probl. Paediat., vol. 13, p. 44-56 (Karger, Basel 1974).

Rabinowicz, T.: Morphological features of the Developing Brain. In: M.A.B. Brazier and F. Coceani. IBRO Symposium 2. Infantile Convulsions, p. 1-23 (Raven Press, New York 1976).

Yakovlev, P.I. and Rakic, P.: Patterns of Decussation of Bulbar Pyramids and Distribution of Pyramidal Tracts on two sides of the Spinal Cord. Transactions of the Amer. Neurol. Assoc. p. 366-367 (1966).

CHAPTER FOUR

FACTORS CONTRIBUTION TO ABNORMAL NEURONAL DEVELOPMENT IN THE CEREBRAL CORTEX OF THE HUMAN INFANT

Dominick P. Purpura, M.D.

INTRODUCTION

Normal neuronal operations depend upon the orderly development of type-specific geometrical features of different classes of neurons and the establishment of appropriate synaptic relations. Analysis of aberrant neurogenetic processes must proceed from knowledge of normal temporo-spatial patterns of cortical development and identification of factors that have a profound impact on physiologic properties of neuronal organizations. These considerations have guided our studies in recent years towards two major objectives: definition of the time course of neuronal differentiation in the cerebral cortex of preterm and older infants, and specification of processes of particular importance for the attainment of optimal neuronal and synaptic functions.

The major morphogenetic event in cortical neuronal differentiation is the elaboration of dendritic systems (Purpura, 1961). This follows from the fact that dendrites of most cortical neurons provide over 95% of the postsynaptic targets for presynaptic inputs. Furthermore the number, distribution and functional types of synapses on different dendritic systems have long been viewed as essential criteria for evaluating integrative operations of multipolar neurons (Purpura, 1967). These are sufficient reasons for a special emphasis on processes of dendritic differentiation and dendritic spine development in the analysis of normal and abnormal cortical development in the human infant (Purpura, 1974, 1975a, 1975b, 1975c, 1976).

A second area of inquiry summarized here is concerned with the geometric properties of neurons and the possible consequences of distortions in neuronal shape. The importance of neuronal geometry in developmental studies has recently been emphasized in pathophysiologic processes underlying several human neuronal storage diseases (Purpura and Suzuki, 1976). These and other studies call attention to the usefulness of an attempt to define aberrant neurobehavioral development in terms of a developmental pathobiology of the neuron (Purpura, 1977).

NORMAL AND ABERRANT DENDRITIC DEVELOPMENT IN THE VISUAL
CORTEX OF THE PRETERM INFANT

The developmental features of dendritic systems of visual cortex neurons in
'normal' and very sick preterm infants (Purpura, 1975c, 1976) may be summarized as follows. The visual cortex bordering the calcarine sulcus of the 25-
week-old preterm infant has densely packed neurons that are not readily
identifiable in terms of specific laminae. Pyramidal neurons throughout the
cortex exhibit cell bodies with prominent apical dendritic shafts of variable
length. Small and medium pyramids have tortuous apical dendritic shafts
some of which terminate in complex growth processes (fig. 1A). Basilar dén-
drites are either absent on superficial pyramids or are represented by a few
fine protoplasmic processes. "Presumptive" basilar dendrites are present on
large pyramidal cells of the cortical depths (fig. 1B). The apical dendritic
shafts of these elements are irregular in contour and exhibit a few filopodium-
like processes (fig. 1B).

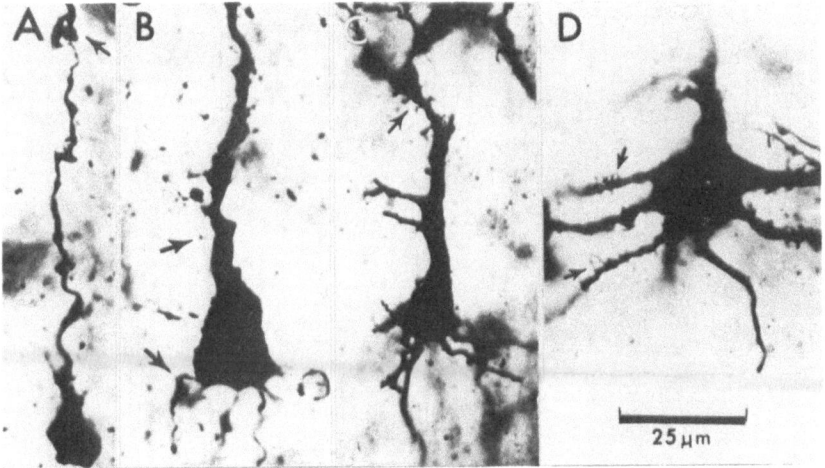

Fig. 1. Examples of rapid Golgi preparations of neurons in calcarine (visual) cortex. A and B,
from a 25-week-old preterm infant. A, small pyramidal cell exhibits tortuous apical dendrite. The
cell body is devoid of basilar processes. A terminal growth cone is seen at arrow. B, large
pyramidal neuron has a few thin perisomatic processes (arrow). The apical dendrite shaft is
irregular and exhibits a rare filopodium-like process (arrow). C and D, from a 33-week-old-
preterm infant. C, superficial pyramidal neuron has many dendritic spines and fine dendritic
growth processes and secondary branches. D, cell body and proximal segments of apical and
basilar dendrites of a Meynert cell. Note conspicuous dendritic spines (arrow). The primary axon
descends in the middle of the figure. (From: Purpura, 1975c.)

DEPTH(μm)

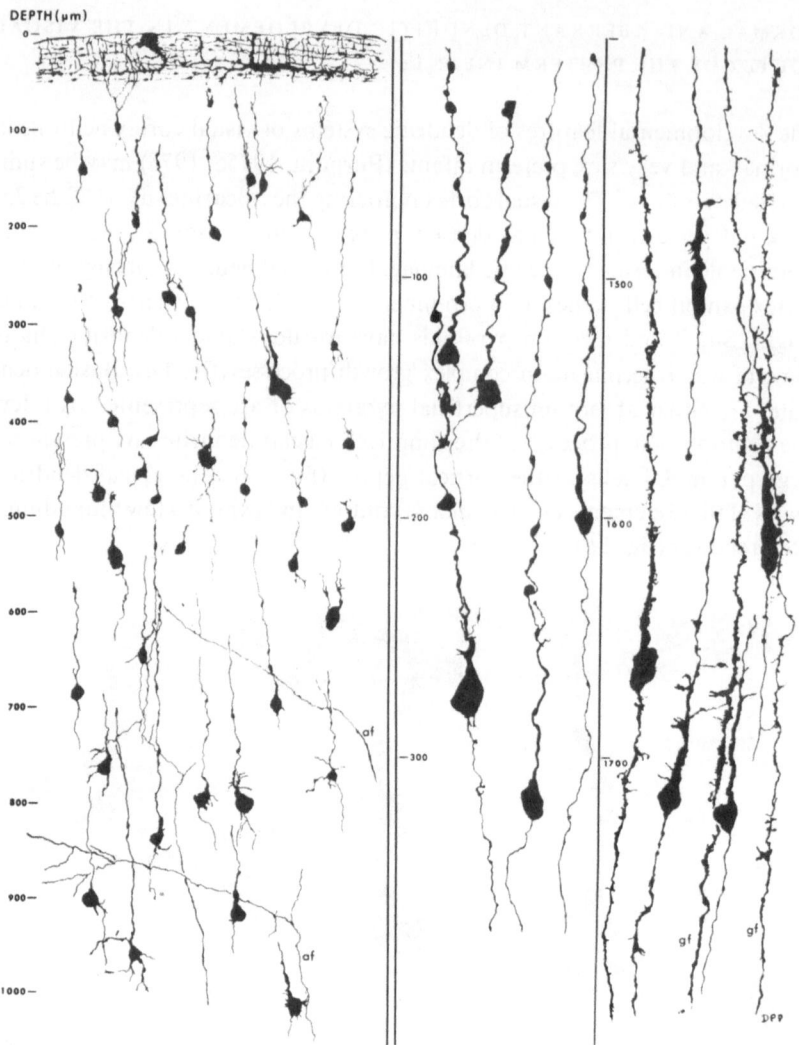

Fig. 2. Camera lucida composite drawings of neurons in calcarine (visual) cortex of a 25-week-old preterm infant. Scale indicates depth in micrometers. Rapid Golgi preparations. Low-magnification drawings at left provide an overview of the maturational features of pyramidal and non-pyramidal cells at different depths. Pyramidal neuron apical dendrites are short, and basilar dendrites are essentially absent. Stellate cells (depth 400-800 μm) are primitive in development. (Right) Superficial pyramidal neurons are shown at higher magnification with apical dendritic growth processes (compare with Fig. 1A). At far right radial glial fibers (gf) are shown in relation to bipolar neurons, with some early dendritic growth processes. These cells are in the upper region of the intermediate zone-cortical plate border. (From: Purpura, 1975c.)

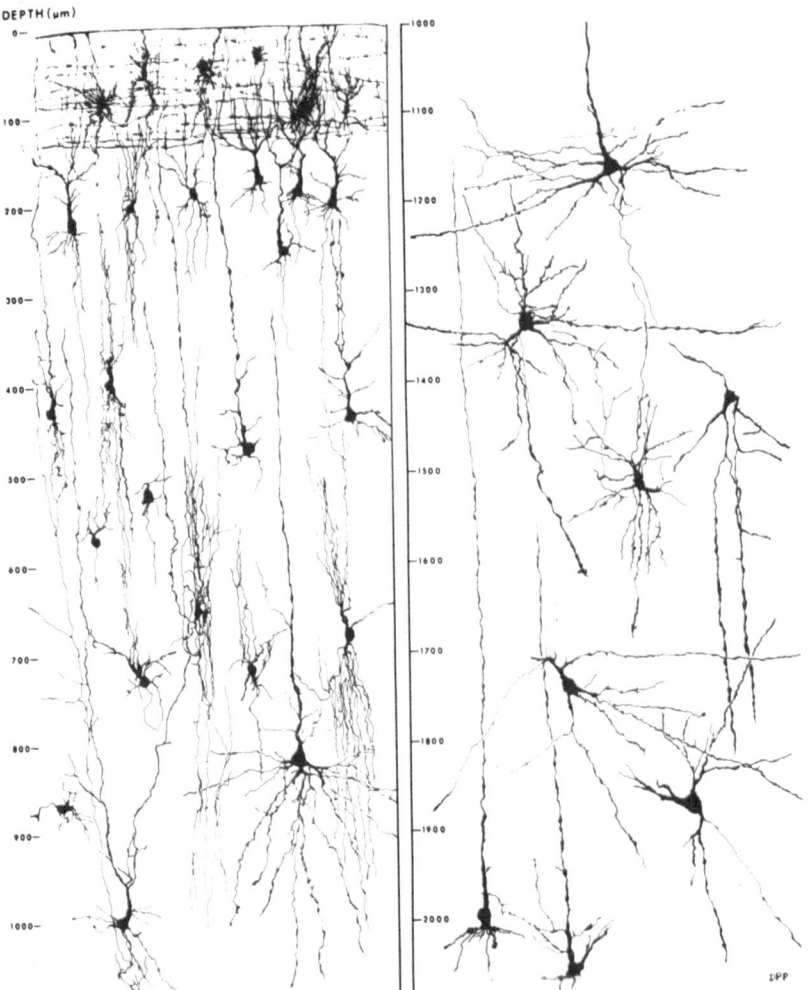

Fig. 3. Camera lucida drawings of neurons in calcarine cortex of a 33-week-old preterm infant. (This infant was born at 29 weeks and survived 4 weeks.) Rapid Golgi preparations. Owing to increase in cortical thickness at this stage the left and right side panels of the composite should be considered as a single continuous "column", 2.0 mm in depth. Superficial pyramidal neurons exhibit considerable apical and basilar dendritic growth. Deep pyramids are also well developed and apical dendrites extend upward through several layers. Stellate cells at 500 to 600 μm show considerable maturation, and large stellate cells in the border zone (1.3-1.7 mm), are well developed; some retain "embryonic" processes directed toward the white matter. Note two immature pyramidal neurons at 2.0 mm with long varicose apical processes. A major feature of the visual cortex at this age is seen in the prominence of cells with double dendritic tufts issuing from upper and lower poles. Compare with Fig. 2. The visual cortex has virtually doubled in thickness from 25 weeks to 33 weeks. Dendritic spines on many neurons are not illustrated at the magnification employed for the drawings. (From: Purpura, 1975c.)

Radial glial fibers are observed throughout the intermediate zone and many extend into the lower margins of the cortical plate (Sidman and Rakic, 1973). In the latter locations neurons with elongated cell bodies and prominent lumpy apical processes are encountered in parallel "relation" to radial glial fibers (fig. 2, far right).

The outer half of the molecular layer of the visual cortex of the 25-week-old preterm infant contains many Cajal-Retzius cells with characteristic tangential processes interrupted by thin radial fibers directed toward the pial surface. Small neurons, probably different varieties of stellate and other non-pyramidal neurons, are almost totally devoid of perisomatic processes. Of particular importance is it to note that neither the apical dendritic shafts arising from pyramidal cell bodies nor cells within the region of termination of geniculocortical afferents exhibit dendritic spines in the 25-week-old preterm infant. The general organization of a "cortical column" of visual cortex neurons in the 6-month-old preterm infant is illustrated in fig. 2.

In contrast to the poor development of dendritic systems and the absence of dendritic spines on visual cortex neurons in the 25-week-old preterm infant, pyramidal neurons in the 33-week-old infant are relatively well developed with respect to these dendritic features (fig. 3). By 33 weeks, there is a significant increase in cortical thickness. The depth of the Cajal-Retzius plexus of the molecular layer has increased. Superficial pyramidal neurons exhibit active dendritic differentiation as evidenced by the number of basilar and apical dendrites with lumpy enlargements, spicules, fine varicosities and filopodium-like processes (Morest, 1969, 1970, Purpura, 1975a). A small superficial pyramidal neuron located at 200 µm toward the right of the cortical column in the overview drawing (fig. 3) is shown in the photomicrograph of fig. 1C. Several varieties of stellate cells in midcortical regions are also prominently seen. Many "giant" stellate neurons bordering the white matter exhibit extensive dendritic systems.

The visual cortex of the 8-month-old preterm infant exhibits many types of neurons designated by Cajal as "cellules à double bouquet dendritique", as well as moderately well developed Meynert cells. The latter neurons (fig. 1D) have thick basilar dendrites. Conspicuous spines are present on proximal apical and basilar dendrites of large and small pyramidal neurons.

The observations summarized in figs. 1-3 and elsewhere (Purpura, 1975c, 1976) provide evidence that neurons of the primary visual cortex in the human infant undergo maximal phases of dendritic differentiation and development during the 6th-8th month of gestation. Of particular significance is the fact that dendritic spines are virtually absent in the 6-month-old preterm infant

DEPTH (µm)

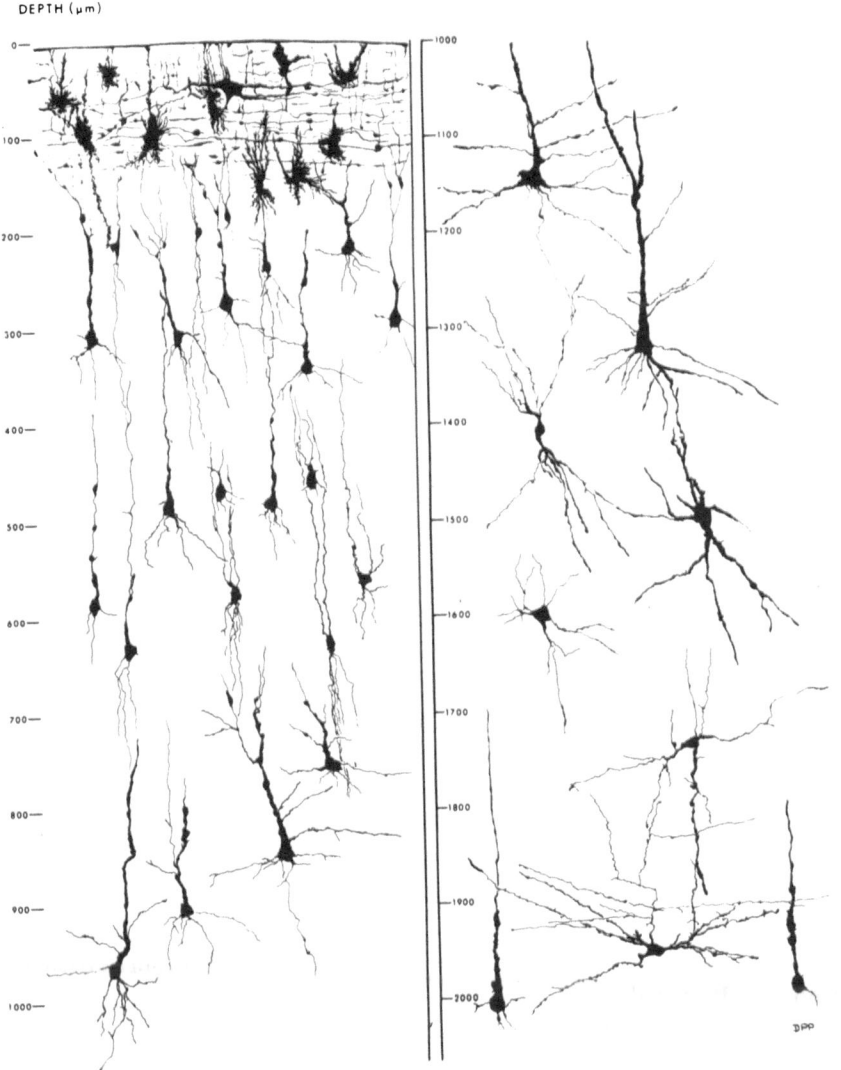

Fig. 4. Camera lucida drawings of neurons in calcarine cortex of *Ne* (34 weeks, c.a.) Rapid Golgi preparations. The molecular layer contains large numbers of glia. Superficial pyramids have poorly developed basilar dendrites and thin apical shafts. None of the dendrites have spines. Stellate cells in the cortical depths (1.4-1.8 mm) are relatively well developed in contrast to most medium and large pyramidal neurons. Double-tufted Cajal cells are present but do not have elaborate dendritic plumes. Cortical neurons are not as well developed for 34-weeks as in the 33-week-old preterm infant whose cortical neurons are illustrated in fig. 3. (From: Purpura, 1976.)

visual cortex but are abundant by 33 weeks. Hence, in a relatively narrow time frame (approximately 2 months) two important developmental events occur: dendritic differentiation and dendritic spine development. It has been suggested from these and earlier ontogenetic studies of laboratory animals (Purpura, 1961, Purpura et al., 1964) that elaboration of dendritic systems and axospinodendritic synapses, as reflected in the appearance of dendritic spines, are the critical morphogenetic events that correlate with the early electrographic maturation of primary visual evoked potentials (VEPs) (Purpura, 1975c).

It is instructive to consider the extent to which electrographic characteristics of VEPs in very sick preterm infants reflect underlying pathologic alterations in cortical neuronal development. Preliminary studies aimed at providing such information have been encouraging, as indicated elsewhere (Purpura, 1976). Two examples of these studies suffice to emphasize the impact of high risk factors on cortical maturation processes in the preterm infant.

One infant in this series (*Ne*) was born at 28 weeks gestation with a birth weight of one kilogram. The clinical course included respiratory distress syndrome, bilateral pneumothoraces, hyperbilirubinemia, hypocalcemia, anemia, necrotizing enterocolitis, coagulopathy and heart murmur. VEPs recorded 9 days before expiration, at 5-6 weeks of age (34 weeks, conceptional age) had electrographic characteristics similar to those of much younger infants (27-30 weeks) (Purpura, 1976). Examination of the morphological features of visual cortex neurons in this infant confirmed the impression of maturational "delay" of neuronal development. Thus when neurons in *Ne* (34 weeks, c.a.) (fig. 4) are compared to the relatively normal appearance of cortical neurons at 34 weeks (fig. 3) it is seen that superficial small and medium pyramids have thin apical dendrites but are virtually devoid of basilar dendrites. Basilar dendrites are occasionally seen on deeper-lying medium pyramids. Apical dendrites frequently exhibit lumpy enlargements and few thin tangential or oblique branches are present. A most important observation in *Ne* is that dendrites do not possess true spines, although filopodium-like processes are detectable on some apical dendritic shafts. It is also of interest to note that the molecular layer in *Ne* is less well organized and contains large numbers of glia. Overall, apical dendrites of medium and large pyramids are shorter and thinner and contain fewer tangential branches (fig. 4), than in the 'normal' 34-week-old preterm infant (fig. 3). Maturational "delay" in *Ne* is thus suggested by three morphologic features: minimal basilar dendritic growth, thin poorly branched apical dendritic shafts, and ab-

DEPTH (μm)

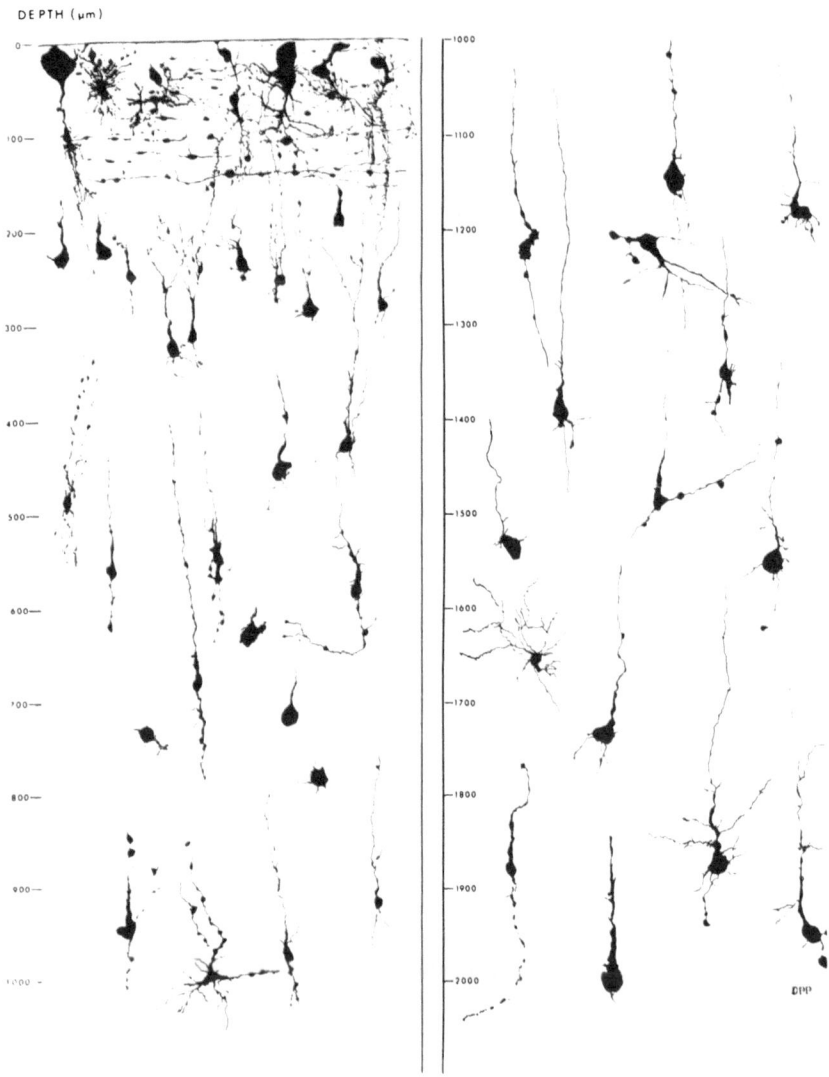

Fig. 5. Camera lucida drawings of neurons in calcarine cortex of *Gl* (34 weeks, c.a.) Rapid Golgi preparations. Apical dendrites of medium and large pyramidal neurons are generally short and thin. Several cells are shown with primitive trailing processes at mid-cortical and deep locations. Basilar dendrites are generally absent, and no cells exhibit dendritic spines. A double-tufted Cajal cell is shown at 500 μm. Compare with fig. 3. (from: Purpura, 1976.)

sence of dendritic spines. These observations suggest that dendritic development attained at the time of birth in *Ne* (28 weeks, c.a.) remained essentially unchanged during the subsequent 5-6 weeks postnatally during which time the infant experienced severe multiple cardiorespiratory and metabolic disorders.

Another preterm infant illustrates consequences of antenatal perturbations. Infant *Gl*, was born after 34 weeks gestation with a birth weight of 2040 g. Premature rupture of membranes occurred 3 weeks prior to delivery. The infant was febrile at birth and icteric at day one. Abnormal posturing was noted and continued into the second day along with bleeding diathesis, abdominal distention, and apneic spells. At 3 days the infant was cyanotic, had a decreased blood pressure, and continued to pass blood clots. Death occurred on the fifth day as a result of intracranial hemorrhage. The VEPs recorded the day before death consisted of grossly abnormal biphasic broad positive-negative potentials widely distributed over the scalp (Purpura, 1976).

Rapid Golgi preparations of visual cortex neurons in *Gl* are shown in the camera lucida drawings of fig. 5. Since the gestational age of *Gl* was similar to that of the infant whose cortex is considered 'normal'-for-dates (fig. 3) the morphologic characteristics of neurons can be compared in terms of their maturational status. The molecular layer of the visual cortex in *Gl* is markedly disorganized and contains poorly developed small and medium pyramidal cells. A curious finding is that many pyramidal neurons at different depths exhibit trailing processes that resemble the primitive trailing processes of postmigratory elements (Morest, 1969, 1970). It is as if these processes have not been resorbed after the neuron has attained its position in cortex. Indeed the persistence of trailing processes of neurons and the general lack of cortical organization in *Gl* (fig. 5) is suggestive of a failure of cortical neuronal maturation with minimal synaptogenesis.

The foregoing two examples emphasize profound alterations in neuronal morphogenesis in association with severe antenatal and perinatal disturbances that place the preterm infant at risk. Obviously the fact that the morphologic studies were possible at all indicates the severity of the risk factors. What is important to stress however, is the correlation between the abnormalities in VEPs and the underlying maturational delay and disturbance in neuronal development (Purpura, 1976). This suggests that significant alterations in the temporal pattern of functional maturation of VEPs may reflect interference with normal neuronal morphogenetic sequences. One conclusion that follows from present considerations is that severe cardiorespiratory distress, infections and metabolic and genetic disturbances of infancy that occur during the period of dendritic differentiation and dendritic spine differentiation can be expected to have profound effects on the development of neurobehavioral competency. Further evidence in support of the view that normal dendritic spine development is a critical factor in the genesis of normal cognitive functions will now be considered.

DENDRITIC SPINE DEVELOPMENT AND DENDRITIC SPINE DYSGENESIS IN MENTAL RETARDATION

The significance of dendritic spine development for the functional maturation of cortical synaptic organizations has been noted above and elsewhere

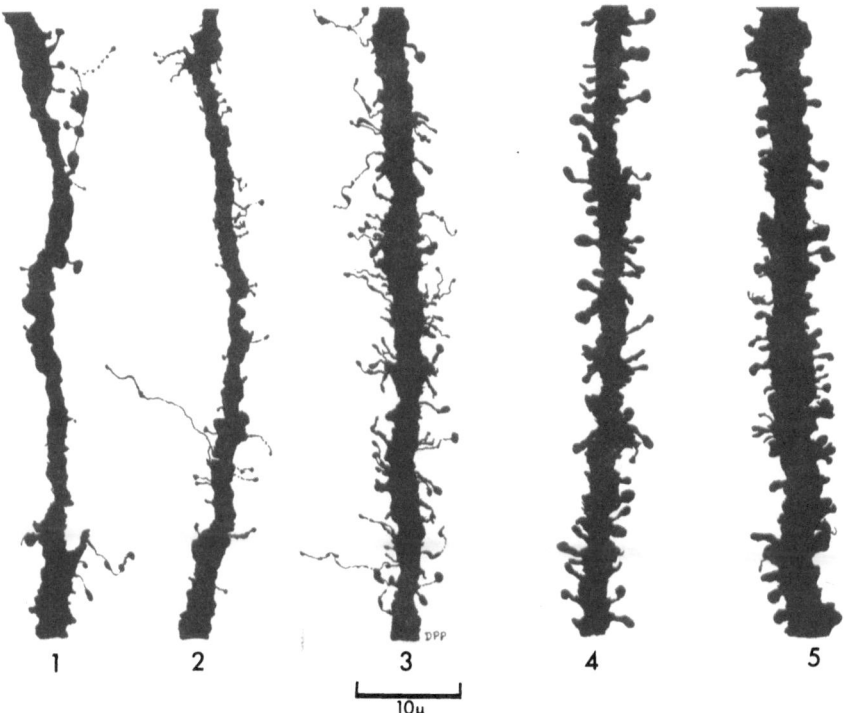

1   2   3   4   5

|_____|
   10μ

Fig. 6. Camera lucida drawings of proximal apical dendritic segments of motor cortex pyramidal neurons at different developmental stages. *1*, 18-week-old fetus; *2*, 26-week-old fetus; *3*, 33-week-old preterm infant; *4*, normal 6-month-old infant; *5*, normal 7-year-old child (accident case). Early phase of dendritic spine differentiation is associated with the development of long, thin filipodium-like spines and relatively few stubby and mushroom-shaped spines. The latter two types are prominent on proximal apical dendritic segments in the postnatal period and into early childhood. (From: Purpura, 1975b.)

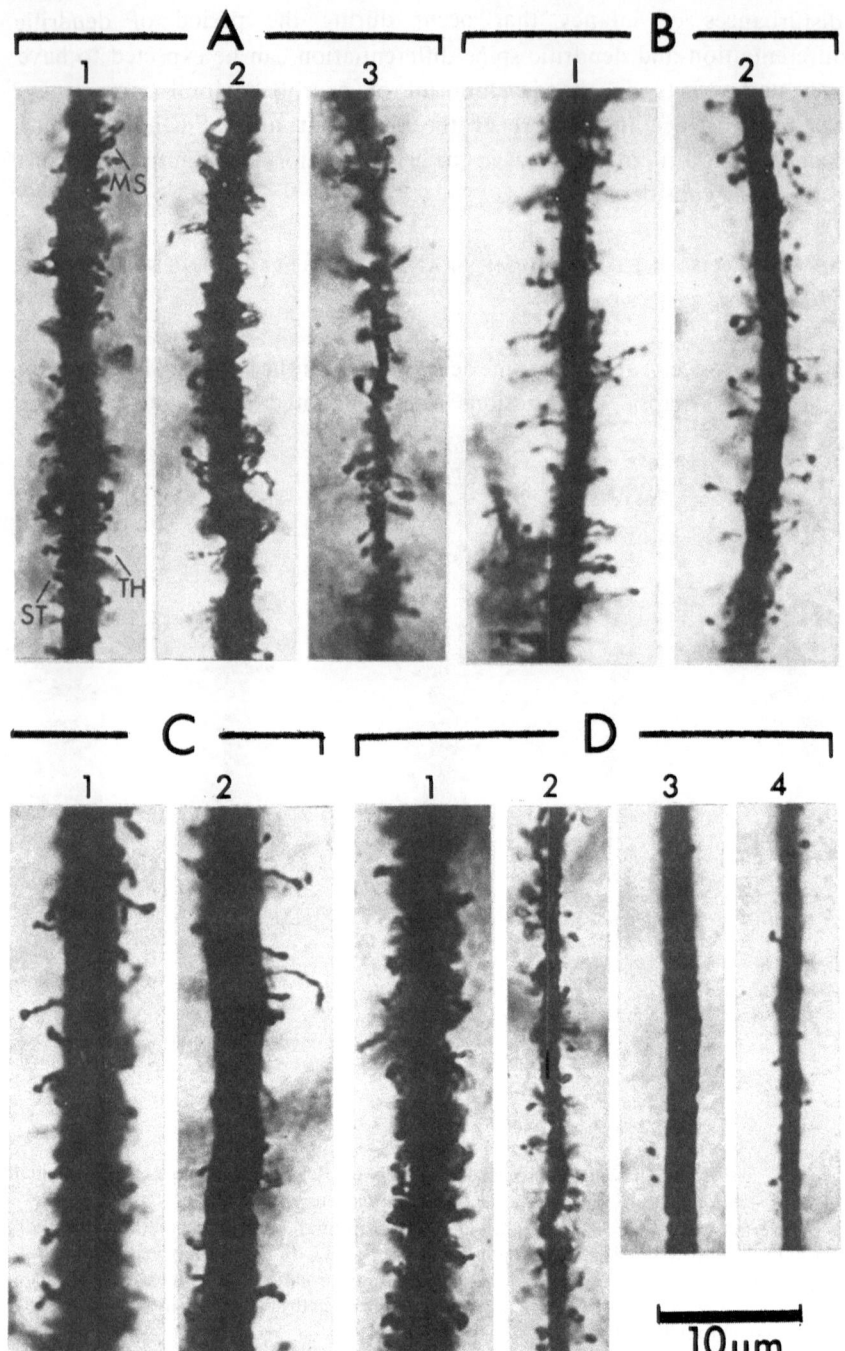

(Purpura, 1975b). It is necessary to consider briefly, the developmental sequence of dendritic spine morphogenesis in order to appreciate findings on dendritic spine alterations in subjects with profound unclassified mental retardation (Purpura, 1974).

It is instructive to examine the changes in proximal segments of apical dendritic shafts of motor cortex pyramidal neurons at different developmental stages (fig. 6) (Purpura, 1975b). Dendrites exhibit dramatic alterations in their associated fine processes during the latter half of gestation. In the 18-week-old human fetus a few long, thin filopodium-like processes with variable terminal expansions are detectable (fig. 6, *1*). By 26 weeks, c.a., more long thin processes appear (fig. 6, *2*) and these predominate by 33 weeks, c.a. (fig. 6, *3*). At this stage shorter and thicker spines are detectable in small numbers. During the postnatal period there is a striking change in the characteristics of dendritic fine processes on proximal segments of pyramidal neuron apical dendrites of motor cortex. This is evident in the changing proportion of long-thin spines to short-thick spines (Purpura, 1975b). Very thin and long spine-like processes seen in the preterm infant (fig. 6, *3*) are significantly reduced in the 6-month-old infant (fig. 6, *4*). At the same time there is an increase in the number of shorter, thicker spines, a trend that seems to persist throughout childhood (fig. 6, *5*).

Marin-Padilla (1972, 1974, 1976) first demonstrated that in infants with trisomic chromosomal aberrations known to be associated with mental retardation, dendritic spines were either very long, thin and tortuous, or short, thin and barely detectable in Golgi preparations. Variable spine loss and other degenerative processes of dendrites were also encountered in the

---

Fig. 7. Rapid Golgi preparations of dendrites of motor cortex neurons in normal and profoundly retarded subjects. *A.* Normal 6-month-old infant with negative neurologic history, postoperative death. *1, 2,* Proximal apical dendritic segments of medium sized layer V pyramidal neurons. Three basic types of dendritic spines are identified: thin (TH), stubby (ST), and mushroom-shaped (MS) spines. *3,* Basilar dendritic segment with a predominance of TH spines. *B 1,2,* Proximal apical dendritic segments of medium sized pyramids in frontal cortex of a 10-month-old retarde. Brain biopsy case. Abnormally long, thin spines predominate; many appear entangled. There is a marked reduction in MS and ST spines. *C 1,2,* Proximal apical dendritic segments of medium pyramidal neurons in motor cortex of a 3-year-old retardate. Note variability in extent of spine loss and distribution of abnormally long, thin spines. *D 1,2,* Proximal and distal segments of apical dendrites, from a normal 7-year-old child, accident case. *3,4,* Examples of apical dendritic segments from a 12-year-old profoundly retarded child. A few TH spines are seen, but otherwise there is almost complete absence of spines. (From: Purpura, 1974.)

studies of Marin-Padilla (1976). However, it is now clear from the author's Golgi studies of dendritic spines in a series of children with unclassified mental retardation (normal karyotypes) that dendrites of cortical neurons may exhibit many of the abnormalities described in trisomic chromosomal disorders. Examples of the morphologic features of dendritic spines in these cases are illustrated in the photomicrographs of fig. 7B, C, D3, D4. Dendrites from a normal 6-month-old infant and a 7-year-old child (accidental death) are also shown for comparison. The morphologic classification of dendritic spines employed by Peters and Kaiserman-Abramof (1970) has been found useful in identifying dendritic spine features in these cases.

The normal distribution of thin (TH), stubby (ST), and mushroom-shaped (MS) spines on dendritic segments from a 6-month-old infant with a normal brain is shown in the photomicrograph of fig. 7A. In fig. 7B dendritic segments from frontal cortex neurons of a 10-month-old retarded child (biopsy specimen) illustrate the marked abnormalities of the dendritic spines in the retardate. These abnormalities are seen in the reduction in short, thick spines and the presence of very long, thin tortuous filopodium-like processes with single or multiple terminal heads.

Additional examples of dendritic spine abnormalities from two other cases of profound mental retardation of unknown etiology are shown in fig. 7C, 7D3, D4. Proximal dendritic segments from a 3-year-old child with seizures and mental retardation show loss of spines and prominence of long, thin-necked spines with large terminal heads. Dendritic segments of cortical neurons from a 12-year-old profoundly retarded child (with a "developmental age" of about one year) are shown in fig. 7D3, D4). Only a few very thin-necked short spines remain on these segments.

The fact that spines may be completely stripped from dendrites in long-standing cases of profound mental retardation (fig. 7D3, D4) suggests that dendritic spine abnormalities observed in young retardates (fig. 7B, 7C) may reflect a progressive degenerative condition involving dendrites and dendritic

---

⟶

Fig. 8. Rapid Golgi preparations of dendritic shafts of cortical neurons obtained postmortem from the retarded child who underwent brain biopsy one year earlier. The abnormal dendritic spines found in this child at the time of biopsy are shown in Fig. 7B. At postmortem dendritic shafts in A, B and C exhibit a few small, thin spines. D, a bundle of dendrites is almost completely devoid of spines. E, dendritic systems are atrophic on cortical neurons. During the year from biopsy (Fig. 7B) until death there was a dramatic progression in dendritic spine abnormalities.

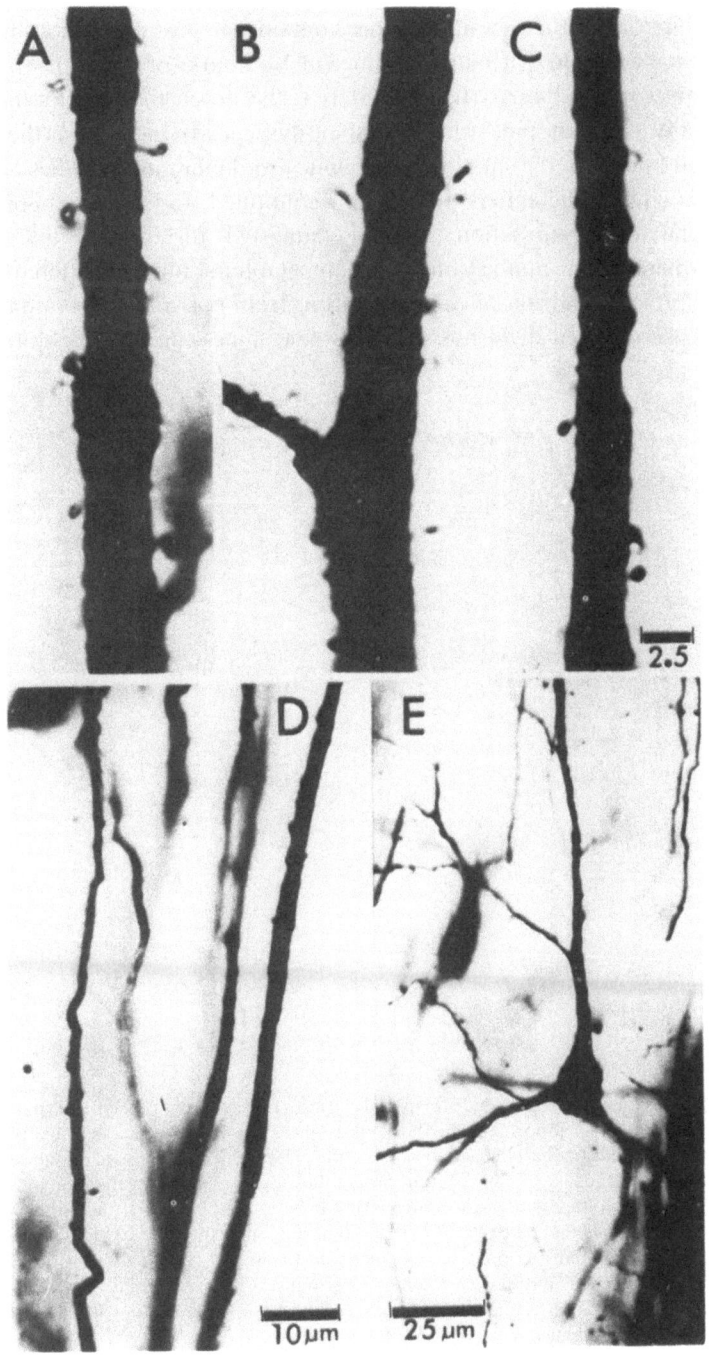

spines. Support for the view has now been obtained in post-mortem studies of dendrites of the child originally examined at 10-months of age at the time of diagnostic cortical biopsy (fig. 7B). It is to be noted that the finding of abnormally long thin spines (dendritic spine dysgenesis) (fig. 7B) was the only abnormal finding in the entire series of clinical and laboratory studies of this retarded child. Thereafter, the infant continued to show progressive neurobehavioral retardation and was admitted to a residential institution where he succumbed one year later. Golgi studies of different segments of apical dendrites of cortical neurons from post-mortem brain tissue in this case revealed dendrites with very few spines (fig. 8). The dramatic

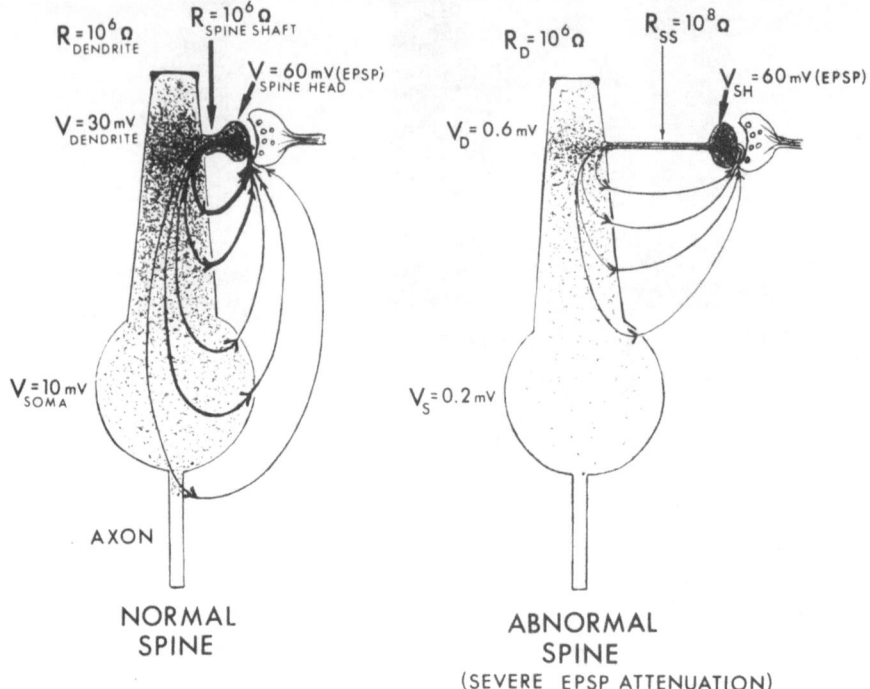

Fig. 9. Possible consequences of dendritic spine dysgenesis for excitatory postsynaptic potentials (EPSPs) generated at an abnormal, long thin spine (right) as compared with a "normal" stubby spine on a thick dendritic shaft (left). In both cases the presynaptic terminals are presumed to be normal and release the same quanta of transmitter following terminal depolarization by a spike impulse. The EPSP voltage at the spine head is 60 mV in both cases. However, due to the 100-fold increase in resistance of the spine stem (R_{ss}) of the abnormal spine there is a 100-fold attenuation of the EPSP at the dendritic input site as compared to the 50% attenuation in the normal situation. The overt effect of this severe attenuation is that very little of the EPSP generated at the spine head (V_{sh}) is seen at the soma-initial segment region. Approximate values of resistances and EPSPs adapted and modified from Rall (1974).

changes in dendritic spines which occurred in the one year period following biopsy (compare fig. 7B and fig. 8) are consistent with the deterioration that was evident clinically. The data illustrated in fig. 7B and fig. 8 represent the first serial study of Golgi impregnated neurons of the human cerebral cortex in a case of profound and progressive mental deficiency.

The morphologic data shown in figs. 7 and 8 point to abnormalities in dendritic spines, ranging from the failure of proximal segment apical dendritic spines to acquire 'mature' characteristics (i.e., ontogenetic changes from long thin spines to thick and stubby spines) to complete loss of spines. It cannot be concluded from Golgi studies alone that the complete stripping of spines from dendrites is evidence of total loss of synapses effected by these postsynaptic elements. While this seems likely it is conceivable that the loss of a dendritic spine might allow the presynaptic element to contact the postsynaptic dendritic shaft directly, bypassing the intermediary spine process. A consequence of this would be that the dendritic branch input site would 'see' the full value of the postsynaptic potential generated by ionic currents at the subsynaptic membrane following transmitter activation of postsynaptic receptor sites.

According to Rall (1974) one function of dendritic spines is to modulate synaptic input to dendrites of different dimensions. In essence, thick dendrites generally have short-thick spines and thin dendrites have long-thin spines. In Rall's view there is an "impedance match" between the spine-stem resistance and the dendritic branch input resistance. According to this hypothesis spine-neck resistance becomes an important variable in determining the extent to which postsynaptic potentials generated at the spine head electrotonically influence the membrane potential of the dendritic input branch and spike generating sites of the pyramidal neuron. For example, the 'normal' relationships between spine neck resistance of a thick spine on a thick dendritic branch might allow for about one-half of the EPSP generated at the spine head to be 'seen' at the dendritic branch input site (fig. 9, left). Assuming reasonable electronic decrement of this EPSP a significant fraction of the potential change is still detectable at the initial axonal segment, the low-threshold spike-generating site of the neuron. In contrast, under conditions of dendritic spine dysgenesis the long, thin neck of an abnormal spine would very greatly attenuate EPSPs. Considerably less of the EPSP generated at the spine head would be 'seen' at the dendritic branch input site due to the very high neck resistance of the long, thin spine (fig. 9, right). Even if this attenuation was less than 10% of 'normal' the combined effect of hundreds or thousands of spine synapses per neuron would significantly distort temporo-spatial patterns of excitatory input. On the other hand, if dendritic spines are eventually lost without concomitant loss of the presynaptic element the maintenance of synaptic transmission at those synapses (i.e. presynaptic element in direct synaptic relation with the dendritic shaft) could result in extraordinary excitatory drives, unattenuated and unmodulated by dendritic spines. This might explain the apparent 'paradox' of loss of dendritic spines and progressive severity of seizure manifestations in profoundly retarded young subjects. Clearly detailed electron microscope studies will be necessary to determine whether the loss of dendritic spines observed in terminal cases of profound mental retardation (fig. 7D3, D4; fig. 8) is associated with a parallel loss of dendritic synapses.

NEURONAL GEOMETRY DISTORTION AND ABERRANT SYNAPSE FORMATION IN HUMAN CEREBRAL CORTEX

There are several classes of developmental disorders in which neuronal abnormalities are expressed by increases in membrane surface area and the formation of aberrant 'dendritic' spine synapses (Purpura and Suzuki, 1976). These disorders include the various neuronal storage diseases that result from specific lysosomal hydrolase deficiencies (Desnick, Thorpe and Fiddler, 1976). It has long been known that neurons in these inherited metabolic disorders frequently exhibit 'ballooned' cell bodies and torpedo-like swellings of axons (Blackwood et al., 1963). These distortions are associated with progressive accumulation of uncatabolized substrates in the form of pathologic cytosomes, e.g., membranous cytoplasmic bodies (Terry and

---

Fig. 10. A-C, Rapid Golgi preparations of neurons in cerebral biopsy tissue obtained from a case with G_{M2}-gangliosidosis, AB variant. A, medium sized layer III pyramidal neuron with extensive well developed apical and basilar dendrites. A large thick process arises from the lateral basal pole of the cell and expands into a fusiform structure (meganeurite). B and C, small layer III pyramids. Meganeurites are seen arising from base of the cell bodies. Insets (b' and c') show that meganeurites are covered with spines. D, pyramidal neuron with a spine-bearing meganeurite (d') from a terminal case of unclassified neurolipidosis. E-H, superficial pyramidal neurons in classical Tay-Sachs disease (infantile G_{M2}-gangliosidosis, B variant); postmortem material. Dendrites are atrophic and cell bodies show some distention. Meganeurites of variable size and shape emerge from connecting stalks to cell bodies (asterisks). Axons (a) arise from the inferior pole of the meganeurites. The meganeurite in H has a secondary "neurite" that bears two filopodium-like processes (arrow). I-L, pyramidal neurons from a postmortem case of Hurler's disease (mucopolysaccharidosis, type I). Dendrites are atrophic and cell bodies are swollen. Meganeurites emerge from the basal pole of the cell. In cells I and J meganeurites exhibit spines (at arrow in J). Magnification bars in μm. (From: Purpura and Suzuki 1976.)

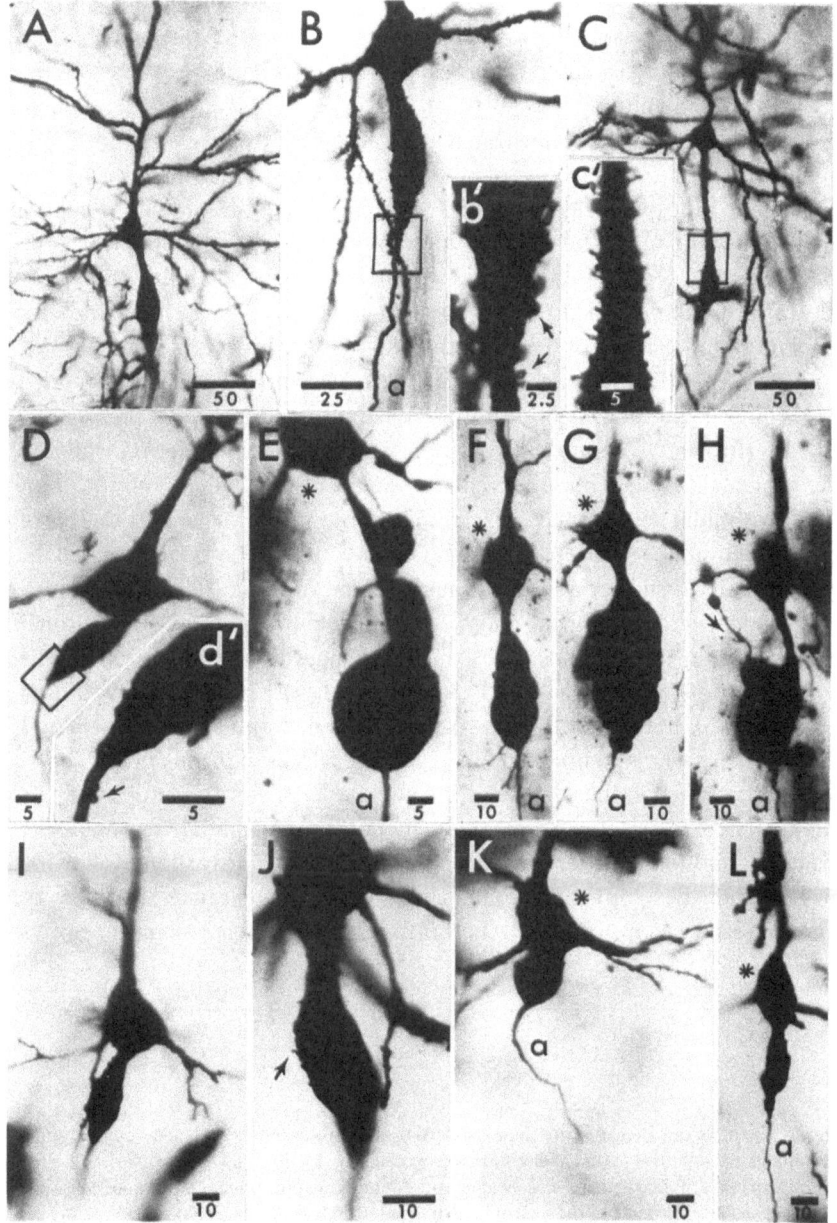

Weiss, 1963). Recent Golgi and electron microscope studies of neurons in these storage disorders have provided evidence that such distortions in neuronal geometry represent the addition of new dendritic-like surface membrane at aberrant sites on affected neurons. Examples of the morphologic features of human cortical neurons in different neuronal storage diseases are shown in fig. 10.

The pyramidal neurons illustrated in fig. 10A-C were revealed by the Golgi method applied to small pieces of middle frontal gyrus obtained at the time of diagnostic biopsy carried out on a 14-month-old infant with seizures, myoclonus to sound and mild motor retardation. Studies established the diagnosis of G_{M2}-gangliosidosis. AB variant (de Bacque, et al., 1975). Each of the three pyramidal neurons in fig. 10A-C shows a large thick process emerging from the base of the cell body. This massive process has been termed a *meganeurite*. (Purpura and Suzuki, 1976). Meganeurites may be studded with spines some of which are in postsynaptic relationship to presynaptic endings (fig. 11). A point of major importance, as shown in fig. 10, is that in the gangliosidoses such as classical Tay-Sachs disease, small and medium-sized pyramidal neurons have bizarre and pleomorphic meganeurites. It is of interest that in many instances the total surface area of a meganeurite may be greater than the surface area of the neuron from which it arises (fig. 10 E-H).

Meganeurite development may be responsible for the onset of neuronal dysfunction in the gangliosidoses since meganeurites form form at a time when neurons are otherwise well preserved (cf. fig. 10A-C) (Purpura and Suzuki, 1976). Meganeurites markedly increase membrane surface area and thus decrease whole neuron (input) resistance. Because of this postsynaptic potentials elicited in the somadendritic membrane of storage neurons will be attenuated to a greater degree than in normal pyramidal neurons. The location and magnitude of neuronal geometry distortion following meganeurite development combine to negate integrative features of somadendritic synaptic inputs. Aberrant synapses that form on meganeurites

Fig. 11. Example of a meganeurite spine (at arrow) in synaptic contact with an axonal terminal containing presynaptic vesicles. (Cerebral biopsy tissue, G_{M2}-gangliosidosis, AB variant). Note prominent postsynaptic density. The meganeurite contains membranous cytoplasmic bodies and other organelles. Elements of the neuropil are normal in appearance. × 24,000. (From: Purpura and Suzuki, 1976.)

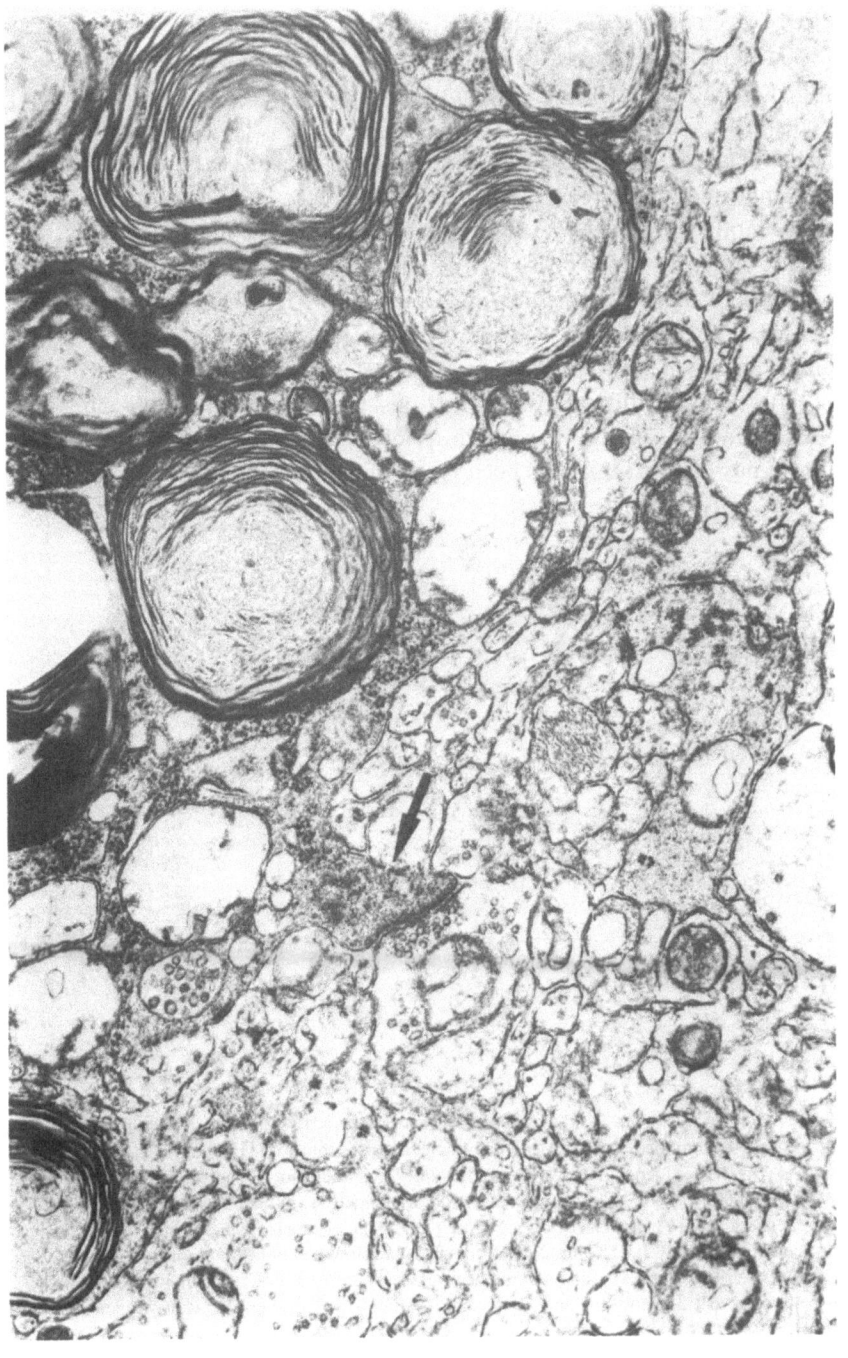

74 DOMINICK P. PURPURA

can be expected to produce additional pathophysiologic effects since these are in a particularly favorable location, close to the initial axonal segment, to influence neuronal excitability to a much greater extent than 'normal' remote axodendritic synapses.

The formation of meganeurites in pyramidal neurons can be expected to have important consequences apart from proposed alterations in electrical and synaptic properties. When a pyramidal neuron undergoes a massive increase in size it must do so at the expense of surrounding neuropil. A composite of camera lucida drawings of medium-sized pyramidal neurons in classical Tay-Sachs disease serves to illustrate this point (fig. 12). By compromising pre- and postsynaptic elements of the neuropil meganeurites may effectively 'deafferent' cortical neurons. Hence the dendritic atrophy and loss of spines in cortical neurons in terminal stages of classical Tay-Sachs disease may be due, in part, to occupation of neuropil space by meganeurites and their Tay-Sachs structures.

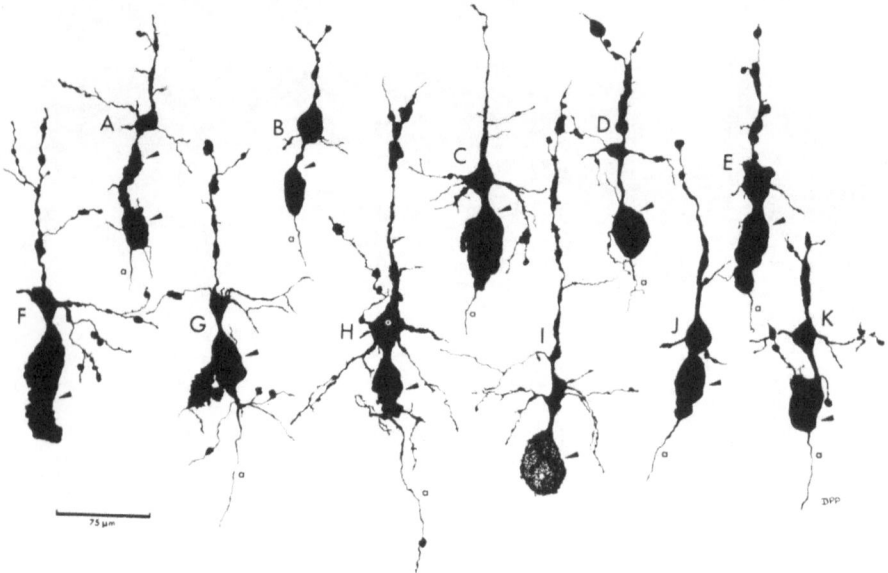

Fig. 12. Composite camera lucida drawings of layer II and III small and medium pyramidal neurons in a terminal case of classical Tay-Sachs disease (infantile G_{M2}-gangliosidosis, B variant). Arrow heads identify primary and secondary meganeurites. Dendrites are atrophic and devoid of spines. Some cell bodies are markedly distended (a, axons). Cells *A, C* and *G* have prominent secondary meganeurites. Secondary neurites emerging from meganeurites are seen in most cells, particularly *D, G, H* and *K*. Cells *C* and *K* are shown in photomicrographs of Fig. 10*G* and *H*, respectively. Note that in most instances the surface area of the meganeurite exceeds the total somadendritic membrane surface area of the neuron that gives rise to the meganeurite. (From: Purpura and Suzuki, 1976.)

The formation of massive, spine-bearing structures (meganeurites) in neurons in human ganglioside storage diseases suggests that the region of the neuron that gives rise to the meganeurite and the meganeurite itself are sites of continued neuronal growth in otherwise mature cells. Evidently meganeurite development involves restricted reactivation of embryonic growth me-

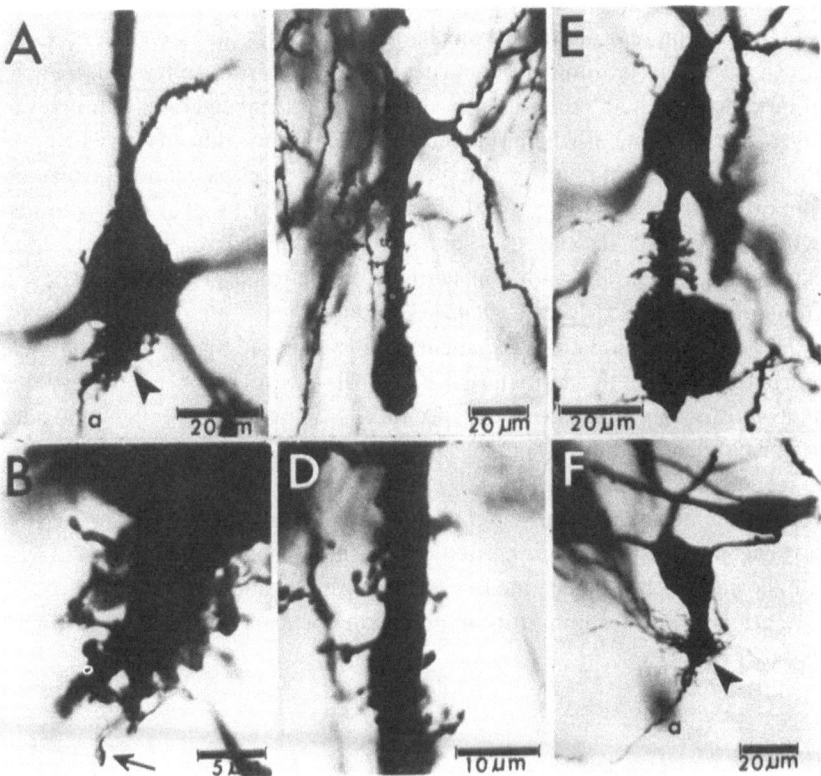

Fig. 13. Rapid Golgi preparations of cortical neurons in feline G_{M_1}-gangliosidosis. A, Medium-sized pyramidal cell of the entorhinal cortex, meganeurite identified, at arrowhead. B, enlargement of meganeurite in A to show tangle of neurites, one of which has an expanded terminal head (at arrow). C, A neuron exhibits a long meganeurite that gives rise to multiple neuritic growth processes, seen at higher magnification in D. E, Another small pyramidal neuron of the cerebral cortex has a meganeurite that ends in a huge globular process. Outgrowth of neurites is prominent in the connecting stalk of the meganeurite. F, Granule cell of the fascia dentata of the hippocampus. Meganeurite, at arrowhead, exhibits extensive neuritic outgrowth. (a, axon). (From: Purpura and Baker, 1977.)

chanisms of the neuron. Some support for this view has come from Golgi studies of mature cortical neurons in a feline model of human G_{M1}-gangliosidosis (Purpura and Baker, 1977). Since outgrowth of neurites is a major morphologic sign of embryonic neurons attempts were made to identify neuritic processes in meganeurites of cortical neurons in the feline mutant.

Aberrant growth processes were observed in several varieties of cortical neurons in the case of feline G_{M1}-gangliosidosis. Of particular importance was the finding that many neurons had meganeurites that gave rise to G_{M1}-gangliosidosis large number of neurites (fig. 13). Neurites seen as outgrowths of meganeurites had 'medusa' like' characteristics and emerged as tangles of fine processes some of which showed expanded terminal heads (fig. 13A and B). A wide variety of neurites and other growth processes (spinules, spicules, filopodia-like elements) occurred on larger meganeurites of small pyramidal neurons (fig. 13C, D and E). Neurites also exhibited beading and tufted expansions, which are common features of neuronal growth processes in Golgi preparations (Morest, 1969, 1970; Purpura, 1975a).

The demonstration that meganeurites in feline G_{M1}-gangliosidosis have morphologic growth characteristics typical of differentiating embryonic neurons raises the question of the specific G_{M1}-gangliosidosis of gangliosides in normal neuronal development and synaptogenesis (Fishman and Brady, 1976). While the findings summarized in fig. 13 provide evidence linking G_{M1}-ganglioside accumulation Fishman neurite formation in well-differentiated cortical neurons in an inherited metabolic disorder it remains G_{M1}-gangliosidosis be determined whether gangliosides influence process formation in normal immature mammalian cortical neurons (Purpura and Baker, 1977).

SUMMARY AND CONCLUSIONS

Golgi studies of neurons in immature human cerebral cortex provide an approach to the analysis of high-risk factors in cortical development that emphasizes several aspects of neuronal developmental pathobiology: The consequences of aberrant dendritic differentiation and dendritic spine development and the effects of ganglioside accumulation on the geometry and synaptic organization of neurons in lysosomal storage diseases.

Cortical neuronal development after the 24th week c.a., is characterized largely by dendritic growth and dendritic spine differentiation. Basilar dendritic proliferation is the most significant component of this developmental phase. Dendritic spines become prominent between dendritic weeks, first as long, thin processes. By 6 months postnatally, spine typology is established on cortical pyramidal neurons. retardation of dendritic growth and differentiation and failure of spine maturation are prominent features of severe prolonged cardiorespiratory distress, antenatal infection and metabolic disturbances in preterm infants 28-33 weeks, c.a.

Dendritic spine dysgenesis is identified as a major micropathologic factor in the etiology of profound unclassified mental retardation with or without seizures. Cortical neurons in this condition show marked reduction in dendritic spine density and prominence of long, thin spines with large terminal heads.

Attenuation or loss of dendritic dendritic spines and reduction in dendritic membrane surface area is the usual response of cortical neurons to major developmental perturbations. In marked contrast ganglioside storage disorders are associated with marked increases in membrane surface area due to the formation of massive dendritic-like structures (meganeurites) between the cell body and axon of pyramidal neurons. Meganeurites are frequently covered with spines and spine-synapses. Evidence is presented that meganeurites in a feline model of human G_{M1}-gangliosidosis give rise to extensive outgrowths of neurites. Meganeurites of well differentiated neurons in some of the ganglioside storage disorders appear to retain embryonic growth characteristics.

The foregoing studies of immature human cerebral cortex provide an approach to the analysis of cortical development that emphasizes alterations in dendritic systems and cell geometry as major factors in the neuronal pathobiology of neurobehavioral disorders of infancy and childhood.

REFERENCES

Blackwood, W., Meyer, A., McMenemey, W.H., Normal, R.M. and Russell, D.S. (Eds.): Greenfield's Neuropathology. (Williams and Wilkins, Baltimore 1963), pp 415-430.

de Bacque, C.M., Suzuki, K., Rapin, I, Johnson, A.B., Whethers, D.L. and Suzuki, K.: G_{M2}-gangliosidosis, AB variant: clinico-pathological study if a case. Acta neuropath. (Bull.) 33: 207-226 (1975).

Desnick, R.J., Thorpe, S.R. and Fiddler, M.B.: Toward enzyme therapy for lysosomal storage disease. Physiol. Rev. 56: 57-99 (1976).

Fishman, P.H. and Brady, R.O.: Biosynthesis and function of gangliosides. Science 194: 906-915 (1976).

Marin-Padilla, M.: Structural abnormalities of the cerebral cortex in human chromosomal aberrations. A Golgi study. Brain Res. 44: 625-629 (1972).

Marin-Padilla, M.: Structural organization of the cerebral cortex (motor area) in human

78 DOMINICK P. PURPURA

chromosomal aberrations. A Golgi study. 1. D_1 (13-15) trisomy. Patau syndrome. Brain Res. 66: 373-391 (1974).

Marin-Padilla. M.: Structural organization of the cerebral cortex (motor area) in human syndrome. A Golgi study. J. Comp. Neurol. 167: 63-81 (1976).

Morest. K.: The differentiation of cerebral dendrites. A study of the post-migratory neuroblast in the medial nucleus of the trapezoid body. Z. Anat. Entwicklungsgeh. 128: 271-289 (1969).

Morest, K.: A study of neurogenesis in the forebrain of opossum pouch young. Z. Anat. Entwicklungsgeh. 130: 265-305 (1970).

Peters, A. and Kaiserman-Abramof, I.R.: The small pyramidal neuron of the rat cerebral cortex. The perikaryon, dendrites and spines. Am. J. Anat. 127: 321-356 (1970).

Purpura. D.P.: Analysis of axodendritic synaptic organizations in immature cerebral cortex. Ann. N.Y. Acad. Sci. 94: 604-654 (1961).

Purpura. D.P.: Comparative physiology of dendrites; in Quarton, Melnechuk and Schmitt. The Neurosciences: A Study Program, pp. 308-323 (Rockefeller Univ. Press, New York 1967).

Purpura. D.P.: Dendritic spine "dysgenesis" and mental retardation. Science 186: 1126-1128 (1974).

Purpura. D.P.: Normal and aberrant development in the cerebral cortex of human fetus and young infant; in Buchwald and Brazier. Brain Mechanisms in Mental Retardation. pp. 141-169 (Academic Press. New York 1975a).

Purpura. D.P.: Dendritic differentiation in human cerebral cortex: Normal and aberrant developmental patterns; in Kreutzberg, Physiology and Pathology of Dendrites. Advances in Neurology 12: pp. 91-116 (Raven Press. New York 1975b).

Purpura. D.P.: Morphogenesis of visual cortex in the preterm infant; in Brazier. Growth and Development of the Brain. pp. 33-49 (Raven Press. New York. 1975c).

Purpura. D.P.: Structure-dysfunction relations in the visual cortex of preterm infants; in Brazier and Coceani. Brain Dysfunction in Infantile Febrile Convulsions. pp. 223-240 (Raven Press. New York 1976).

Purpura. D.P.: Developmental pathobiology of cortical neurons in immature human brain; in Gluck. Intrauterine Asphyxia and the Developing Fetal Brain. (Year Book Medical Publishers. Chicago. Ill. 1977) in press.

Purpura. D.P. and Baker. H.J.: Neurite induction in mature cortical neurons in feline G_{M1}-ganglioside storage disease (1977) in press.

Purpura. D.P.. Shofer. R.J.. Housepian. E.M. and Noback. C.R.: Comparative ontogenesis of structure-function relations in cerebral and cerebellar cortex. Prog. Brain Res 4: 187-221 (1964).

Purpura. D.P. and Suzuki. K.: Distortion of neuronal geometry and formation of aberrant synapses in neuronal storage disease. Brain Res. 116: 1-21 (1976).

Rall. W.: Dendritic spines. synaptic potency and neuronal plasticity; in Woody. Brown, Crow and Knipsel, Cellular Mechanisms Subserving Changes in Neuronal Activity. (Brain Inf. Service. UCLA. Los Angeles. 1974).

Sidman. R. and Rakic. P.: Neuronal migration with special reference to developing human brain: A review. Brain Res. 66: 1-36 (1973).

Terry, R.D. and Weiss. M.: Studies in Tay-Sachs disease II. Ultrastructure of cerebrum. J. Neuropath. exp. Neurol. 22: 18-55 (1963).

CHAPTER FIVE

ASSESSMENT AND SIGNIFICANCE OF BEHAVIOURAL STATES

Heinz F.R. Prechtl

1. THE STATE CONCEPT

The human newborn infant is equipped with a large set of interrelated abilities and competences, not only physiologically but also in his behaviour. A central aspect of this biologic organization is the existence of the so-called behavioural states. They are a convenient categorization of many conditions which are relatively stable over time and which can be easily recognized if they occur again. The main contributions came from two sources: first, from observational studies (Wolff, 1959, 1966), and second, from polygraphic recordings which had their origin in an extension of EEG recording to other physiologic variables such as respiration, EKG, EMG of the chin muscle, and oculogram. The latter contribution was particularly promoted by the new interest in the study of the ontogeny of sleep states in the mid-1950's. Many of these later investigations focused exclusively on sleep cycles, neglecting the awake states and sometimes leading to erroneous conclusions.

A difficult, and still unsolved problem due to lack of general agreement, is the categorization of the neonate's behavioural states. If the number of arbitrarily selected criteria is too large, too many states or "indeterminate states" will frequently be found, because not all criteria are present at every moment. If the number of selected criteria is too small, the necessary discrimination is lost. Furthermore, the identification of states depends on the time window which is employed. Twenty seconds, thirty seconds, one minute or three minutes, respectively, have been used by different investigators which make it difficult to compare results. Whatever the time window, the selected state criteria should be applied consistently for all states. This has not always been the case. Parmelee et al. (1967) and later Anders (1974) used five criteria for the sleep state identification. In order to escape the problem of designating too many epochs as "indeterminate sleep", four, then three, out of five criteria were used to identify a sleep state. Other criteria were employed for awake states. An attempt to standardize terminology, criteria and recording techniques for polygrams of newborn infants (Anders, Emde and Parmelee 1971) was unfortunately doomed to fail because of lack of a common concept, and unfamilarity not only with the principles of signal analysis but also with the requirements of technical equipment.

An alternative approach is a taxonomy of behavioural states based on criteria applicable both to direct observation and to a polygraphic record (Prechtl and Beintema 1964; Prechtl et al. 1968). Four criteria are sufficient to distinguish reliably five different behavioural states. The criteria are consistently applied to all states, namely: eyes open or closed, respiration regular or irregular, and absence or presence of continuous gross movements and of vocalization. Table I defines each state as a mutually exclusive, four dimensional vector (Prechtl 1974).

Table 1. Vectors of behavioural states.

	Eyes open	Respiration regular	Gross movements	Vocalization
state 1	−1	+1	−1	−1
state 2	−1	−1	0	−1
state 3	+1	+1	−1	−1
state 4	+1	−1	+1	−1
state 5	0	−1	+1	+1

signs: +1 = true; −1 = false; 0 = true or false

2. CLINICAL SIGNIFICANCE OF POLYGRAPHIC ANALYSIS OF STATES

From neurologic examinations of newborn infants (Prechtl and Beintema 1964), it appeared that some infants with neurologic deviations had unusual sequences of behavioural states during the course of the examination. The question arose as to whether their reactivity to stimulation is impaired or whether they have a dysregulation of their state cycles. This problem has been solved by long polygraphic recordings followed by visual and/or computer analysis (Prechtl et al. 1969). There are several aspects of states which are important.

a) The state profile
The first step in the analysis of a polygram is the determination of the state

STATE-PROFILE

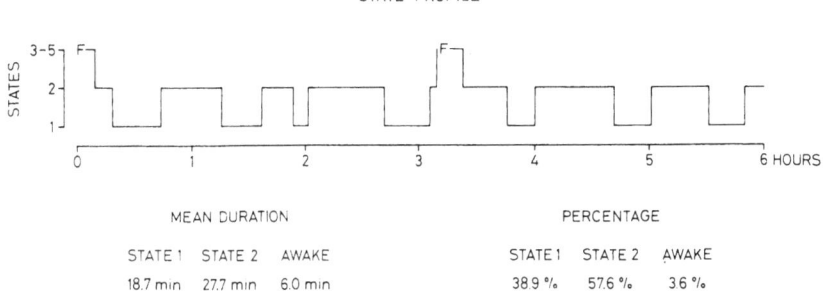

Fig. 1. Example of a state profile from a six-hour-recording of an eight-day-old full-term infant. F stands for feeding. Note variations of duration of state epochs. (Mean and percentage from this record.)

transitions. This can be done by visual inspection of the recording using the state criteria. From these data a state profile (fig. 1) can be easily constructed. The next step is the computation of the mean duration and percentage of each state during the whole record. This method is simple and does not require automatic analysis.

It is important, that a change of state is accepted only if it lasts three minutes or longer. It is also crucial to analyse recordings of several hours duration. An analysis of mean duration and percentage of states derived from only one or two sleep-cycles will in most cases lead to an unrepresentative state assessment. Intra-individual variability in state durations of normal newborns between feeds is typically so large that an arbitrary selection of two contingent states will constitute an inaccurate index of a basic biologic process. Therefore, we carry out six hour recordings as a routine procedure. Only then are sufficient data available for statistical analysis.

In a series of studies mean durations and percentage distributions of the various states have been investigated in healthy low-risk infants and in a variety of abnormal babies or infants belonging to groups at-risk for brain dysfunction.

In the healthy infants an important fact was found. State 1 is a stable state, hardly influenced by stressful conditions. A comparison of the states of 20 infants during their first hours after birth with the states on their fifth day did not reveal a difference in state 1 but the infants were more awake on the first day and had therefore less state 2 than on the fifth day (Theorell et al. 1973). This exchange between state 2 and the awake states has been also found when normal newborn infants were sleeping in a baby chair with an angle of 30-40°

from the horizontal. The same infants were more awake at the expense of state 2 than during their sleep in a horizontal supine position (Casaer et al. 1973). Stress induced by circumcision led to a similar phenomenon, again using the same infants as their own controls and comparing three 1-hour periods surrounding the "stress" of circumcision (Anders and Chalemion 1974). State 2 and wakefulness seem to be easily exchangeable, while the more complex and homeostatically controlled state 1 is less vulnerable in contrast to what would have been expected on theoretical grounds.

Similar results come from studies on abnormal infants. In six hour recordings Down's syndrome infants spent $25^{\circ}_{\ 0}$ in awake states as compared to $10^{\circ}_{\ 0}$ in healthy infants of the same age, again exclusively at the expense of state 2 (Prechtl et al. 1973). In the same investigation it was found that neonates with hyperbilirubinemia (10 mg $^{\circ}_{\ 0}$ or more unconjugated bilirubin) spend more time in state 2 and are correspondingly less awake.

State 1 does not remain unaffected in all conditions. Anti-epileptic medication given for neonatal convulsions significantly increased state 1 and reduced state 2. It is possible that the prolonged highly regular respiration, and in addition the disappearance of startles, indicates an abnormal state rather than a physiologic state 1 (Prechtl et al. 1973). In infants of diabetic mothers, Schulte et al. (1969) reported a normal sleep cycle length and an increase in the relative amount of REM sleep with a significant reduction in the duration of state 1 (non-REM-sleep). A significant reduction in mean duration and percentage of state 1 has been observed in infants of mothers with hypertension and toxaemia (Huisjes et al. 1975). This reduction of state 1 was only correlated with low oestrogen values in the mothers' urine during the last few weeks of pregnancy. It did not show any relation to pre- or full-term birth, nor to adequate and small-for-dates birth weight. All infants were recorded at about their expected term.

In an extensive study of states in the neonatal period in infants with severe brain malformations and/or chromosomal anomalies, Monod and Guidasci (1976) reported that in general "newborn babies with brain malformation appeared to be poor sleepers. Amount of wakefulness is higher than in normal neonates. Some babies were nearly insomniac." A special merit of this study is the comparison of the results obtained by different state criteria. The great discrepancy between the findings illustrates strikingly the dilemma of the different state categorizations which make data from different authors incomparable.

Despite these shortcomings, the assessment of state (mean durations and percentages) measured by hand from transition to transition is simple and reliable within different techniques.

Recently Anders (1974) suggested scoring the whole polygram visually in 30 second epochs, then putting these subjective estimates on punch cards and carrying out digital computer analysis. The shortcomings of this technique are obvious. No objectivity is gained and the assessment of the average durations by the binary auto-correlation (Globus 1970), which Anders advocates is invalid for relatively short recordings between two feeds.

b. The organization of states
Besides the more elementary quantitative aspects of states as discussed in the previous chapter, the complex organization of the various state criteria and their concomitants is an important expression of the neural regulatory mechanisms of states as well as of their dysfunctions. This has been clearly pointed out by Monod et al. (1967) who measured by hand the relationships between various parameters such as EEG types, EMG, regularity of respiration and absence of motility in healthy infants. They also studied the deviations of these relationships in pathologic newborns. Prechtl (1968) introduced direct computer analysis of analog polygraphic signals by analog-to-digital conversion and processing into compiled polygrams (fig. 2). This step towards an objective statistical description of many variables and their parameters in a consecutive three minute epoch analysis eliminated the need for subjective scoring systems. Not only was a sophisticated form of data reduction obtained in this way, but it has permitted a complex mathematical analysis of the relationship between the variables (Prechtl et al. 1969), fulfilling the recent but superfluous request of Anders (1975, p. 57). Multiple regression and discriminant analysis distinguished significantly between healthy and neurologically impaired groups of infants. A shortcoming of this method is the application of mathematical models, designated for independent data, to serially correlated data (the consecutive three-minute epochs). Groups of infants were compared on non-parametric difference tests but the individual case could not be reliably interpreted. A break-through in the polygraphic analysis was made by Scholten (1976) who replaced the epoch analysis by a running window analysis and obtained continuous signals of respiratory rate, heart rate and EMG-activity which could then be cross-correlated. By "running window" is meant: counting the events such as breaths and heart beats within a 20 second epoch. Then this window is shifted by 20 msec. and a count of these events is made again and so on. Considering

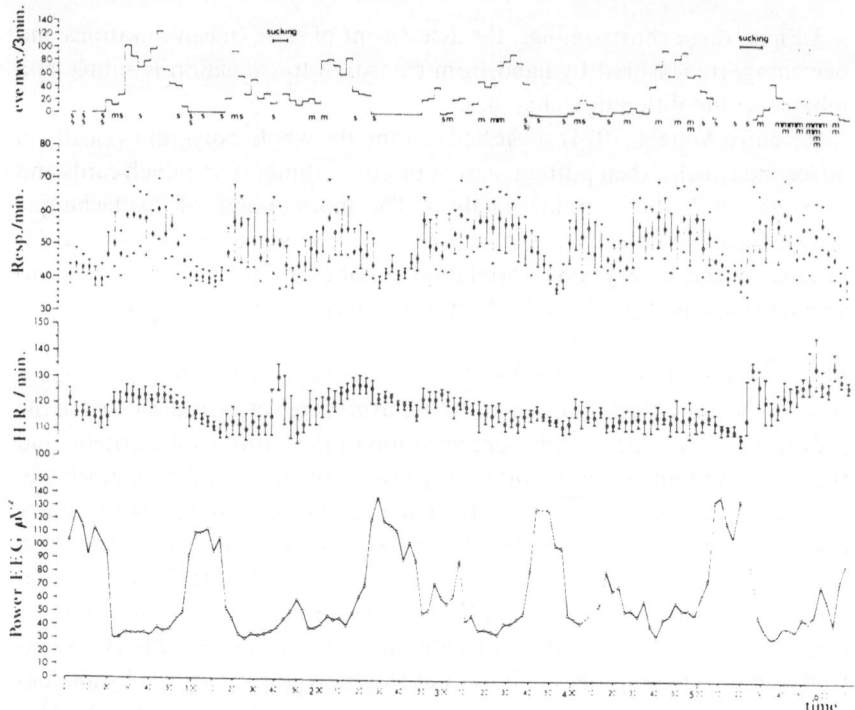

Fig. 2. Plot of a computer analysed compiled polygram with epoch analysis per consecutive three-minute epochs. Respiration and heart rate are given as medians and interquartile ranges of breath/breath heart beat/heart beat intervals. Rapid eye movements were counted per three minutes and EEG is indicated as the mean square voltage per three minutes. Normal full-term infant, eight days old. (From Prechtl 1968.)

the relatively slow processes, this approach is almost a continuous estimation of respiratory rate and heart rate. The averaged EMG activity of several muscles is assessed with the same window. An example of the results obtained in this way is presented in fig. 3.

From these data the following values are calculated, for example per ten minute time windows:
1) The means and variances of breathing rate, heart rate and EMG-activity.
2) The cross-correlation between each pair of the three signals.
3) The value of the phase shift at the extreme of the cross-correlation coefficient.

The ten minute window is then shifted by five minutes, the computations are repeated and the results plotted. There is thus an overlap of five minutes in

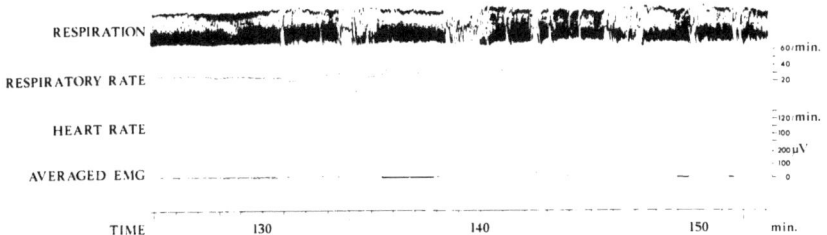

Fig. 3. Example of a pen write-out of the respiratory-signal, the counted respiratory rate and heart rate as well as the averaged EMG employing a running window of 20 seconds. (From Scholten 1976, by kind permission.)

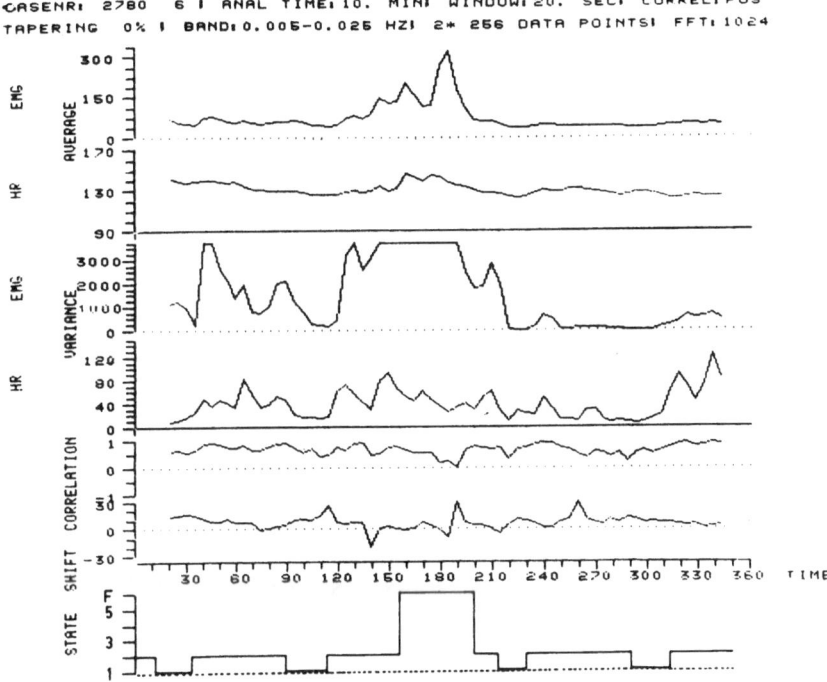

Fig. 4. Computer plot of a compiled profile of EMG and heart rate averages and their variances plotted every five minutes. Below cross-correlation coefficients and phase of their extremes. At the bottom, state profile. F indicates feeding. five-days old healthy full-term newborn. (From Scholten 1976, by kind permission.)

each of the estimated values. Examples of the results obtained with this time window are given in figs. 4 to 6. Shorter windows can also be used which increase the resolution but blur the slow trends. The technique can cater for other variables such as the number of rapid eye movements and EEG power in relation to the previously mentioned signals.

The positive correlation between motility (EMG-average) and heart rate is illustrated in fig. 4. Every movement (increased EMG) is accompanied by a rise in heart rate. Furthermore, a trend in the heart rate can be clearly seen. From the onset of the recording, which was a few minutes after the end of a feeding, there is a steady decline in the heart rate which repeats itself after the feed in the middle of the recording towards the end at 360 minutes. During the feeding, EMG-activity and heart rate lose their positive correlation.

The relationship between EMG-activity and respiratory rate is given in fig. 5. A negative correlation exists during state 2 but not during state 1. Gross movements and especially stretches lead to apnoeas and consequently a reduction in the rate of breathing. During state 1 only startles occur which are followed by a brief pause in respiration. Therefore the variance of the re-

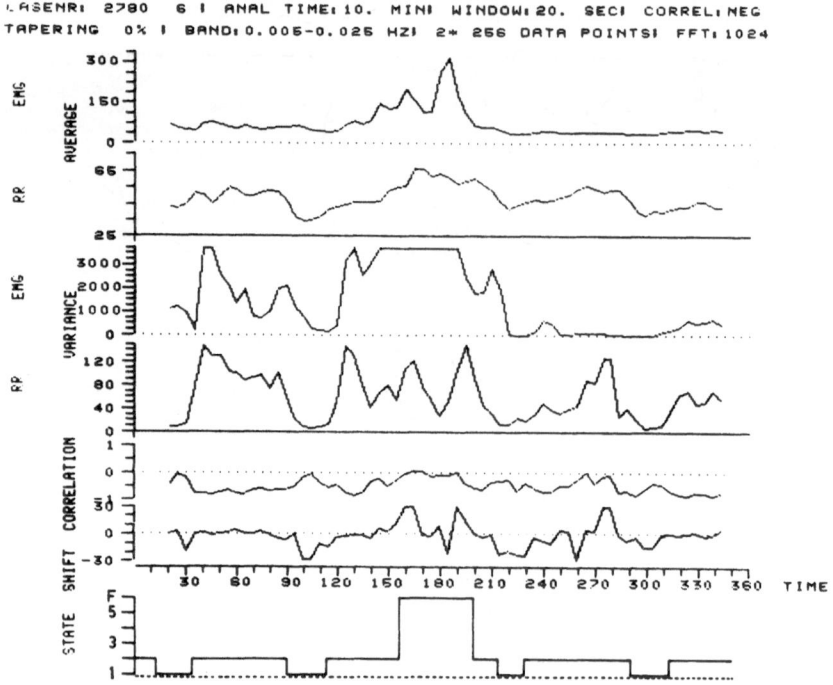

Fig. 5. Same graphical presentation as in figure 4 but for EMG and respiratory rate respectively. (From Scholten 1976, by kind permission.)

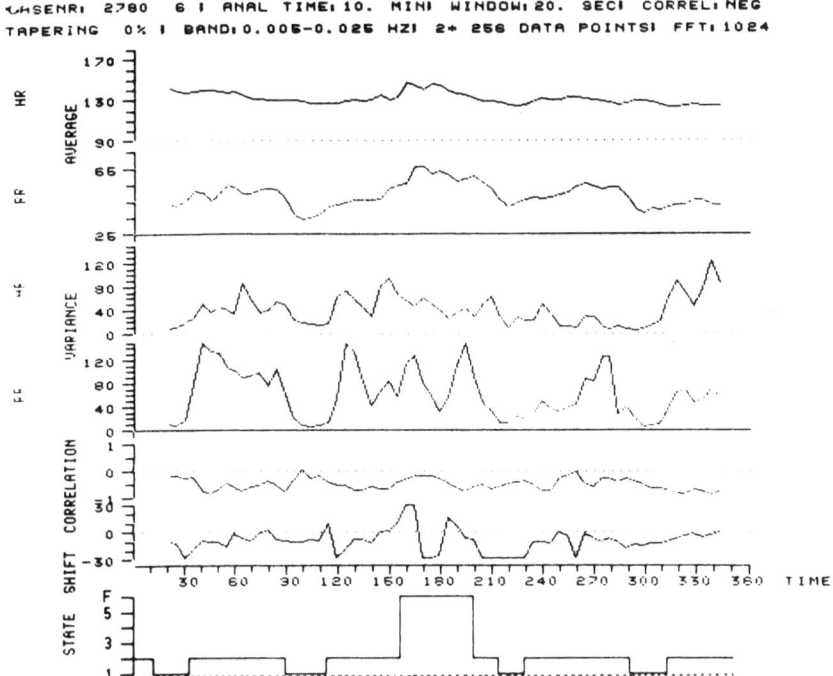

Fig. 6. Same graphical presentation as in figure 4, but for heart rate and respiratory rate respectively. (From Scholten 1976, by kind permission.)

spiratory rate remains low. The tonic activity during state 1 has no marked effect on the respiration.

The last pair of variables to be shown are heart rate and respiratory rate (fig. 6). Their correlation is negative and can be explained by the previously described relationship of each of the two variables with motility.

In all three figures the cross-correlation above a value of 0.2 are significantly different from zero at a 5% level of confidence. When the variance is simultaneously very low, correlation coefficients higher than 0.2 should be ignored.

With this newly developed technique a comprehensive analysis of six-hour recordings is possible. It shows the dynamics of the relationships between different signals in relation to state cycling and is thus a qualitative and quantitative description of the organization of behavioural states. It also shows trends between feeds and trends within single state epochs. Dysfunction of the regulatory mechanisms can be easily detected in changes

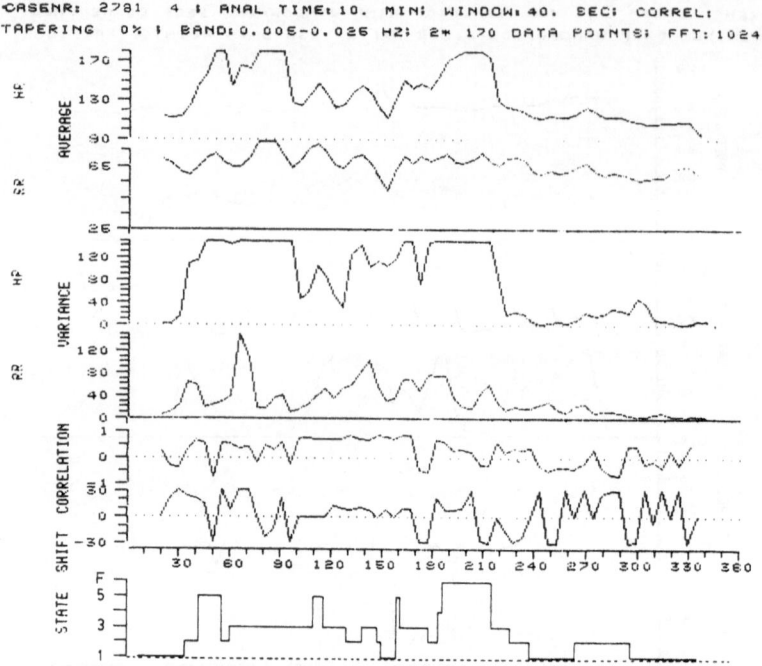

Fig. 7. Values for heart rate and respiratory rate from a four-days-old abnormal full-term infant. (From Scholten, unpublished, by kind permission.)

of the cross-correlations and the phase relations. An example of altered organization in an infant with neurologic impairment, who was asphyxiated at birth, is given in fig. 7. The relation between heart rate and respiratory is no longer consistently negative (as in fig. 6) but occasionally reaches significant positive values indicating a dissociation between the time course of heart rate and respiratory rate. The advantage of this method above the statistical treatment of the three-minute epoch analysis is the possibility to interpret the statistical findings for the individual case. Future research will be directed to the age-dependent aspects in healthy pre- and full-term infants and to a variety of abnormal infants, employing different time windows for the cross-correlation.

3. CONCLUSIONS

The investigation of behavioural states in infants has supplied empirical data on the time course and cycling of states in healthy infants. Deviations were found in infants with a variety of abnormal conditions. The analysis of the regulation and organization of state criteria and their concomitants leads us to a better understanding of the regulatory mechanisms and to a diagnosis of their dysfunctions. It is obvious that infants with abnormalities in their behavioural states may in many cases create difficulties for the caregiver, who has to develop an intimate mutual relationship with the young infant.

SUMMARY

In recent years it has become evident that for many aspects of physiologic and behavioural research on newborn infants the consideration of sleep and awake states is of utmost importance. Due to the differences in selecting criteria various scales of behavioural states have been designed.

States can be assessed by direct observation or by recording of respiration, EKG, EEG, eye movements and EMG with electronic equipment in a so-called polygram. There are advantages and disadvantages in each of these techniques.

In the newborn infant sleep and awake states occur in regular cycles but may be disturbed in babies with brain dysfunction. An important aspect is the relationship between the various recorded signals within and between the behavioural states. In this case automatic analysis can be of great help as an extension of hand-analysis.

The spontaneously generated behavioural states of the infant are of great significance for the development of the interaction between infant and mother or caregiver. Abnormalities of the behavioural states may have profound effects on social behaviour.

REFERENCES

Anders, T.F.: The infant sleep profile. Neuropädiatrie 5: 425-442 (1974).
Anders, T.F.: Maturation of sleep patterns in the newborn infant. In: E.D. Weitzman (Ed.) Advances in Sleep Research. Vol. 2, 43-66 (Spectrum Publications, Inc. Flushing, N.Y., 1975).
Anders, T.F. and Chalemion, R.J.: The effects of circumcision on sleep-wake states in human neonates. Psychosomatic Medicine 36: 174-179 (1974).
Anders, T.F., Emde, R. and Parmelee, A.: A manual of standardized terminology, techniques and criteria for scoring states of sleep and wakefulness in newborn infants. In: Brain Information Service, UCLA, School of Health Service (Los Angeles, 1971).
Casaer, P., O'Brien, M.J. and Prechtl, H.F.R.: Postural behaviour in human newborns. Agressologie 14, no. B.: 49-56 (1973).
Globus, G.: Quantification of the REM sleep cycle as a rhythm. Psychophysiol. 7: 248-253 (1970).

90   HEINZ F. R. PRECHTL

Huisjes, H.J., Okken, A., Prechtl, H.F.R. and Touwen, B.C.L.: Neurological and pediatric findings in newborns of mothers with hypertensive disease in pregnancy. In: Z.K. Stembera, K. Polacek, V. Sabata (Eds.) Perinatal Medicine, 531-535 (Georg Thieme Verlag, Stuttgart, 1975).

Monod, N., Eliet-Flescher, J. and Dreyfus-Brisac, C.: Le sommeil du nouveau-né et du prématuré. III. Les troubles de l'organisation du sommeil chez le nouveau-né pathologique: Analyse des études polygraphiques. Biol. Neonat. 11: 216-247 (1967).

Monod, N. and Guidasci, S.: Sleep and brain malformation in the neonatal period. Neuropädiatrie 7: 229-249 (1976).

Parmelee, A.H., Wenner, W.H., Akiyama, Y., Stern, E. and Flescher, J.: Electroencephalography and brain maturation. In: A. Minkowski, Regional development of the brain in early life, 459-476 (Blackwell Scientific Publications, Oxford, 1967).

Prechtl, H.F.R.: Polygraphic studies of the full-term infant. II. Computer analysis of recorded data. In: M.C.O. Bax & R.C. MacKeith (Eds.) Studies in infancy. Clinics in Developmental Medicine no. 27: 26-40 (Heinemann, London, 1968).

Prechtl, H.F.R.: The behavioural states of the newborn infant (a review). Brain Research 76: 185-212 (1974).

Prechtl, H.F.R., Akiyama, Y., Zinkin, P. and Kerr Grant, D.: Polygraphic studies of the full-term newborn. I. Technical aspects and qualitative analysis. In: M.C.O. Bax & R.C. MacKeith (Eds.) Studies in infancy. Clinics in Developmental Medicine no. 27: 1-25 (Heinemann, London 1968).

Prechtl, H.F.R. and Beintema, D.J.: The neurological examination of the full term newborn infant, Heinemann, London, 1964.

Prechtl, H.F.R., Theorell, K. and Blair, A.W.: Behavioural state cycles in abnormal infants. Dev. Med. Child Neurol. 15: 606-615 (1973).

Prechtl, H.F.R., Weinmann, H. and Akiyama, Y.: Organization of physiological parameters in normal and neurologically abnormal infants: Comprehensive computer analysis of polygraphic data. Neuropädiatrie 1: 101-129 (1969).

Scholten, C.A.: Computeranalyse in polygrammen. Een methodologisch onderzoek. Thesis, University Groningen (1976).

Schulte, F.J., Lasson, U., Parl, U., Nolte, R. and Jürgens, U.: Brain and behavioural maturation in newborn infants of diabetic mothers. Neuropädiatrie 1: 36-43 (1969).

Theorell, K., Prechtl, H.F.R., Blair, A.W. and Lind, J.: Behavioural state cycles of normal newborn infants. Dev. Med. Child Neurol. 15: 597-605 (1973).

Wolff, P.H.: Observations on newborn infants. Psychosom. Med. 221: 110-118 (1959).

Wolff, P.H.: The causes, controls and organization of behaviour in the neonate. In: Psychological Issues, Vol. 5, No. 1, Monogr. 17. (International University Press, New York, 1966).

CHAPTER SIX

SLEEP BEHAVIOR: GENETIC REGULATION*

JEAN-LOUIS VALATX, M.D.

For twenty years, phenomenology and biochemical mechanisms have been described in the majority of mammals (2, 15, 21, 26, 30). In each species, results have been obtained in randomly selected individuals representing an average of a non homogenous population. In spite of intraspecies variability, significant changes have been shown between different species (1, 30). Are these differences due to environmental factors or the result of alterations in the functioning of the central nervous system?

To answer this question, a genetic approach can be a good model. We have chosen the mouse (Mus musculus) because it is the best adapted species for this study. For fifty years, numerous pure inbred strains of mice have been isolated by selecting somatic traits such as coat color, plasmatic enzyme level, metabolic inborn errors, various neurologic mutations, etc. (13, 26). Now the laboratory mouse can be used as a prototype in behavioral genetics to separate the different components of a given behavior and determine the respective influence of environmental and hereditary factors (9, 20, 23).

The observation of low intra-strain variability and high inter-strain variability in sleep patterns allows a genetic approach, a "genetic dissection" (7) of the sleep-waking cycle in order to attempt to understand the neurophysiological mechanisms of the observed variations (12, 26).

The results we present here indicate that the origin of these sleep variations might be located "in utero" at the moment of the formation of the nervous system. The proposed model allows the study of fetal development and eventual alterations of nervous structures involved in the production and regulation of sleep-waking cycle and also their role in the development of the central nervous system.

* This work was partly supported by DRME (grant 74232), INSERM (U52) and CNRS (ATP
   We are greatly indebted to Mrs Paut for her skilful technical assistance. We thank Don Bailey,
The Jackson Laboratory for his help with the RI strains.

METHODOLOGY

Animals. All the subjects were male, 12-15 week old mice from four inbred strains: C57BRcd/Orl (BR), C57BL6/Orl (B6), BALBc/Orl (C), CBA/Orl, F1 hybrids: BR,CBA and BR,C. Then, to study eventual linkage with the histocompatibility genes (H loci) used as markers, Recombinant Inbred (RI) strains of Bailey were studied. Production of these strains is described by Bailey (5): "The RI strains were derived from the cross of two unrelated and highly inbred progenitor strains, C and B6, and then maintained independently from the F2 generation under a full-sib mating regimen. This procedure progressively fixed the chance recombination of alleles as inbreeding proceeded and as full homozygotosis was approached. The resulting battery of strains can be looked upon as a finite but replicable recombinant population." For each H locus a characteristic strain distribution pattern (SDP) is established by determining which allele (B6 or C) is present in these RI strains.

We have studied parental strains: C57BL6/By, BALBc/By, F1 hybrids and seven RI strains: CXBD, CXBE, CXBG, CXBH, CXBI, CXBJ and CXBK.

Sleep recordings. Under nembutal anesthesia, at least 8 mice from each strain were chronically implanted with cerebral and muscular electrodes. Each mouse was put on synthetic litter in a glass jar with standard food and tap water ad libitum. Environmental conditions were constant insofar as possible: ambient temperature was $24 \pm 1°C$, lighting schedule consisted of 12 hours of light (7.00-19.00) and 12 hrs of darkness (19.00-7.00).

After a ten day period of adaptation to recording conditions, continuous recording (24 hrs. per day) for a week was carried out. By visual analysis, all the recordings were scored by 30-second epochs (26).

Statistical analysis of RI data, variance and Student-Newman-Keuls multiple range test, was performed at The Jackson Laboratory, Bar Harbor, Maine, USA.

RESULTS

Sleep waking cycle in inbred strains and F1 hybrids

Results are shown in table 1 and can be summarized as follows:

In inbred strains, 1) Total, night and day sleep durations significantly differed from strain to strain.

Table 1. Sleep patterns in inbred strains and F1 hybrids (duration: minutes ± standard error of the mean).

	Slow Wave Sleep Duration			Paradoxical Sleep Duration			PS/SWS		
	24 hr	Night	Day	24 hr	Night	Day	24 hr	Night	Day
CBA	746±7.10	322±5.14	424±3.75	66.5±1.62	24.0±0.77	42.5±1.24	8.91	7.47	10.03
CBA.BR F1	741±10.7	331±2.25	410±6.97	65.0±2.10	26.5±1.40	38.5±1.84	8.77	8.00	9.39
BR.CBA F1	754±11.9	316±9.76	438±6.76	65.5±2.26	23.0±1.56	42.5±2.01	8.68	7.27	9.70
BR	639±7.70	180±8.72	459±3.20	73.0±2.50	19.0±1.35	54.0±1.49	11.42	10.55	11.76
BR.C F1	692±8.65	277±3.50	415±3.60	67.0±2.0	20.0±1.20	47.0±1.40	9.68	7.22	11.32
C.BR F1	735±7.42	342±4.34	393±3.32	80.0±2.5	36.0±1.30	44.0±1.20	10.88	10.52	11.19
C	723±9.34	318±6.49	405±8.74	54.0±1.8	31.0±1.50	23.0±1.15	7.46	9.74	5.67

Table 2. Sleep patterns in recombinant inbred strains (duration: minutes ± standard error of the mean).

	Slow Wave Sleep Duration			Paradoxical Sleep Duration		
	24 hr	Night	Day	24 hr	Night	Day
B6	741.0±8.20	322.5±5.26	418.5±5.63	75.23±1.73	19.73±0.81	55.50±1.10
B6.C F1	753±12.9	345.8±10.5	408.1±5.14	71.57±1.62	33.34±1.46	38.33±0.81
C.B6 F1	735.8±11.8	329.0±10.3	406.0±5.70	68.04±1.39	29.92±1.93	38.12±0.88
C	726.6±9.77	311.3±0.68	415.3±9.6	56.80±2.46	30.53±1.33	26.27±1.58
CXBD	689.1±10.6	270.4±13.6	418.7±4.8	66.82±1.71	22.06±1.08	44.76±1.07
CXBE	805.0±7.58	370.6±5.39	434.4±5.4	70.59±1.72	30.03±0.93	40.57±0.92
CXBG	681.5±7.87	311.2±4.12	370.3±6.5	70.30±1.28	37.00±1.37	33.30±1.07
CXBH	630.1±10.8	275.5±8.65	355.6±7.0	50.70±1.42	19.20±1.20	31.50±1.50
CXBI	779.8±8.75	353.1±5.1	426.7±7.58	77.20±1.80	39.26±1.33	37.96±1.33
CXBJ	563.9±15.4	150.2±12.5	413.7±7.75	74.70±2.53	12.36±1.48	62.33±2.85
CXBK	750.2±7.83	350.3±6.41	399.9±8.76	87.00±1.91	37.43±1.74	49.6±1.64

Table 3. Strain Distribution Pattern of sleep in Recombinant Inbred Strains of mice compared to Strain Distribution Pattern of three histocompatibility genes. Brackets indicate intermediate duration between B6 and C mice. Thus, there are at least 2 possibilities of linkage for night PS duration (H-37 and H-38). Study of congenic lines will allow to say what is the right linkage.

	CXBD	CXBE	CXBG	CXBH	CXBI	CXBJ	CXBK
24 hr PS duration	( )	B6	B6	C	B6	B6	B6
H-2	B6	B6	B6	C	B6	B6	B6
Night PS duration	B6	C	C	( )	C	B6	C
H-37	B6	C	C	B6	C	B6	C
H-38	B6	C	C	C	C	B6	C

2) Slow wave sleep (SWS) duration varied independently from paradoxical sleep (PS) duration.

3) Strains could be classified in three groups according to their PS duration: the two C57 strains had the highest amount of PS, Balb/c the lowest and CBA the intermediate one. Considering slow wave sleep duration, it was possible to do another classification.

4) Day-night variations were another parameter to separate strains. C57 strains had the greatest difference while Balb/c had no significant difference between day and night sleep duration.

In F1 hybrids, results differed according to crossings.

In CBAI, BR hybrids, animals from reciprocal crosses (CBxBR and BRxCBA) are identical for all the sleep parameters (night, day PS and SWS durations) and not significantly different from CBA strain.

In C,BR hybrids, reciprocal crosses (CxBR and BRxC) were not statistically different for total PS and SWS durations. However, paradoxical sleep duration was not different from the BR strain while slow wave sleep duration was not different from the C strain.

For day-night differences, maternal effects seemed to be important in C,BR hybrids while in CBA,CR hybrids they were not apparent (see table 1).

Sleep waking cycle in Recombinant Inbred Strains

Variance analysis and Student-Newman-Keuls multiple range test allowed us to rank by order the RI strains with regard to parental strains and F1 hybrids (table 2). We have observed the most reliable differences for PS duration (total and night durations).

This analysis allowed us to classify each RI strain as not different from one or another parental strain. Thus, a strain distribution pattern (SDP) could be established for total and night PS durations (table 2). Comparison with H loci SDP indicated that total and night paradoxical sleep durations were possibly linked to H 2 locus and non H 2 locus (H 37, H 38) respectively (table 3).

DISCUSSION

These results from inbred and recombinant strains give a new insight into sleep physiology. The first finding is the existence and stability of the differences in sleep duration between strains of mice. The second point is that these differences are transmissible according to Mendelian laws of heredity.

Thus, sleep behavior can be considered as a complex somatic trait which may be subdivided into several units: duration of each stage of sleep, circadian rhythm and also phasic activity of paradoxical sleep. Cespuglio et al. (1975) showed that total number of eye movements per minute of PS and their patterns of occurrence were different from strain to strain and had a hereditary component (10).

The strain distribution pattern of recombinant inbred strains indicated that the total paradoxical sleep duration may be linked with the H2 locus, the major histocompatibility complex (MHC) of the mouse. Further studies with specific congenic lines will allow us to confirm or disprove this linkage. This is important because the H2 locus has been located on chromosome 17 next to the T locus region, the genes of which specify cell surface structure in embryos. Several recessive mutations at the T locus are known to provoke selective defects in the neural tube during embryonic development (4, 5, 6). The observed differences in sleep patterns could be the consequence of some discrete alterations of some parts of the brain.

The cause of inter-strain differences may be due to various factors such as 1) *structural alterations* of nervous structures involved in the production of sleep (brainstem nuclei: raphe and locus coeruleus systems) or in the regulation of the sleep cycle (some hypothalamic nuclei, pineal gland, etc.), 2) *alterations of sensitivity* to environmental factors which can modify sleep: light (8, 11), ambient temperature (28), food absorption (22), isolation.

The effects of these factors are mediated by a genetically influenced anatomic support. So, it is difficult to say what are the actual causes of variations in sleep duration of the mouse.

The interesting finding is that these variations are linked to H and T systems which determine cell surface structure and thus, might be originated "in

utero" at the moment of neuronal differentiation and formation of connections between neurons as a consequence of discrete alterations in the cell-
cell recognition of neurons involved in sleep regulation. If this is true, it is
possible to predict which alterations could induce these differences: 1) change
in the number of neurons in a given part of the brain (25); 2) alterations of
neuron surface concerning membrane permeability, number of receptors,
number of connections, etc.; 3) changes in metabolic activity: quantity or
substrate affinity of a given enzyme involved in the neurotransmitter
synthesis.

Knowing nervous structures supposed to be related to sleep mechanisms
and using mice bearing mutations at the T H2 complex, it will be possible to
refine genetic dissection and go further into the understanding of biologic
mechanisms controlling this complex behavior.

That is an important point because paradoxical sleep is the only stage of
sleep in utero and at birth. Hypothetical functions of PS during ontogenic
development would be to set up neuronal circuitry responsible for primary
motor automatisms, functions different from those in the adult, which would
be in relation to adaptation to the environment (3, 14, 15, 17, 18, 19, 27, 29).

SUMMARY

A genetic approach provides a good model for studying sleep mechanisms not
only in adults but also in pre- and post-natal maturation.

Using several inbred strains of mice, their F1 hybrids and backcrosses, it
was possible to show that SWS duration was determined independently of PS
duration.

Furthermore, using histocompatibility genes as markers, a sleep study of
recombinant inbred strains of Bailey indicated 1) several genes are directly or
indirectly involved in the determination of total PS duration; 2) these genes
seemed to be linked to H2, the major histocompatibility complex of the
mouse; 3) these genes were different from those which determine night
PS duration.

Further studies with specific congenic lines of mice will allow us to demonstrate these linkages. The major histocompatibility complex has been
located on chromosome 17 next to the T locus which specifies cell surface
structure on embryos. In the future, this model will be useful in studying total
development of brain structures involved in the production and regulation of
the sleep cycle.

REFERENCES

1. Allison, T. and Van Twyver, H.: The evolution of sleep; Nat. Hist. 79: 169-175 (1974).
2. Allison, T., Van Twyver, H. and Goff, W.R.: Electrophysiological studies of the echidna, tachyglossus aculeatus. I. Waking and sleep; Arch. Ital. Biol.110: 145-184 (1972).
3. Astic, L.: Le sommeil avant et après la naissance; Thèse de Doctorat d'Etat es Sciences (Lyon 1976).
4. Bach, F.H. and Van Rood, J.J.: The major histocompatibility complex. Genetics and biology; New England J. Med. 295: 806-813 (1976).
5. Bailey, D.W.: Genetics of histocompatibility in mice. I. New loci and congenic lines; Immunogenetics 2: 249-256 (1975).
6. Bennett, D.: The T-locus of the mouse; Cell 6: 441-454 (1975).
7. Benzer, S.: Genetic dissection of behavior; Scientific American 229: 24-37 (1973).
8. Borbely, A.A., Huston, J.P. and Waser, P.T.: Control of sleep states in the rat by short light dark cycles; Brain Res. 95: 89-102 (1975).
9. Bovet, D., Bovet-Nitti, F. and Oliverio, A.: Genetic aspects of learning and memory in mice; Science 163: 139-149 (1969).
10. Cespuglio, R., Musolino, R., Debilly, G., Jouvet, M. and Valatx, J.L.: Organisation différente des mouvements oculaires rapides du sommeil paradoxal chez deux souches de souris; C. R. Acad. Sci. (Paris), 280: 2681-2684 (1975).
11. Fishman, R. and Roffwarg, H.P.: REM sleep inhibition in the albino rat; Exp. Neurol. 36: 106-178 (1972).
12. Friedman, J.K.: A diallel analysis of the genetic underpinnings of mouse sleep; Physiol. Behav. 12: 169-175 (1974).
13. Green, E.L.: Biology of the laboratory mouse; 2nd Edn (Mc Graw-Hill, New York 1966).
14. Greenberg, R. and Pearlman, C.: Cutting the REM nerve: an approach to the adaptive role of REM sleep; Perspect. Biol. Med. 17: 513-521 (1974).
15. Hennevin, E. et Leconte, P.: La fonction du sommeil paradoxial: faits et hypothèses; Année Psychol. 2: 489-519 (1972).
16. Jouvet, M.: The role of monoamine and acetylcholine containing neurons in the regulation of the sleep-waking cycle; Ergebn. Physiol. 64: 165-307 (1972).
17. Jouvet, M.: Essai sur le rêve; Arch. Ital. Biol. 111: 564-576 (1973).
18. Jouvet-Mounier, D., Astic, L. and Lacote, D.: Ontogenesis of the states of sleep in rat, cat, and guinea pig during the first post natal month; Develop. Psychobiol. 2: 216-239 (1970).
19. Kitahama, K., Valatx, J.L. and Jouvet, M.: Apprentissage d'un labyrinthe en Y chez deux souches de souris. Effets de la privation instrumentale et pharmacologique du sommeil; Brain Res. 108: 75-86 (1976).
20. Lindzey, G. and Thiessen, D.D.: Contributions to behavior-genetic analysis. The mouse as a prototype; Appleton-Century-Crofts, New York 1970).
21. Moruzzi, G.: The sleep-waking cycle; Ergebn. Physiol. 64: 2-164 (1972).
22. Mouret, J. and Bobillier, P.: Diurnal rhythms of sleep in the rat: augmentation of paradoxical sleep following alterations of the feeding schedule; Int. J. Neurosci. 2: 265-270 (1971).
23. Oliverio, A.: Genetic and biochemical analysis of behavior in mice; in Kerkut, G.A. and Phillis, J.W., Progress in Neurobiology, Vol. 3 pp. 193-215 (Pergamon Press, Oxford, New York 1974).
24. Ross, R.A., Judd, A.B., Pickel, V.M., Joh, T.H. and Reis, D.J.: Strain dependent variations in number of midbrain dopaminergic neurons; Nature (in press).
25. Sidman, R.L., Green, M.C. and Appel, S.M.: Catalog of the neurological mutants of the mouse (Harvard University Press 1965).
26. Valatx, J.L. et Bugat, R.: Facteurs génétics dans le déterminisme du cycle veille-sommeil chez la souris; Brain Res. 69: 315-330 (1974).

27. Valatx, J.L., Jouvet, D. et Jouvet. M.: Evolution électroencéphalographique des différents états de sommeil chez le chaton. Electroenceph. Clin. Neurophysiol. 17: 218-233 (1964).

28. Valatx, J.L. Roussel, B. et Cure. M.: Sommeil et température cérébrale du rat au cours de l'exposition chronique en ambiance chaude; Brain Res. 55: 107-122 (1973).

29. Valatx, J.L. et Nowaczyk, T.: Essai de suppression pharamacologique du sommeil paradoxal chez le rat nouveau-né; Rev. E.E.G. Neurophysiol (in press).

30. Zepelin, H. and Rechthshaffen. A.: Mammalian sleep. longevity and energy metabolism. Brain Behav. Evol. 10: 425-440 (1974).

CHAPTER SEVEN

DEVELOPMENTAL PLASTICITY IN THE VISUAL CORTEX*

MICHEL IMBERT**

A better understanding of the function of sensory neurons has been achieved in the last decade by the determination of the stimulus parameters which are critical for triggering these neurons. Numerous experiments, following the pioneering studies by Hubel and Wiesel (10, 11) have clearly demonstrated that the neurons in the primary visual cortex of the cat (and of the monkey) are preferentially activated by rather complex visual stimuli, usually edges, slits or bars of particular and precise orientation, moving across their receptive field. The orientation of the edge which maximally activates each neuron is different from one cell to the next, but every orientation is equally represented. The majority of these neurons are binocularly driven, but are differentially influenced by the two eyes (2, 11, 16, 17).

A consensus seems to have been reached concerning the possibility that the complex neural interconnections, which give such a highly specific visual response, can be modified. An alteration in the visual environment of an immature kitten seems sufficient to alter the response of cortical cells profoundly (3, 4, 6, 7, 8, 9, 13, 18, 19, 20, 21, 22, 23, 24, 24, 26). Selective visual experience is effective in this respect only when it occurs during a limited critical period of postnatal development and when the cat has had no previous visual experience.

There are two alternative interpretations of these results:

1. Changes observed are mediated by instructive environmental processes which are normally required for the development of specialized cortical cells. According to Hirsch and Spinelli (8, 9) "the instructional hypothesis can

* This work was supported by grants from the CNRS (RCP 348) and the INSE⟨M (ATP6-74-27) and a contract of the DGRST (no 74-7-800).
Miss Paulette Saillour rendered valuable technical assistance.
** Most of the results of these experiments have been published in detail elsewhere (5,15) in collaboration with P. Buisseret.

account for the presence of units whose receptive field characteristics closely
match the stimuli presented during the animal's development."

2. Changes result from the selective modification of a pre-wired structure
existing at birth and involve the degeneration of classes of non used synaptic
afferents. This selective hypothesis involves two complementary aspects: first,
that mechanisms for modification of receptive field characteristics are available, second, that specific visual neurons, with adult receptive field arrange-
ment, are present before modification. This last aspect is still controversial.
According to Hubel and Wiesel (12) "cortical cells of visually inexperienced
kittens strongly resembled cells of mature cats in their responses to patterned
stimuli," and "clear receptive field orientation and directional preferences to
movement are seen in cortical cells of newborn visually inexperienced kittens." On the other hand, according to Barlow and Pettigrew (1) "diffuse
binocular connections and the mechanism for directional movement selectivity appear to be innately determined, but the mechanism for disparity and
orientational selectivity requires visual experience." Pettigrew (20) has recently reaffirmed this conclusion.

Experiments were undertaken in order to:
– Compare receptive field properties of visual cortical neurons in two groups
  of kittens at different ages. The first group was normally reared (NR), the
  second group was reared in complete darkness (DR) from the first or
  second day of age up to the day of the acute experiment.
– Evaluate the degree of plasticity of cortical units in 5-6 week old kittens,
  reared with or without visual experience, in response to a "conditioning
  exposure" while recording from a given neuron.

The activity of single cells was recorded extra-cellularly metallic electrodes
stereotaxically located in area 17, in kittens which were anesthetized with
Nembutal or Penthotal given intraperitoneally and curarized. The characteristics of the receptive field (size, position, shape) were mapped by manually
projecting small spots or slits of light upon a wide tangent screen diffusely
illuminated at mesopic level. On completion of this preliminary, rather qualitative examination, the neurons were studied using computer programmed
visual stimuli. The stimulus was a stationary or a moving visual pattern usually a spot, a slit or a bar of any size, contrast, orientation or speed. These
different parameters being precisely controlled by a computer, which is also
used to form the PSTH.

Four types of cells were identified:
    TYPE 1: visually non-responsive units. They have almost no spontaneous

activity and cannot be stimulated by any peripheral stimulation. They are however identifiable because they can be mechanically activated.

TYPE 2: non-specific units. They are usually activated by a circular stimulus moving in any direction across their receptive field.

TYPE 3: immature units. They are preferably activated when a rectilinear stimulus moves in a direction orthogonal to the optimal orientation of their roughly rectangular and rather large ($10°-8°$) receptive field. However this orientation is rather imprecise since the cell can be activated by orientation up to $45°$ on both side of the optimal one. But in all cases there exists a direction in which the stimulus is ineffective.

TYPE 4: specific units. They have all the characteristics of the simple or complex cells in an adult cat. Their receptive field is rectangular ($4°-2°$) and smaller in size than the one of an immature cell.

In normally reared kittens, four periods after birth could be distinguished:

1. Between the 8th and the 11th day, the cells were visually non-responsive.

2. During the second period, between the 12th and the 17th day, visual units are already evident, with spatially localized fields and definite characteristics of the trigger feature. They can be classified according to the types defined above. (See table 1.)   Examples of non-specific and of specific cells

Table 1. Properties of different types of visually responsive units and different ages.

	Age group 12-17		Age group 18-28		Age group 29-42	
	NR	DR	NR	DR	NR	DR
Non specific	23%	40%	9%	56%	—	94%
Immature	54%	37%	49%	30%	24%	4%
Specific	23%	23%	42%	14%	76%	2%

Age group in days: NR = normally reared,   DR = dark reared.

recorded in a 15 day old kitten are given in figure 1. In A, the cell was driven by both eyes but was preferentially activated by the left eye. This cell is classified as non-specific: the receptive field of a relative large size, is not precisely limited; it is not possible to separate *on* and *off* regions, and, furthermore, the cell is activated by a moving spot in any direction as illustrated by the arrows in the central diagram. A specific cell is illustrated in figure 1B. This cell is activated only in a one way, horizontal direction, by a vertically oriented moving bar (left histogram). There is no response for the movement of the

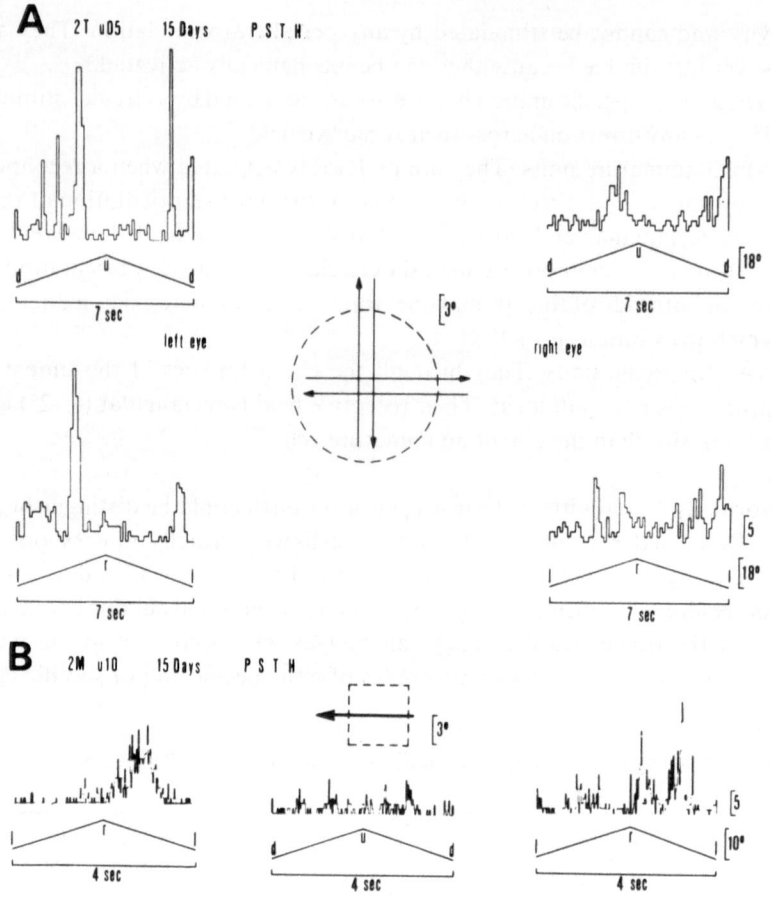

Fig. 1. PSTH recorded in two 15 day old NR kittens. In the center: diagram of receptive field and direction of movement of the activating stimulus.

A. Non-specific cell binocularly activated with a left dominance. On the left, histograms formed by a moving spot presented to the left eye: on the upper left, from down (d) to up (u) and up to down. Lower left from left (l) to right (r) and right to left. On the right: histograms formed by a moving spot presented to the right eye. Upper right: vertical movements. Lower right: horizontal movements.

B. Specific cell monocularly activated by the contralateral eye. Histograms formed by a vertically oriented slit moving across the receptive field. On the left, movement from left (l) to right (r) and from right to left. In the center, movement from down (d) to up (u) and up to down. On the right, movement from right to left and left to right. On the right, the slit is longer than on the left.

stimulus in a vertical direction (central histogram). A longer vertically oriented bar moving horizontally (right histogram) gives a bi-modal response, which probably reveals an inhibitory influence from the side regions of the receptive field.

As soon as a visual response appears, about 25% of the units were found clearly and positively activated by selectively oriented stimuli; and a tendency to monocular dominance was also regularly observed.

3. Between the 18th and the 28th day, all these types of visual units (non specific, immature and specific) continue to be found. However the number of specific cells increases while the number of non-specific ones decreases (see table 1). A specific cell (simple cell) in a 23 day old kitten is illustrated in figure 2.

4. Finally, between the 29th and the 42nd day, the cells behaved almost like adult cells.

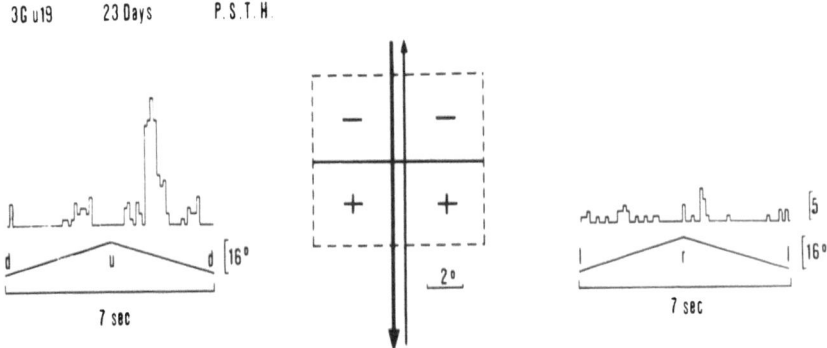

Fig. 2. Specific cell monocularly driven. PSTH recorded in a 23 days old NR kitten. In the center: diagram of the receptive field, + ON region, − OFF region, and direction of movement of the slit. The thickness of the arrow indicates the strength of the response. On the left, histogram formed by a horizontal slit moving from down (d) to up (u) and up to down. On the right, histogram formed by the movement from the left (l) to the right (r) and the right to the left by the same slit, vertically oriented.

In dark-reared kittens, the distribution and the four types of cells observed during the second period (12th-17th day) was similar to the distribution found in normally reared kittens; in particular the percentage of the specific units (23) is the same in both cases. Later, during the third period, between the 18th and the 28th day, the number of specific cells decreased. Two examples of specific cells in a 19 day old DR kitten are illustrated in figure 3. The number of non-specific cells increased (see table 1).

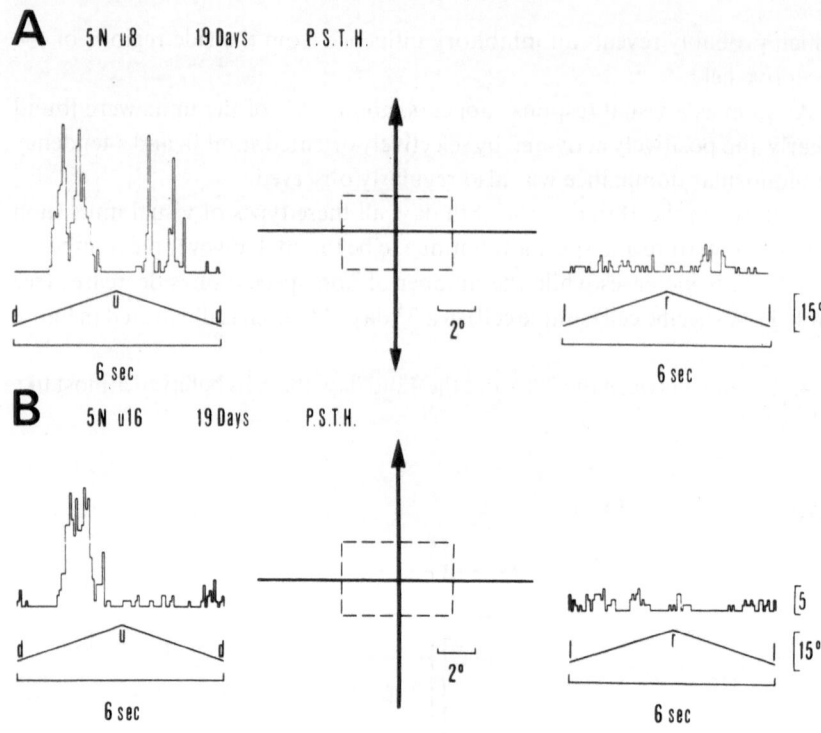

Fig. 3. PSTH of two specific cells in a 19 day old DR kitten.
A. In the center, diagram of the receptive field and direction of movement of the activating stimulus. The thickness of the arrow indicates the strength of the response. On the left, histogram formed by a horizontally oriented slit moving from down (d) to up (u) and from up to down. On the right, histogram formed by the same but vertically oriented slit moving from right (r) to left (l) and left to right.
B. The same for another cell. In A, a specific cell monocularly activated by the contralateral eye. The cell responds to a two way vertical direction. In B, a binocular specific cell, mostly activated by the contralateral eye in a one-way direction.

Later on, during the last period, between the 29th and the 42nd day, the non specific units became by far the most frequent. The figure 4 is a quantitative summary of the results described above.

First of all, it is remarkable that when the first clearly visual responses are recorded, approximately 25% of the visual cells are specific, whatever the rearing conditions of the animal. After that period, in the NR kittens the specific cells tend to increase, whereas, on the DR kittens, the characteristics of the early specific units disappear until all the units are non-specific.

In a previous work (5) it was shown that the distribution of cells according

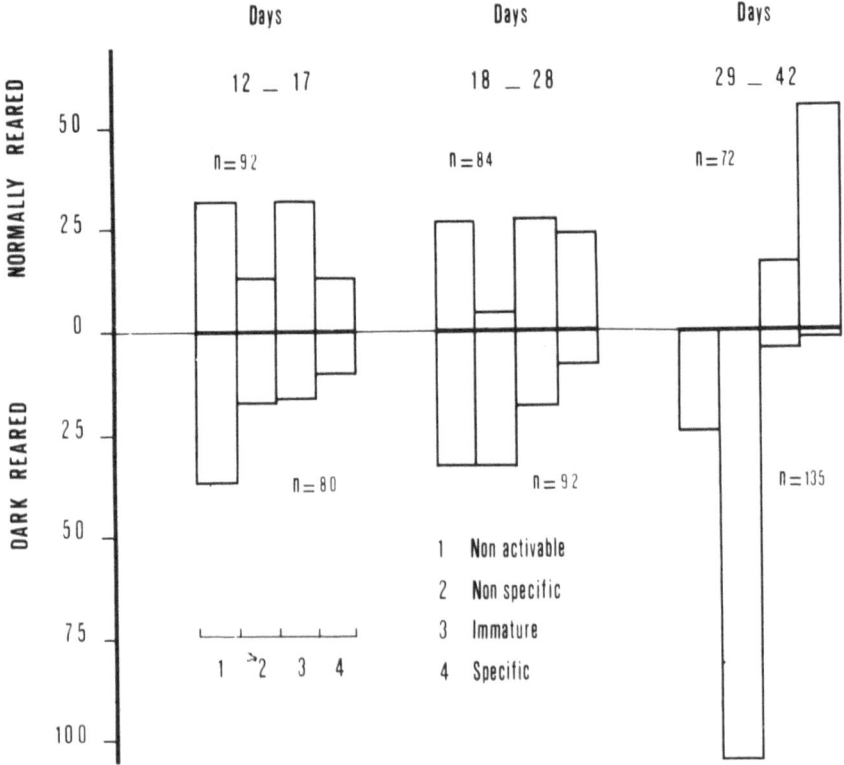

Fig. 4. Distribution of the different types of cells in 3 age groups in the NR kittens (upper part), and in the DR kittens (lower part).

to their ocular dominance was not different in either group, normally or dark reared, and was similar to that previously described for "adult" cells. On the whole we confirm this statement. However up to the end of the third week the specific cells, in the normally or dark reared kittens, are generally monocularly driven, in contrast with the specific ones of the adult cortex.

Recently, Blakemore and Mitchell (4) have shown that no more than one hour of selective exposure is sufficient to modify the preferred orientation of the visual cortical neurons. Pettigrew, Olson and Barlow (18) have also reported reversal of eye dominance "in response to conditioning stimulation applied while the neurons are under observation." These investigations suggest that it should be possible – during an acute experiment – to follow the process by which a given neuron gains its specific properties.

The following experiments were performed with DR kittens, 5 to 6 weeks

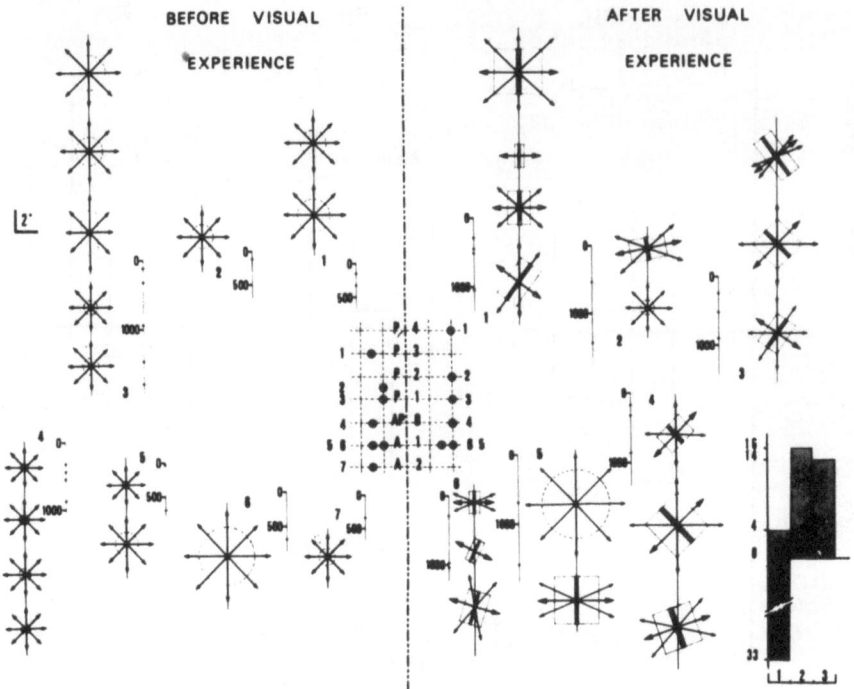

Fig. 5. Diagrammatic reconstruction of a 2 stage acute experiment. A 5 week old dark-reared kitten. The left half is before visual experience and the right half is after visual experience. At the center are the stereotaxic coordinates of the cortical penetrations. To either side of this the individual RF are plotted in the sequence in which they were recorded. The characteristics of the RF were plotted using the dominant eye, but most of the cells were binocularly driven. In the extreme lower right of cells are classified as in Fig. 1: 1) non-specific, 2) immature and 3) specific cells, before visual experience (lower part) and after (upper part).

old. After having chosen a binocular non-specific neuron, typical in these kittens, as a function of the stability of the recording, a stationary grating made of alternatively white and black stripes of unequal widths (covering $50° \times 50°$) was binocularly presented for a duration of 400ms at the rate of one per second. Adaptation and stabilization of the retinal image were avoided not only by a repetitive stimulation but also by a slight shift of the grating at each presentation in such a way that the edges of the bars never fell upon the same part of the retina during two successive presentations. This conditioning procedure was interrupted every 10 minutes to form a PSTH. The purpose of this experiment was to try to force a non specific cell to respond preferentially

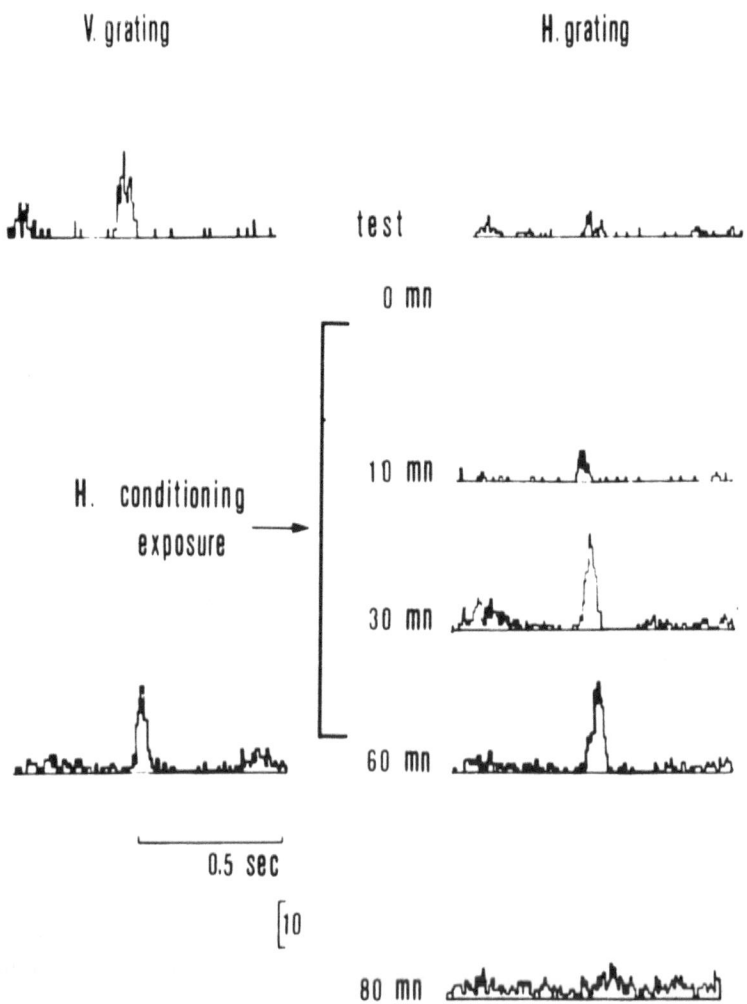

Fig. 6. Conditioning exposure of a cortical cell in a normally reared kitten. Test before conditioning: on the left, response obtained by reptitive stimulation with a vertical grating (V, grating): on the right, response obtained by repetitive stimulation with an horizontal grating (H, grating), during conditioning with an horizontal grating (H, conditioning exposure): 10, 30 and 60 min after the beginning. At 60 min after termination of the conditioning, the cell still presented response with a V grating. The newly acquired response with an H grating disappears 20 min after the end of the conditioning exposure.

to an elongated pattern with the same orientation as that of the conditioning grating.

In spite of a very long period of conditioning exposure, up to 8 hours, we failed to detect any acquired selectivity. No preferential responding was observed after exposure for an oriented edge and there was no indication of directional selectiveness for a moving stimulus.

One would therefore suppose that there is an irreversible effect due to the complete absence of light during the first weeks of the postnatal life preventing organization of specific response to orientation. In order to test such an effect we have performed some two stage experiments with the same kitten under conditions of endotracheal intubation.

First, cortical cells of one hemisphere were studied in order to test their non-specificity. After this experiment was finished, the kitten recovered in darkness. When it was completely awake, it was taken out of the dark-room and was permitted to run freely in the laboratory. By the fifth to the sixth hour, it appeared as normal as NR kittens of the same age. It was then put in the dark room for 6 more hours before the second experiment was performed under the same conditions as the first one. The cortical cells of the unused hemisphere were characterized: most of the units were sufficient to provide some orientation selectivity.

One such experiment is illustrated in the figure 5. The histogram in the lower right hand side shows the distribution of the neurons according to their response specificity before (lower part) and after (upper part) visual experience. Six hours of light were sufficient to provide some orientation selectivity. In fact, the number of immature cells was larger in this kitten than in the normally reared kittens of the same age.

The fact that it is possible to specify with a few hours of light may indicate that the experimental conditions used for "conditioning exposure" were not suitable for revealing the possibility of plastic changes in the properties of cortical cells. In order to test this possibility the same kind of "conditioning exposure" was performed on 5-6 week old kittens reared in a normal visual environment. The purpose was to try to change the prime orientation of a specific unit.

In the case illustrated in figure 6 a stationary vertical grating caused predominantly *off* responses; the orientation used corresponded to the preferred orientation of the cell studied. The grating was then rotated to 90, and a new PSTH was performed. In this particular case only a small response was recorded. The conditioning exposure of the horizontal grating (H-conditioning) was then begun. After 10 minutes, there was a clear increase in the response of the unit, which was maximal after a half hour. During and at the end of the conditioning exposure the test stimulus with a vertical grating gave a response identical to that before exposure. The cell was then activated by two perpendicular orientations: the initial preferred orientation and the orientation acquired during the conditioning exposure: this acquisition was transient and disappeared 20 minutes after the end of the conditioning exposure.

On the main issue. our findings are that as soon as neurons become visually activated about $25^0_{,0}$ of the recorded visual units are definitely specific in terms of orientation selectivity. These neurons are present in the earliest stages. even in the absence of any visual experience. However. in order for specificity to be maintained and to keep developing. the kitten must be allowed visual experience by the third week of life.

Our "conditioning exposure" data suggest that under our experimental conditions it was not possible to induce specific responding in the dark-reared kittens. whereas it was possible to unmask characteristics which. under normal conditions would not be apparent in the normally reared kittens. The transient change of orientation selectivity seems to indicate that. in these young animals. neurons acquire their orientation selectivity while losing the possibility of responding to other orientations. This loss is not complete at that time. since a conditioning exposure can elicit it (14).

The difference in effectiveness of the "conditioning exposure" in the two groups of kittens may indicate that the development of specific characteristics requires more than passive visual functioning. In fact. an active visuomotor interaction with the environment is necessary.

SUMMARY

The functional organisation of visual cortical cells was studied in two groups of kittens between 8 days and 6 weeks of age: one group normally reared. the other reared in total darkness. The following results were obtained. 1) The cells properties of 6 week old normally reared kittens were similar to those of adult cats. 2) In 8 day old kittens. 25 percent of the visual units had the same properties as in adult cats. The other cells found were 'immature' or 'non specific.' their proportion decreasing gradually until disappearance at the age of 6 weeks. Correlatively the proportion of specific units increase. 3) The cells of 6 week old dark-reared kittens were totally 'non specific'. The receptive fields showed neither directional nor orientational properties. Six hours of active experience followed by 6 hours of darkness. are sufficient to induce specific properties. 4) In 15 day old dark-reared kittens. visual cells were found identical with those of normally reared kittens. The initial specific units disappeared before the age of 6 weeks. 5) The distribution of cells according to the ocular dominance was not different in either group and was similar to that previously described for "adult" cats. 6) "Conditioning exposure" with an oriented grating induced changes in orientational sensitivity on normally reared kittens but not in dark-reared kittens.

REFERENCES

1. Barlow, H.B. and Pettigrew, J.D.: Lack of specificity of neurons in the visual cortex of young kittens. J. Physiol. (Lond.) 218: 98-100 (1971).
2. Bishop, P.O., Henry, G.H. and Smith, C.J.: Binocular interaction fields of single units in the cat striate cortex. J. Physiol. (Lond.) 216: 39-68 (1971).
3. Blakemore, C. and Cooper, G.F.: Development of the brain depends on the visual environment. Nature 228: 477-478 (1970).
4. Blakemore, C. and Mitchell, D.E.: Visual cortex: modification by very brief exposure to the visual environment. Nature 241: 467-468 (1973).
5. Buisseret, P. and Imbert, M.: Visual cortical cells: their developmental properties in normal and dark reared kittens. J. Physiol. (Lond.) 255: 511-525 (1976).
6. Cynader, M., Berman, N. and Hein, A.: Cats reared in stroboscopic illumination: Effects on receptive fields in visual cortex. Proc. Nat. Acad. Sci. USA, 70, no. 5, 1353-1354 (1973).
7. Ganz, L., Fitch, M. and Satterberg, J.A.: The selective effect of visual deprivation on receptive field shape determined neurophysiologically. Exp. Neurol. 22: 614-637 (1968).
8. Hirsch, H.V.B. and Spinelli, D.N.: Visual experience modifies distribution of horizontally and vertically oriented receptive fields in cats. Science 168: 869-871 (1970).
9. Hirsch, H.V.B. and Spinelli, D.N.: Modification of the distribution of receptive field orientation in cats by selective visual exposure during development. Exp. Brain Res. 13: 509-527 (1971).
10. Hubel, D.H. and Wiesel, T.N.: Receptive fields of single neurons in the cat's striate cortex. J. Physiol. 148: 574-591 (1959).
11. Hubel, D.H. and Wiesel, T.N.: Receptive fields, binocular interaction and functional architecture in the cat's visual cortex. J. Physiol. (Lond.) 160: 106-154 (1962).
12. Hubel, D.H. and Wiesel, T.N.: Receptive fields of cells in striate cortex of very young, visually inexperienced kittens. J. Neurophysiol. 26: 994-1002 (1963).
13. Hubel, D.H. and Wiesel, T.N.: Binocular interaction in striate cortex of kittens reared with artificial squint. J. Neurophysiol. 28: 1041-1059 (1965).
14. Hubel, D.H. and Wiesel, T.N.: The period of susceptibility to the physiological effects of unilateral eye closure in kittens. J. Physiol. (Lond.) 206: 419-436 (1970).
15. Imbert, M. and Buisseret, P.: Receptive field characteristics and plastic properties of visual cortical cells in kittens reared with or without visual experience. Exp. Brain Res. 22: 25-36 (1975).
16. Pettigrew, J.D., Nikara, T. and Bishop, P.O.: Responses to moving slits by single units in cat striate cortex. Exp. Brain Res. 6: 373-390 (1968a).
17. Pettigrew, J.D., Nikara, T. and Bishop, P.O.: Binocular interaction on single units in cat striate cortex: simultaneous stimulation by single moving slit with receptive fields in correspondence. Exp. Brain Res. 6: 391-410 (1968b).
18. Pettigrew, J.D., Olson, C. and Barlow, H.B.: Kitten visual cortex: short-term, stimulus induced changes in connectivity. Science 180: 1202-1203 (1973).

19. Pettigrew, J.D., Olson, C. and Hirsch, H.V.B.: Cortical effect of selective visual experience: degeneration or reorganization? Brain Res. 51: 345-351 (1973).
20. Pettigrew, J.D.: The effect of visual experience on the development of stimulus specificity by kitten cortical neurons. J. Physiol. (Lond.) 237: 49-74 (1974).
21. Shlaer, R.: Shift in binocular disparity causes compensatory change in the cortical structure of kittens. Science 173: 638-641 (1971).
22. Van Sluyters, R.C. and Blakemore, C.: Experimental creation of unusual neuronal properties in visual cortex of kitten. Nature 246: 21-28 (1973).
23. Vital-Durand, F. and Jeannerod, M.: Role of visual experience in the development of optokinetic response in kittens. Exp. Brain Res. 20: 297-302 (1974).
24. Wiesel, T.N. and Hubel, D.H.: Effects of visual deprivation on morphology and physiology of cells in the cat's lateral geniculate body. J. Neurophysiol. 26: 978-993 (1963).
25. Wiesel, T.N. and Hubel, D.H.: Comparison of the effects of unilateral and bilateral eye closure on cortical units responses in kittens. J. Neurophysiol. 28: 1029-1040 (1965a).
26. Wiesel, T.N. and Hubel, D.H.: Extent of recovery from the effects of visual deprivation in kittens. J. Neurophysiol. 28: 1060-1072 (1965b).

CHAPTER EIGHT

LANGUAGE REPRESENTATION AND BRAIN DEVELOPMENT*

Henry Hecaen

INTRODUCTION

This chapter deals only with the disorders of language due to cortical lesions in childhood, excluding the developmental aphasias and dyslexias, as well as dysarthria and dysphonia due to subcortical lesions.

One may say that the acquired aphasia of childhood refers only to disturbances of language due to cerebral lesions which have occurred after language acquisition.

The clinical pattern of acquired aphasia of childhood has several essential differences from that of adults (Bernhardt, 1897; Guttmann, 1942; Branco-Lefevre, 1950; Basser, 1962; Lenneberg, 1967; Alajouanine et Lhermitte, 1965; Hecaen, 1976). As early as 1885 Bernhardt observed that acquired aphasia in childhood was relatively frequent, transient, and predominantly of an expressive nature. Freud (1897) clearly distinguished acquired aphasia from developmental language retardation, emphasizing that it occurred with much greater frequency after right hemispheric lesions than did acquired aphasia in adults. Subsequently only single case reports are found in the literature. In 1942 Guttmann presented the first detailed and systematic analysis of childhood aphasia, describing 16 of his own cases. Most subsequent authors, with occasional exceptions, have tended to agree on the main clinical features. Regardless of lesion localization, the spontaneous speech is nonfluent. Mutism is common initially. Subsequently, there is a reduced initiative for speech, hesitations, dysarthria and an impoverishment of language with a reduced vocabulary. The reported frequency and severity of comprehension defects vary with the study. Guttmann (1942) and Branco-Lefevre (1950) believe disorders of comprehension are rare; Alajouanine and Lhermitte

* Travail de l'Unité de Recherches Neuropsychologiques et Neurolinguistiques (U-111) de l'I.N.S.E.R.M., Laboratoire de Pathologie du Langage de l'E.H.E.S.S., E.R.A. No 274 au C.N.R.S., 2ter, rue d'Alésia – 75014 PARIS.

(1965) and Collignon et al. (1968) report they can be found in 1/3 of childhood aphasia cases and may be quite severe.

Recovery from aphasia also presents characteristics special to the developing brain. Motor aphasia, the usual variety, may have a good prognosis, recuperation often being complete in four weeks (Guttmann, 1942). The prognosis in mixed motor and sensory aphasia remains more guarded, however. In his studies of cerebral dominance for language in the child Lenneberg (1967) attempted to define a critical period for the acquisition of language, a period corresponding to the development of cerebral dominance. Age of onset of cerebral injury was used as a guide. If aphasia were acquired prior to age 3, recovery was rapid and complete. Before 10 years of age, a true aphasia would develop, but slow recovery was the rule. If the child were between 11 and 14 when brain damage occurred, recovery was less likely. Lenneberg concluded that a period roughly between 2-3 years of age and puberty was a critical period. During this time a state of cerebral plasticity existed; following this period the hemispheres have achieved their final specialization.

Recently, the conclusions of Lenneberg have been disputed by Krashen (1973). This author lowered the end of the period during which transfer between the hemispheres is still possible to 5 years of age. He pointed out that in the series of Basser (1962) there were no instances of aphasia with a right-sided lesion occuring after age 5.

Krashen compared the frequency of right-sided lesions in several series of childhood aphasias where the lesion occurred before or after 5 years with the frequency of right-sided lesions producing aphasia in adults. He noted that in the older children the percentage of right-sided lesions producing aphasia was similar to that observed in adults. Krashen finds support for this argument as well in the results of studies of dichotic listening, particularly those of Kimura (1963) and Knox and Kimura (1972). The superiority of the right ear, that is the predominance of the left hemisphere in the reception of verbal material (digits) appears toward the age of 6, and that of the left ear (right hemisphere) for familiar sounds toward age 5. Since this work, Nagafuchi (1970) and Ingram (1974) have found a right ear superiority for verbal sounds (words) as early as 3 years of age.

In 1976, we proceeded to revise our observations concerning cortical lesions in the child (from this analysis, two new cases have been observed). Then we compared our results with those of the literature to study the problem of the ontogenesis of the hemispheric dominance.

POPULATION, METHOD AND RESULTS

We have studied each child with a cortical lesion observed at the Centre
Neurochirurgical Sainte Anne (Dr. Mazars) during the past 15 years. The
youngest was 3½ years old, and we set the upper limit at 15 years. The total
number of cases observed was 28. There were unilateral left-sided lesions in 18
cases, right-sided lesions in 7, and 3 cases with bilateral lesions.

The lateralization of the lesion was confirmed by surgery and or the results
of clinical, radiologic and electroencephalographic testing. However, in the
cases due to trauma we were unable to completely exclude a lesion of the other
hemisphere.

Tables 1, 2 and 3, give the age at the onset of brain pathology, sex, manual
preference, etiology, localization, associated neurologic and neuropsychologic symptoms, and the clinical picture and evolution of the language disorder. In 20 of 28 children with cortical lesions a language disorder was found.
With respect to unilateral lesions, a disorder of language was present in 16 of
18 cases with left hemispheric lesions, and in 2 of 7 cases (of which one was
left-handed) with right-sided lesion. This corresponds to an incidence of 88%
with left, and 28% with right hemispheric lesions. These percentages correspond approximately to those in the series of Basser, viz: 86% and 46%
respectively.

The frequency and evolution of the various disorders of verbal behavior
observed with left-sided lesions are presented in table 4.

Clinically, two characteristics emerged from this study: a positive feature;
the frequency of mutism, loss of ability to initiate speech, and, more generally,
loss of ability to communicate; and a negative feature: the fact that paraphasias are rare and logorrhea is nonexistent. Articulatory disorders were the
rule; and disorders of auditory comprehension occurred in more than one-
third of the cases. The finding of poor auditory comprehension was similar to
that of Alajouanine and Lhermitte (1965), but in contradiction with the observations of Guttmann (1942) and Branco-Lefevre (1950) who stressed the
rarity of this defect. In those studies in which auditory comprehension was
impaired, the comprehension defect often cleared rapidly after the acute
stage. Problems with naming were more frequent and more persistent, however. Disorders of writing appeared to be the most frequent and longest
lasting of all language defects. Among the associated neuropsychologic symptoms, acalculia was the most common.

We have also studied the relation between different aspects of the language
disorder and lesion localization. Mutism appeared chiefly in association with

Table 1. Left-sided lesions.

Name and Sex	Age	H	Etiology	Coma	Loc.	Symptoms- neurol., neuro- psychol.	Mu- ism	Artic. dis.	Compr. dis.	Nam- ing dis.	Para- pha- sia	Read- ing dis.	Writ- ing dis.	Evolution
L.S. M	6	R	Abcess	0	R	R. Hemipl. Astereognosia Constr. Apr.	16 d	+	+	+	0	?	?	15 d Improvement
L.M. F	6	R	Tr.	3 d	?	R. Hemipl. Hemianesth. Acalculia	8 d	±	0	0	0	+	+	1 m Dis. Writing Reading
I.P. F	7	R	Angioma Hemat.	0	T	0	0	0	±	+	+	0	±	R 1½ m
L.P. M	8	R	Abcess	0	T	R. Hemipl. Hemianesth.	0	+	0	±	0	0	+	1 m Improvement
M.C. F	8	R	Angioma	0	PRT	R. Hemipl. Hemianops.	28 d	+	0	+	0	+	+	2 m lack of words Agraphia Alexia
R.D. M	8	R	Tr.	.7 d	?	R. Brachial. Monopares. Acalculia	12 d	±	+	+	0	0	±	6 y Dis. Writing ±
C.S. E.	8	R	Hemat. Angioma	0	T	R. Sup. Quad.	0	0	±	+	0	0	±	2 m Improvement
G.J.P. M	8	R	Hemat.	0	F post.	R. Hemipl. BLF Apr. Acalculia	1 m 5 d	+	0	0	0	0	±	2 m Articul. Dis.
V.C. F	8	R	Tumor	0	T	R. Hemianop. Acalculia	0	0	0	0	0	0	0	1 y
G.E. M	9	L	Tr.	8 d	?	R. Hemipares. Acalculia	8 d	±	0	0	0	0	+	4 m Improvement
L.J.P. M	11	R	Tr.	0	T + sub corti- cal.	R. Hemipl. Astereognosia Acalculia	3 m	+	+	+	0	+	+	6 y Writing ±

Table 1. Left-sided lesions (cont.).

Name and Sex	Age	H	Etiology	Coma	Loc.	Symptoms-neurol.-neuropsychol.	Mutism	Artic. dis.	Compr. dis.	Num-ing dis.	Para-phasia	Read-ing dis.	Writ-ing dis.	Evolution
C.P. M	12	R	Tr.	14 d	?	R. Brachial. Monopares. Constr. Apr. Acalculia	13 d	0	0	0	0	+	±	3 m R
B.D. M	12	R	Tr.	15 d	?	R. Hemipl. Acalculia	0	+	+	+	0	+	+	18 m No change
G.N. M	13	R	Tumor	0	F	R. Hemipl. Acalculia R. Astereogn.	2 m	+	0	0	0	+	±	3 m Articul. Dis.
LE BP M	14	L	Tr.	19 d	?	R. Brachial. Monopares. Acalculia Finger Agn. BLF Apr.	0	±	+	+	0	+	+	2 y R
D.G. M	14	R	Tr.	7 d	?	R. Hemipl. Acalculia	5 d	±	0	0	0	±	0	1 m R
M.S. F	14	R	Abcess	0	R	R. Hemipl. R. Hemianesth.	0	±	0	0	0	0	0	3 m R
L.A. M	15	R	Tr.	1 d	?	R. Hemianops.	0	0	0	0	0	0	0	R

H. = Handedness; Loc. = Localization; d = day; m = month; y = year; Artic. Dis. = Articulatory Disorders; Compr. Dis. = Comprehension Disorders; Hempl. = Hemiplegia; L = Left; R = Right; Monopar. = Monoparesis; Hemianesth. = Hemianesthesia; U.S.A. = Unilateral Spatial Agnosia; Hemianops. = Hemianopsia; Quad. = Quadranopsia; B.L.F. Apr. = Buccolinguofacial Apraxia; Const. Apr. = Constructional Apraxia; R = Recovery.

Table 2. Right-sided lesions.

Name and Sex	Age	H	Etiology	Coma	Loc.	Symptoms- neurol., neuro-psychol.	Mu-ism	Artic. dis.	Compr. dis.	Nam-ing dis.	Para-pha-sia	Read-ing dis.	Writ-ing dis.	Evolution
W.C. M	3½	R	Tr.	1 d	?	L. Hempl.	6 d	±	0	0	0	?	?	15 d
D.H. M	6	L	Tr.	5 d	?	0	5 d	±	0	0	0	0	+	Writing Dis. 1 y ½
A.F. F	8	L	Tumor	0	R	L. Somato-Sensory Dis.	0	0	0	0	0	0	0	
L.P.J. M	12	R	Tr.	0	?	L. Hempl.	0	0	0	0	0	0	0	
B.V. F	13	R	Hemat. Angioma	2 d	T	L. Sup. Quad. USA Constr. Apr.	0	0	0	0	0	0	Spat-ial Dys-graph	2 m
M.P. F	14	R	Hemat. Aneur.	0	F	0	0	0	0	0	0	0	0	
P.M. M	15	R	Tr.	2 d	F	0	0	0	0	0	0	0	Spat-ial Dys-graph	

H. = Handedness; Loc. = Localization; d = day; m = monthy; y = year; Artic. Dis. = Articulatory Disorders; Compr. Dis. = Comprehension Disorders; Hempl. = Hemiplegia; L = Left; R = Right; Monopar. = Monoparesis; Hemianesth. = Hemianesthesia; U.S.A. = Unilateral Spatial Agnosia; Hemianops. = Hemianopsia; Quad. = Quadranopsia; B.L.F. Apr. = Buccolinguofacial Apraxia; Const. Apr. = Constructional Apraxia; R = Recovery.

Table 3. Bilateral or diffuse lesions.

Name and Sex	Age	H	Etiology	Coma	Loc.	Symptoms-neurol.-neuro-psychol.	Mu-ism	Artic. dis.	Compr. dis.	Num-ing dis.	Para-phasia	Read-ing dis.	Writ-ing dis.	Evolution
G.M. F	7	R	Tr.	21 d	?	R. Hempl. Bil. visual defects Vis. Agnosia Const. Apr.	21 d	0	0	0	0	+	+	3 y Alexia Agraphia
S.G. F	11	L	Tr.	7 d	Diff.	L. Hemiplegia Sensory Dis. Acalculia	12 d	+	0	±	0	0	±	4½ y Writing Dis.
R.B. M	16	R	Tr.	0	F	L. facial paresis	0	0	0	0	0	0	0	

H. = Handedness; Loc. = Localization; d = day; m = monthy; y = year; Artic. Dis. = Articulatory Disorders; Compr. Dis. = Comprehension Disorders; Hempl. = Hemiplegia; L = Left; R = Right; Monopar. = Monoparesis; Hemianesth. = Hemianesthesia; U.S.A. = Unilateral Spatial Agnosia; Hemianops. = Hemianopsia; Quad. = Quadranopsia; B.L.F. Apr. = Buccolinguofacial Apraxia; Const. Apr. = Constructional Apraxia; R = Recovery.

Table 4. Frequency of different aphasic symptoms in 16 cases due to left hemisphere lesions.

	Number of cases	Percentage	Evolution
Mutism	10	62	from 5 days to 3 months
Articulatory disorders	13	81	persistent in 4 cases
Auditory verbal comprehension disorders	7	43	persistent in 1 case
Naming disorders	8	50	persistent in 3 cases
Paraphasia	1	6	disappearance
Reading disorders 15 cases only	9	60	persistent in 3 cases
Writing disorders	13	86	persistent in 7 cases

anterior (frontal or Rolandic) lesions. It was present in 4 of the 5 cases with anterior lesions, and in one of 4 cases with temporal lesions (though in this case there was also involvement of the brain stem). Articulatory disorders were similarly noted in all five anterior cases, and in 2 of the 4 cases of temporal lesions, though again, one of these patients had subcortical involvement. Conversely, disorders of auditory verbal comprehension were found especially with temporal lesions (3 of 4 cases). There were no localizing features in the disorders of naming, writing or reading, at least during the acute period. We should note that this difference in the clinical picture according to localization was present in both the youngest and oldest children in the series.

DISCUSSION

Our results confirm, then, several of the well-established findings in acquired aphasia of childhood, namely their great frequency in cases of unilateral left lesions and their relative frequency in right lesions, the usually rapid and almost complete recovery even in cases of dominant hemisphere lesions, and their unique character in relation to adult aphasia (e.g. absence of logorrhea, rarity of paraphasias). These findings also appear to indicate that language disorders occur with right hemispheric lesions only in very young patients. With regard to left-sided lesions, whatever the age of the children, one notes that the language disturbance takes on a different aspect depending on the intrahemispheric localization of the lesion, in a manner similar to, but not identical with, the regional syndromes as observed in adults. Finally, although the recovery is certainly more striking than in the adult, it is important to stress the persistence, at times permanent, of mild verbal deficits, particularly in writing.

Findings from studies of aphasia in childhood thus support the notion that very young children have hemispheric equipotentiality for language, and that cerebral dominance is established in the course of maturation.

Due largely to Basser's data (1962) it is known that hemispherectomies performed on either the right or the left side produce similar effects.

Permanent dysphasia has never been observed after left hemispherectomy performed for damage having occurred before the acquisition of language.

On the other hand, the bilateral Wada test showed language representation in the right hemisphere in only six of 140 right-handed cases (4%) when the left hemispheric involvement was not early in life, but in the reverse situation, that of early lesions, language representation was on the right side in four of 31 cases (13%) and bilateral in two subjects (6%) (B. Milner, 1974a).

There appears to be a critical period during which each hemisphere may support verbal activity, but this period may be shorter than has been suggested by Lenneberg (1967).

On the other hand two other lines of argument favor hemispheric specialization established much earlier in life, or even innate.

First, psychologic examination of subjects having had early unilateral hemispheric damage reveal a type of deficit in accordance with the lesioned hemisphere, similar to that encountered in adults, i.e. deficits in verbal tasks for left damage, deficits in non-verbal tasks, in particular spatial tasks, from right lesions (McFie, 1961; Fedio and Mirsky, 1969; Woods and Teuber, 1973; Rudel et al., 1974).

A second set of arguments in favor of early specialization derives from anatomical research on the structural asymmetry of the hemispheres in the region of the planum temporale, an asymmetry demonstrated in adults by Geschwind and Levitsky (1968) and in the fetus and newborn by Teszner et al. (1972), Witelson and Pallie (1973) and, above all, by Wada et al. (1975). These latter authors examined 100 fetal and newborn brains and compared them with 100 adult brains. In newborns as in adults, the planum temporale is larger on the left than on the right in approximately 90% of cases. These authors stress that this difference is more marked in adults than in newborns, and that there is very likely a further development of the left planum during maturation.

The existence of this structural difference in embryonic life appears particularly likely if we consider it in relation to observations of behavior, such as those of Eimas et al. (1971), Morse (1972), which show that infants, as early as the first month, perceive phonemes in the same categorical fashion as adults.

It appears, then, that we now have evidence, both behavioral and anatomic, for an extremely early – indeed, even innate – lateralization of functional representation. On the other hand, we have also mentioned those findings in favor either of initial hemispheric equipotentiality or of the possibility of functional displacement from one hemisphere to the other.

Against these contradictory findings, we should point out that the very precise study of verbal capacities of subjects who have been hemispherectomized early revealed differences to Dennis and Kohn (1975) and to Dennis and Whitaker (1976), depending on which hemisphere was removed. When only the right hemisphere is still functioning we certainly cannot speak of resembling aphasia but one notes a deficit in the syntactical capacities as well as in comprehension, production or the handling of syntactical relations although phonologic and semantic capacities have been normally developed.

It is also important to emphasize that language representation can well be in the left hemisphere, in spite of an early and significant lesion of that side. Using the Wada test, Milner (1974) demonstrated that language could be represented in the left hemisphere even with considerable involvement of that hemisphere, if the posterior temporoparietal region was not destroyed. Besides the factor of age, the site of the lesions, therefore, appears of equal importance in the displacement of functional representation.

The relationship of these various findings to the effects of early hemispheric lesions and to the manner of subsequent language recovery is still not sufficient to construct a precise systematization of the ontogeny of cerebral dominance. While there is no doubt that factors of time and site of lesion are important, the role of experience ought not to be neglected. The concept of a critical period for language acquisition should also be reformulated as in the work of Fromkin et al. (1974), based on observations in a young girl deprived of all verbal stimulation from the age of 20 months to 13 years and 9 months of age, after which time some limited language acquisition was possible. Tests of dichotic listening showed that both verbal and non-verbal material were processed by the right hemisphere. The critical period is not just a period of hemispheric equipotentiality, but one in which those stimuli necessary to set in motion a preformed area in the left hemisphere should appear.

The necessity for adequate stimulation in order to insure the functioning of pre-existing mechanisms seems to emerge from a comparison of the faculty which exists in infants for discriminating phonetic contrasts such as /r/ versus /1/ in a speech context (Eimas, 1975) and the lack or diminution of this faculty in adult speakers of a language which does not have this phonemic contrast

(Miyawaki et al.). For Eimas, based on findings from experiments with adults, this mechanism could consist of detectors of phonetic features. This mechanism would depend on a biologic structure peculiar to the human infant and require little exposition to adequate stimulation in order to be operational. If ever the exposition did not occur at all during the critical period, it would be liable to lose its sensitivity altogether.

The apparent contradiction between the findings in acquired aphasia in children and arguments in favor of an early, probably innate, hemispheric specialization, can be resolved if we accept the existence of a critical period during which specific stimuli are required for the development of the functional potentialities of the preformed area. On the other hand, a less specialized region will assure function either at an inferior level, or at the expense of other functions which it normally subserves.

The hypothesis of a displacement of language representation to the other hemisphere can apply, it would seem, to only a very limited period time. Subsequently, the situation will be different. As we have seen, there is both clinical and anatomic evidence to show that there is not only a functional but a structural asymmetry between the hemispheres, present at the earliest age and moreover, becoming more accentuated in adulthood.

REFERENCES

Alajouanine Alajouanine Th. and Lhermitte F.: Acquired aphasia in children. Brain, 1965, 88, 653-662.
Basser L.S.: Hemiplegia of early onset and the faculty of speech with special reference to the effects of hemispherectomy. Brain, 1962, 85, 427-460.
Bernhardt M.: Ueber die Spatische cerebral paralyse un Kindersalter (Hemiplegia Spastisca Infantilis) nebst einem Excurse über "Aphasie bei Kindern." Summarized in Neurologisches Centreblatt. 1886-180-181. Virchow's Archiv., 1885, 102, 1.
Branco Lefevre A.F.: Contribuição para o estudo da psicopatologia, da afasia em crianças. Arq. Neuropsiquiat. (São Paulo), 1950, 8, 345-393.
Dennis M. and Kohn B.: Comprehension of syntax in infantile hemiplegics after cerebral hemidecortication: left hemisphere superiority. Brain and Language, 1975, 2, 472-482.
Dennis M. and Whitaker H.A.: Language acquisition following hemidecortication: Linguistic superiority of the left over the right hemisphere. Brain and Language, 1976, 3, 404-433.
Eimas P.D.: Auditory and phonetic coding of the cues for speech: Discrimination of the (r-l) distinction by young infants. Perception and Psychophysics, 1975, 18, 341-347.
Eimas P.D., Siqueland E., Jusczyk P. and Vigorito J.: Speech perception in infants. Science, 1971, 171, 303-306.
Fedio P. and Mirsky A.F.: Selective intellectual deficits in children with temporal lobe or cephalic epilepsy. Neuropsychologia, 1969, 7, 287-300.
Freud S.: Die Infantile cerebrallähmung. Vienna, 1897.

Fromkin V.A., Krashen S., Curtiss S., Rigler D. and Rigler M.: The development of language in Genie: A case of language acquisition beyond the "critical period." Brain and Language, 1974, 1, 81-108.

Geschwind N. and Levitsky W.: Human brain, left-right asymmetries in temporal speech regions. Science. 1968, 161, 186-187.

Guttmann E.: Aphasia in children. Brain, 1942, 65, 205-219.

Ingram D.: Cerebral speech lateralization in young children. Neuropsychologia, 1975, 13, 103-105.

Kimura D.: Speech lateralization in young children as determined by an auditory test. J. of Comparative and Physiological Psychology, 1963, 56, 899-902.

Knox C. and Kimura D.: Cerebral processing of non verbal sounds in boys and girls. Neuropsychologia. 1970, 8, 227-237.

Krashen S.: Lateralization language learning and the critical period. Some new evidence. Language Learning, 1973, 23, 63-74.

Lenneberg E.: Biological foundations of language. J. Wiley and Sons. New York, London, Sydney. 1967.

McFie J.: Intellectual impairment in children with localized post infantile cerebral lesions. J. Neurol. Neurosurg. Psychiat., 1961, 24, 361-365.

Milner B.: Hemispheric specialization: scope and limits. 75-88 in The Neurosciences. Third Study Program. Edited by F.O. Schmitt and F.G. Worden. The M.I.T. Press, Cambridge, London, 1974a.

Milner B.: Personal communication. 1974b.

Miyawaki W., Strange W., Verbrugge R., Liberman A.M., Jenkins J.J. and Fujimara O.: An effect of linguistic experience: The discrimination of r and l by native speakers of Japanese and English. Perception and Psychophysics. 1975, 18, 331-340.

Morse P.A.: The discrimination of speech and non speech stimuli in early infancy. J. Exp. Child Psychol., 1972, 14, 253-260.

Nagafuchi M.: Cited by D. Ingram. 1974.

Rudel R., Teuber H.L. and Twitchell T.E.: Levels of impairments of sensorimotor early damage. Neuropsychologia, 1974, 12, 95-108.

Teszner D., Tzavaras A., Gruner J. and Hecaen H.: L'asymétrie droite-gauche du planum temporale. A propos de l'étude anatomique de 100 cerveaux. Rev. Neurol., 1972, 126, 444-449-

Wada J., Clark R. and Hamm A.: Cerebral hemispheric asymmetry in humans. Arch. of Neurol., 1975, 32, 239-246.

Witelson S.F. and Pallie W.: Left hemisphere specialization for language in the newborn: Neuroanatomical evidence of asymmetry. Brain, 1973, 96, 641-646.

Woods B.T. and Teuber H.L.: Early onset of complementary specialization of cerebral hemispheres in man. Tr. Am. Neurol. Assoc., 1973, 98, 113-115.

CHAPTER NINE

ZINC AND THE DEVELOPING NERVOUS SYSTEM: TOXIC EFFECTS OF ZINC ON THE CENTRAL NERVOUS SYSTEM OF THE PREWEANLING RABBIT

ARTHUR L. PRENSKY, M.D.* AND LAURA HILLMAN, M.D.**

In 1869 Raulin (34) first noted that zinc was essential for growth of the organism Aspergillus Niger. However, it was not until 1961 that Prasad and his associates (31) described the syndrome of iron deficiency anemia, hepato-splenomegaly, hypogonadism, dwarfism and geophagia that is now recognized to be associated with zinc deficiency in children. Prior to this Blamberg et al. (2) had noted that zinc deficiency in hens reduced chick production and increased the number of embryonic abnormalities found in those animals born alive. Subsequently, the teratogenic effect of zinc depletion on the nervous system has been the subject of study by Hurley and her associates (21, 22, 44) and Warkany and Petering (47). When pregnant dams are subjected to a zinc deficient diet a high incidence of CNS malformations are found in living offspring including exencephaly, hydrocephalus, malformations of the mid-brain, absence of the optic nerves and microphthalmia. While it has never been clearly established that zinc deficiency in man results in an increased number of congenital malformations of the central nervous system, it has been noted that anencephaly is found in an increased incidence in areas of Egypt and Iran where zinc deficiency is common (5, 38).

The subject of zinc deficiency in humans has been reviewed on numerous occasions since 1961 (3, 16, 30), and there is now reasonably conclusive evidence that supplementation of zinc deficient children and young adults produces an acceleration in growth, weight and bone age (35). Surprisingly, these zinc deficient children have not been examined in detail for behavioral or intellectual abnormalities. However, Henkin and his associates (19, 20) have noted that zinc depletion does result in a dysfunction of taste and smell with resultant anorexia. It has been suggested on several occasions that significant deficiencies in body zinc content may not be limited to populations of the

* Supported in part by NIH Grant T32 NS 7027 and a grant from the Allen P. & Josephine B. Green Foundation, Mexico, Missouri.
** Supported in part by NIH Grant 1R01 HDO 998.

Middle East but occur in impoverished areas throughout the world in populations suffering from protein-calorie malnutrition, or can even be seen in infants and children of middle class families whose diets are restricted in zinc content (17, 24, 36, 43). Once again, there has been no attempt in these populations to define a neurologic syndrome resulting from zinc depletion.

Experimental zinc deficiency has been produced in suckling rats and is associated with decreased weight gain and with biochemical changes in the structure of the central nervous system which can, in part, be distinguished from those found in similarly undernourished controls or animals whose dams are fed an ad-lib diet (10, 11). When compared to pair-fed controls zinc depleted animals show a greater reduction in brain weight relative to body weight and a parallel decrease in the DNA content of the forebrain (37). The reduction in protein per cell in zinc-deficient pups also exceeds that found in pair-fed controls during the first and second weeks of development but not thereafter (11). However, differences in the RNA content between these two groups were slight. The lipid content of the zinc deficient forebrain is only two-thirds that found in pair-fed controls (37). Administration of labeled precursors shows that zinc deficient animals exhibit a decrease in the uptake of ^{3}H-thymidine and ^{3}H-UTP in brain DNA and RNA respectively (10, 37), and ^{35}S in brain protein (37). These changes are associated with an altered ratio in rat brain monosomes to polysomes (10, 12), and is consistent with the known zinc requirement of DNA polymerase (29, 40, 42), thymidine kinase (33) and RNA polymerase (45) in other cell systems as well as the increase in ribonuclease found in other tissues of zinc deficient rats (32). Prenatal (15) or early postnatal (4, 25) zinc deficiency results in impaired avoidance reactions and decreased learning, as measured by maze tests, in adult animals.

The effects of ingestion of large amounts of zinc orally was investigated in rats (18) and dogs (6) as early as 1927. These initial studies produced negative results presumably because only a small amount of zinc was added to the diet. Later investigations of rats using larger quantities of zinc are summarized by Van Reen (46). They indicate that if zinc constitutes 0.5 to 1% of the diet, anorexia, growth failure and a copper-responsive anemia occurs (14). Neurologic abnormalities have not been described. Zinc toxicity in man generally results in acute gastrointestinal or pulmonary disease. However neurologic complications were reported in a single instance (28) and consisted of slight ataxia and lethargy. Recently, a family has been reported with an inherited abnormality in zinc metabolism resulting in elevated plasma levels of zinc that are between three and four times that found in the normal population (41). No one in this family is reported to have neurologic disease.

However, it was not clear that a large proportion of this zinc was in a freely diffusible form. It does appear that the mammalian central nervous system is surprisingly resistant to large quantities of orally ingested zinc.

The present studies were undertaken to determine if parenterally administered zinc could be shown to have a toxic effect on developing rabbit brain. Both the subcutaneous and intraocular administration of zinc chloride appear to have effects that could be potentially damaging to the central nervous system of the developing rabbit.

METHOD

Pregnant New Zealand white rabbits were purchased from Isaacs Lab Stock in Litchfield, Illinois. Litters were altered at birth to contain five to seven animals. Nursing dams were fed an ad-lib diet. One litter of animals (average weight 250 grams) received subcutaneous injections of 16 mg. of Zn Cl_2 (1.4 × 10^{-4}M of zinc) on five occasions between days 11 and 17 of postnatal life. There was a high initial mortality unassociated with convulsions or pathologic evidence of infection or shock. Surviving animals were sacrificed on day 17 of life after a neurologic examination and electroencephalogram had been performed. They were compared to controls injected subcutaneously with hypertonic sodium chloride.

Other animals were injected intraocularly with 2 mg. Zn Cl_2 (10 µl) on the 23rd and 24th, or 24th and 25th days of life (a total dose of 3.68 × 10^{-5}M of zinc was given) and sacrificed 18 to 36 hours later. The injections were made into the right eye. The volume of the eye was calculated by water displacement as 0.8 cc. The final concentration of Zn Cl_2 in the eye therefore is 1 × 10^{-2}M. Other animals were injected in the same manner with a cumulative dose of 2 × 10^{-8}M of lead in the form of lead acetate or with normal saline. Two hours following the last injection of metal 10 µc of proline, L-(2, 3, 4, 5, $^{-3}$H), specific activity 60.0 Ci/mmol, was injected into both the right and left eyes. At the time of sacrifice the right eye, the left and right superior colliculi and left frontal lobes of three animals in each group were frozen immediately and saved for biochemical investigation. Tissue and sera from other animals were saved for determination of zinc levels. Small pieces of the right optic nerve and left optic tract were fixed in phosphate-buffered gluteraldehyde for electron-microscopy. The eyes, optic nerves and tracts and brain of one animal in each group were removed as a unit and fixed in 10% formalin for pathologic investigation.

Routine pathologic studies were performed on formalin fixed tissue stained with hematoxylin and eosin. Electronmicroscopic studies of the optic nerve and tract were performed by Dr. James Nelson using a Phillips 100 electronmicroscope.

BIOCHEMICAL PROCEDURES

Zinc determinations in tissue (27) and sera were performed by Dr. Laura Hillman using a Varian 1250 A atomic absorption spectrophotometer. Amino acids in cerebrum and cerebellum were determined with a Technicon Model TSM amino acid analyzer after extraction of frozen tissue in 5% perchloric acid at 0°C. Routine biochemical studies of sera were performed in the clinical chemistry laboratories.

Lipids were extracted by the method of Folch et al. (9) from samples of mixed gray and white matter of systemically injected animals. Tissues from animals injected intraocularly with tritiated proline were extracted by homogenization in 5% TCA at 0°C. The homogenate was then centrifuged and the precipitate washed twice with diethyl ether and once with ethanol/ether (3:1, V/V) and then dissolved in 1N NaOH at 37°C. The supernatant was washed three times with ether to remove TCA. A portion of the supernatant was dissolved in Eastman 3a70. An aliquot of the sodium hydroxide extract was dissolved in NCS buffer and mixed with lipid scintillation fluid containing 4 cc of Omniflour per liter of toluene. Samples were counted on a Packard Tricarb Liquid Scintillation counter and the counts corrected for absorption by the addition of an internal standard of ³H-proline. Proteins were determined by the method of Lowry et al. (26). The radioactivity in the TCA extract is calculated per mg. of protein in the corresponding NaOH extract.

RESULTS

The subcutaneous injection of zinc chloride in developing rabbits over a six-day period results in a neurologic syndrome characterized by lethargy, slight ataxia, decreased strength in the hind limbs which are slightly hypotonic, and an impaired righting reflex. The electroencephalogram of zinc injected animals is considerably slower than that of controls (fig. 1). Pathologic changes in the nervous system are minimal and consist only of a mild gliosis in cerebral gray and white matter with some pairing of astrocytic nuclei. There is no

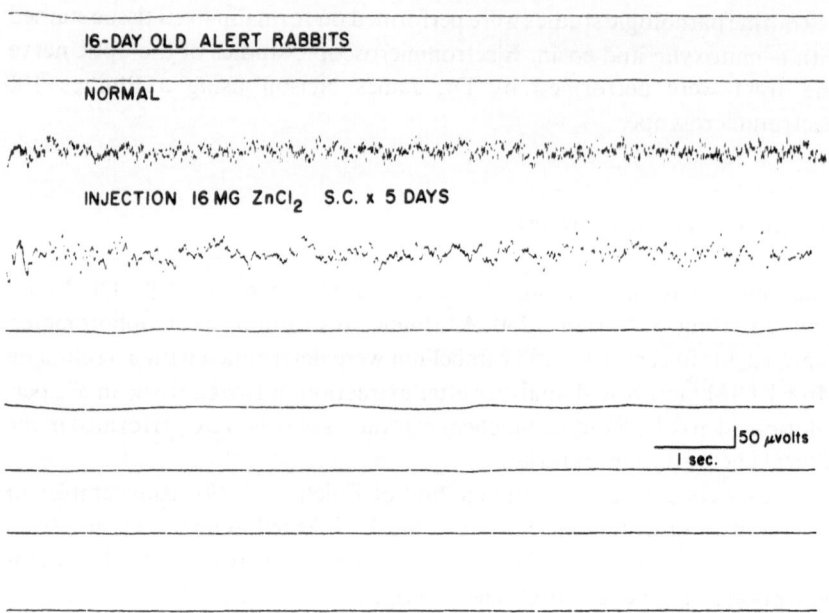

Fig. 1. EEG obtained from 16 day old rabbits injected with either normal saline or 16 mg day of zinc chloride for 5 days. The EEG of the zinc injected animal is considerably slower exhibiting both alpha and theta activity.

evidence of neuronal loss or vascular damage. No changes are seen in the cerebellum or spinal cord. The kidneys of zinc injected animals reveal evidence of tubular distension and toxic granulation in the cytoplasm of the renal tubular epithelium; the glomeruli appear normal. Control kidney: 21.7 µg Zn/g wet weight, Zn injected kidneys 112 and 73 µg Zn/g wet weight. There is a slight increase in sites of extramedullary hematopoesis in the liver but the parenchymal tissue and the intrahepatic biliary system appear unaffected. The combined sera of these animals revealed a BUN of 23.7, an SGOT of 16.0 and a calcium of 13.4 mg%. The zinc in the sera was elevated to 1898.2 µg% (the range of control values was 100-150 µg%), however zinc levels within the nervous system were normal ranging from 13.4 ± 5.8 µg/gm fresh weight in the brain stem to 16.5 µg in cerebellum and cerebrum. The lipid content of cerebrum and brain stem as a percent fresh weight did not differ significantly from the control animal of the same age (table 1). The levels of amino acids were also similar in the cerebrum and cerebellum of two zinc infected animals when compared to a control of the same age.

Table 1. Lipid content (% fresh weight) of the brains of rabbits injected s.c. with 16 mg of zinc chloride from day 11 to 17.

	Zinc	*Control*
Cerebrum	12.3 ± 0.8	14.9
Brain Stem	6.17 ± 0.19 (2)*	5.32

* Number of animals is given in parentheses.

The intraocular injection of zinc chloride, lead acetate or sodium chloride resulted in no alteration in body weight. When sacrificed, animals in the control group weighed 469.2 ± 20.9 grams, while those in the zinc and lead groups weighed 478.5 ± 28.9 grams, and 472 ± 59.9 grams respectively. The retinal tissue of animals injected with sodium chloride appeared normal, and there was no definite pathologic abnormality in those injected with lead acetate. However, patchy destruction of retinal tissue was noted in animals injected with zinc chloride. Nevertheless, large sections of the retina appeared intact. Examination of formalin fixed sections of the optic nerves and tract by light microscopy showed no abnormalities. However, electronmicroscopy revealed occasional degenerating axons in the optic nerves of both the lead and zinc injected group. The optic tract appeared to be normal.

The intraocular injection of zinc was associated with only a very slight increase in serum ($155 \pm 11 \mu g\%$) and no increase in the forebrain levels of zinc. Similarly, cpm/ml of sera in zinc injected animals was 26,900 which did not differ significantly from the control (24, 960) or the lead injected animal (26,050). The specific activity of proteins (dpm/mg) of the left frontal pole of zinc injected animals was approximately 50% of the controls with a concomitant reduction in the radioactivity of the TCA extract expressed as dpm/mg protein (table 2). The specific activity of protein in the right eye of zinc injected

Table 2. Specific activity of protein (dpm/mg) of the left frontal pole of 26-day-old rabbits injected in the right eye and sacrificed after 36 hours.*

	Protein	*TCA Extract (dpm/mg protein)*	*Ratio*
Control**	228.4 ± 29.1(2)	1373 ± 127.3(2)	0.17
Zinc	117.1 ± 21.9(3)	631.1 ± 62.3(3)	0.19

* Animals were injected in the right eye with 2 mg of $Zn Cl_2$ or 10 μl of normal saline on days 23 and 24 of life. 10 μci of ³H-proline was injected into both eyes on day 24.
** Figures are given as the mean ± SD. Number of animals is given in parentheses.

Table 3. Specific activity of protein (dpm/mg) in the right eye of 26-day-old rabbits injected in the right eye and sacrified after 36 hours.*

	Mean ± S.D.
Control**	987.2
Zinc	402 ± 42.4
Lead	1394 ± 213

* Animals were injected in the right eye with 2 mg of Zn Cl$_2$. 3 μg of lead acetate or 10 μl of normal saline on days 23 and 24 of life. 10 μci of ^{3}H-proline was injected into both eyes on day 24.
** Figures are given as the mean ± S.D.

animals was only about one-third of that found in control or lead injected animals (table 3).

A major difference was seen when comparing the accumulation of label in the left and right superior colliculi. The specific radioactivity of proteins in both the left and right superior colliculi increases rapidly between 18 and 36 hours after injection (tables 4 and 5) to well above the levels found in the anterior portion of the cerebral hemisphere. The specific activity of protein in

Table 4. Specific activity of protein (dpm/mg) of the right superior colliculus of 26-day-old rabbits injected in the right eye.*

	Time of Sacrifice After Injection	
	18 Hours	36 Hours
Control**	429.5 ± 58.8(2)	3124 ± 1470(3)
Zinc	100.3 ± 16.1(2)	5042 ± 2054(3)
Lead	371.0 ± 90.5(2)	3375 ± 528(3)

* Animals were injected in the right eye with 2 mg of Zn Cl$_2$. 3 μg of lead acetate or 10 μl of normal saline on days 24 and 25 of life and sacrificed 18 hours thereafter or days 23 and 24 of life and sacrificed 36 hours thereafter. 10 μci of ^{3}H-proline was injected into both eyes 2 hours following the last injection of metal.
** Mean ± S.D. Number of animals is given in parentheses.

Table 5. Specific activity of protein (dpm/mg) of the left superior colliculus of 26-day-old rabbits injected in the right eye.*

	Time of Sacrifice After Injection	
	18 Hours	36 Hours
Control**	309.1 ± 24.0(2)	1274 ± 571 (3)
Zinc	10.3 ± 4.5(2)	137.7 ± 18.5(3)
Lead	259.7 ± 31.3(2)	2348 ± 716 (2)

* Animals were injected in the right eye with 2 mg of Zn Cl$_2$. 3 μg of lead acetate or 10 μl of normal saline on days 24 and 25 of life and sacrificed 18 hours thereafter or days 23 and 24 of life and sacrificed 36 hours thereafter. 10 μci of ^{3}H-proline was injected into both eyes 2 hours following the last injection of metal.
** Mean ± S.D. Number of animals is given in parentheses.

Table 6. Radioactivity in the colliculi of 26-day-old rabbits injected in the right eye and sacrificed after 36 hours.*

	Protein (dpm/mg)	TCA Extract (dpm/mg protein)	Ratio
Control**			
Right	3124 ± 1470(3)	794 ± 50.3(3)	3.93
Left	1274 ± 571(3)	728 ± 68.4(3)	1.75
Zinc			
Right	5042 ± 2054(3)	965 ± 63 (3)	5.22
Left	135.7 ± 18.5(3)	896 ± 141.4(3)	0.15
Lead			
Right	3375 ± 528(3)	1026 ± 254 (3)	3.29
Left	2348 ± 716(2)	984 ± 363 (3)	2.39

* Animals were injected in the right eye on days 23 and 24 of life with either 2 mg of Zn Cl$_2$, 3 μg of lead acetate or 10 μl of normal saline. 10 μci of ^{3}H-proline was injected into both eyes on day 24.
** Activities are given as the mean ± S.D. Number of animals is given in parentheses.

the zinc injected animal at 18 hours is lower in both left and right colliculi, although the right superior colliculus receives almost all its fibers from the uninjected left eye (13). At 36 hours there is no significant difference between specific activities of control, lead or zinc injected groups in the right colliculus. In the left superior colliculus, however, the specific radioactivity remains extremely low in the zinc injected animal. It is approximately 2.5% of the activity found in the right colliculus of the same animals and less than 10% of the specific activity found in either lead or control groups. There is no significant difference in the specific activity of TCA extracts of either left or right colliculi in any of the three groups (table 6). At 36 hours the specific activity of protein in the left superior colliculus of zinc-injected animals is similar to that in the protein of the left frontal pole as is the ratio of the specific radioactivities in protein to that in the TCA extract.

DISCUSSION

The subcutaneous injection of 16 mg. of zinc chloride on five occasions from day 11 to 17 results in changes in the neurologic examination and the EEG, suggestive of a metabolic encephalopathy. The only pathologic change noted is a mild astrocytosis in cerebral gray and white matter. These changes are unassociated with increased zinc content in brain stem, cerebellum or cerebrum, or with any changes in brain weight or lipids as a percent of fresh weight in cerebrum or brain stem. The amino acid content of cerebrum and cerebellum were also similar to control specimens. The volume of the injected solution, 1 cc, represents a concentration of approximately 110 millimoles/liter. Injection of the same volume of a more concentrated solution of sodium chloride failed to produce clinical, physiologic or pathologic changes. Even though there is evidence of renal tubular damage, the basis of this encephalopathy remains unknown since the BUN was not elevated. However, the serum calcium was elevated and possibly had reached levels which could account for the abnormalities noted. The cytochrome oxidase activity in livers of zinc intoxicated rats is decreased (7, 48) but there is no evidence to indicate that this phenomenon can occur within the central nervous system without an elevation in tissue zinc levels. In vitro low concentrations of zinc (10^{-5}M) inhibit mitochondrial respiration possibly by altering the relationship between NADH and NAD (39), or by a reduction in cytochrome b (23). However, the concentration of zinc in brain is approximately 1.5×10^{-6}M. Nevertheless, further investigations of this model will involve comparing effects of zinc injections upon synthesis relative to oxidative metabolism. With the exception of the hippocampus, the brain appears relatively resistant to the accumulation of zinc after systemic injection even in the presence of a lesion (1). This type of encephalopathy has not been described previously in either rats or dogs and its occurrence in rabbits is presumably related either to a species difference or to the route of administration since in all other instances zinc was ingested orally. It is possible that subcutaneous injections of zinc chloride result in transient elevations of zinc that are greater than those that occur when comparable amounts are administered by the oral route.

The retinal studies which were described pose a number of problems. There appears to be a non-specific reduction in radioactivity in the left colliculus when compared to the right which averages between 35 and 50% in control and lead injected animals. This may represent the non-specific effect of irritation caused by multiple injections or leakage of isotope from an eye that had been injected previously on several occasions. Unfortunately, the left eyes were not available for comparison.

More perplexing is the decrease in the specific activity of protein noted in the zinc injected animals in the right superior colliculus at 18 hours and in the frontal pole of the cerebrum at 36 hours. This occurred in the absence of increased zinc levels in the cerebral hemisphere or sera at 36 hours. While this might suggest that the administration of small quantities of zinc intraocularly has a non-specific effect on protein synthesis, the specific activity of proteins in the right colliculus of zinc injected animals at 36 hours after injection is not significantly different from that found in either the control or lead injected groups. Thus this matter must await investigation with larger numbers of animals.

The extremely low specific activities of protein in the left superior colliculus can in part be attributed to the irritant effects of zinc chloride in the right eye which resulted in patchy retinal damage and a 65% reduction in the specific radioactivity of protein in that eye. However, large areas of retina appeared grossly undamaged and the reduction in specific activity in the proteins of the eye was much less than that found in the left colliculus. The specific activity of protein in the left colliculus of zinc injected animals and the ratio of specific radioactivities in the protein to that in the TCA extracts is almost identical to that found in the left frontal pole of the cerebrum. This suggests that any label which reached the left colliculus probably did so through the systemic circulation which is heavily labeled. It seems reasonable to suggest that axonal transport of labeled protein from intact areas of the retina of the right eye to the left superior colliculus is almost entirely abolished in vivo when the zinc injected into the eye is no greater than 3.68×10^{-5} moles. This is somewhat more than the concentration of zinc found necessary to inhibit axonal transport in vitro by Edström and Mattsson (8).

Since we are observing cerebral abnormalities in the presence of normal brain zinc levels, after the subcutaneous injection of approximately 31 µg of Zn per gram/day, a particularly fruitful area for future investigation in these models appears to be the relationship between zinc and other trace metals in the nervous system relative to the toxic effects following the administration of this element. The extremely high serum levels of zinc which were found may result in functional abnormalities in other organs unassociated with evidence of cell destruction with secondary effects on the CNS or possibly alter the transport and protein binding of other trace metals and thus indirectly influence their brain levels.

SUMMARY

Both excessive and deficient quantities of zinc in the diet lead to decreased appetite, growth failure and anemia. A zinc deficient diet during pregnancy results in a high incidence of congenital malformations, and in the developing animal there is a decrease in brain DNA, RNA and protein with concomitant behavioral abnormalities in adult life. Zinc toxicity has not previously been associated with neurologic deficits. Rabbits injected subcutaneously with 16 mg. of zinc chloride from day 11 to 17 of life developed a neurologic syndrome characterized by decreased spontaneous activity, mild ataxia, weakness of the hind legs and a diminished righting reflex. This was associated with a mild gliosis in cerebral gray and white matter, but unassociated with a reduction in brain weight or lipid content or an increase in cerebral zinc levels (16.5 µg/gm fresh weight). Renal tubules are dilated and toxic granulations seen in the cytoplasm of tubular cells, but the BUN is 23.7. Serum calcium is elevated to 13.4 mg%. Injection of 2 mg. of zinc chloride into the right eye for two days with sacrifice 18 to 36 hours after the intraocular injection of ^{3}H-proline into both eyes results in patchy retinal damage to the right eye and a 65% decrease in the specific activity of ocular proteins. The specific activity of protein in the left colliculus is only 2.5% that in the right, however. It approximates that seen in the left frontal cortex, as does the ratio of the specific activities in the protein fraction to that in the TCA pool, suggesting that concentrations of zinc in the range of 4×10^{-5}M almost completely abolish axonal transport of labeled protein.

REFERENCES

1. Berger, M.L. and O'Leary, J.L.: Zinc distribution in mouse brain subsequent to hippocampal lesions. Arch. Neurol. 32: 295-297 (1975).
2. Blamberg, D.L., Blackwood, U.B., Supplee, W.C. and Combs, C.F.: Effect of zinc deficiency in hens on hatchability and embryonic development. Proc. Soc. Exp. Biol. Med. 104: 217-220 (1960).
3. Burch, R.E. and Sullivan, J.F.: Clinical and nutritional aspects of zinc deficiency and excess. Med. Clin. North Am. 60: 675-685 (1976).
4. Caldwell, D.F., Oberleas, D., Clancy, J.J. and Prasad, A.S.: Behavioral impairment in adult rats following acute zinc deficiency. Proc. Soc. Exp. Biol. Med. 133: 1417-1421 (1970).
5. Damyanov, I. and Dutz, W.: Anencephaly in Shiraz, Iran. Lancet 1:82 (1971).
6. Drinker, K.R., Thompson, P.K. and Marsh, M.: An investigation of the effect of long-continued ingestion of zinc, in the form of zinc oxide, by cats and dogs, together with observations upon the excretion and the storage of zinc. Am. J. Physiol. 80: 31-64 (1927).
7. Duncan, G.D., Gray, L.F. and Daniel, L.J.: Effect of zinc on cytochrome oxidase activity. Proc. Soc. Exp. Biol. Med. 83: 625-627 (1953).
8. Edström, A. and Mattsson, H.: Small amounts of zinc stimulate rapid axonal transport in vitro. Brain Res. 86: 162-167 (1975).
9. Folch, J., Lees, M. and Stanley, G.H.S.: A simple method for the isolation and purification of total lipids from animal tissues. J. Biol. Chem. 226: 497-509 (1957).
10. Fosmire, G.J., Al-Ubaidi, Y.Y., Halas, E. and Sandstead, H.H.: The effect of zinc deprivation on the brain. Adv. Exp. Med. Biol. 48: 329-345 (1974).
11. Fosmire, G.J., Al-Ubaidi, Y.Y. and Sandstead, H.H.: Some effects of postnatal zinc deficiency on developing rat brain. Pediatr. Res. 9: 89-93 (1975).
12. Fosmire, M.A., Fosmire, G.J. and Sandstead, H.H.: Effects of zinc deficiency on polysomes. Fed. Proc. 33: 699 (1974).

13. Giolli, R.A. and Guthrie, M.D.: The primary optic projections in the rabbit: An experimental degeneration study. J. Comp. Neurol. 136: 99-126 (1969).
14. Grant-Frost, D.R. and Underwood, E.J.: Zinc toxicity in the rat and its interrelation with copper. Austr. J. Exp. Biol. Med. Sci. 36: 339-345 (1958).
15. Halas, E.S. and Sandstead, H.H.: Some effects of prenatal zinc deficiency on behavior of the adult rat. Pediatr. Res. 9: 94-97 (1975).
16. Halsted, J.A., Ronaghy, H.A., Abadi, P., Haghshenass, M., Amirhakemi, G.H., Barakat, R.M., and Reinhold, J.G.: Zinc deficiency in man: The Shiraz experiment. Am. J. Med. 53: 277-284 (1972).
17. Hambidge, K.M., Hambidge, C., Jacobs, M. and Baum, J.D.: Low levels of zinc in hair, anorexia, poor growth, and hypogeusia in children. Pediatr. Res. 6: 868-874 (1972).
18. Heller, V.G. and Burke, A.D.: Toxicity of zinc. J. Biol. Chem. 74: 85-93 (1927).
19. Henkin, R.I., Patten, B.M., Re, P.K. and Bronzert, D.A.: A syndrome of acute zinc loss: Cerebellar dysfunction, mental changes, anorexia, and taste and smell dysfunction. Arch. Neurol. 32: 745-751 (1975).
20. Henkin, R.I., Schecter, P.J., Raff, M.S., Bronzert, D.A. and Friedewalt, W.T.: Zinc and taste acuity: A Clinical study including a laser microprobe analysis of the gustatory receptor area; in Pories, W., Clinical Applications of Zinc Metabolism, pp. 204-228 (C.C. Thomas, Springfield, ill. 1975).
21. Hurley, L.S., Gowan, J. and Swenerton, H.: Teratogenic effects of shortterm and transitory zinc deficiency in rats. Teratology 4: 199-204 (1971).
22. Hurley, L.S. and Mutch, P.B.: Prenatal and postnatal development after transitory gestational zinc deficiency in rats. J. Nutr. 103: 649-656 (1973).
23. Kleiner, D. and Von Jagow, G.: On the inhibition of mitochondrial electron transport by Zn + ions. Febs Lett. 20: 229-232 (1972).
24. Lehmann, B.H., Hansen, J.D.L. and Warren, P.J.: The distribution of copper, zinc and manganese in various regions of the brain and in other tissues of children with protein-calorie malnutrition. Br. J. Nutr. 26: 197-202 (1971).
25. Lokken, P.M., Halas, E.S. and Sandstead, H.H.: Influence of zinc deficiency on behavior. Proc. Soc. Exp. Biol. Med. 144: 680-682 (1973).
26. Lowry, O.H., Rosebrough, N.J., Farr, A.L. and Randall, R.J.: Protein measurement with the Folin phenol reagent. J. Biol. Chem. 193: 265-275 (1951).
27. Meret, S. and Henkin, R.: Simultaneous direct estimation by atomic absorption spectophotometry of copper and zinc in serum urine and cerebrospinal fluid. Clin. Chem. 17: 369-373 (1971).
28. Murphy, J.V.: Intoxication following ingestion of elemental zinc. JAMA 212: 2119-2120 (1970).
29. Poiesz, B.J., Battula, N. and Loeb, L.A.: Zinc in reverse transcriptase. Biochem. Biophys. Res. Comm. 56: 959-964 (1974).
30. Prasad, A.S.: Zinc deficiency in man. Am. J. Dis. Child. 130: 359-361 (1976).
31. Prasad, A.S., Halsted, J.A. and Nadimi, M.: Syndrome of iron-deficiency anemia, hepatosplenomegaly, hypogonadism, dwarfism and geophagia. Am. J. Med. 31: 532-546 (1961).
32. Prasad, A.S., Oberleas, D.: Ribonuclease and deoxyribonuclease activities in zinc-deficient tissues. J. Lab. Clin. Med. 82: 461-466 (1973).
33. Prasad, A.S. and Oberleas, D.: Thymidine kinase activity and incorporation of thymidine into DNA in zinc-deficient tissue. J. Lab. Clin. Med. 83: 634-639 (1974).
34. Raulin, J.: Etude cliniques sur la végétation. Ann. Sci. Natur. Botan. Biol. Végétale 11: 93-299 (1869).
35. Ronaghy, H.A., Reinhold, J.G., Mahloudji, M., Ghavami, P., Spivey, M.R. and Halsted, J.A.: Zinc supplementation of malnourished schoolboys in Iran: Increased growth and other effects. Am. J. Clin. Nutr. 27: 112-121 (1974).

36. Sandstead, H.H.: Zinc nutrition in the United States. Am. J. Clin. Nutr. 26: 1251-1260 (1973).
37. Sandstead, H.H., Gillespie, D.D. and Brady, R.N.: Zinc deficiency: Effect on brain of the suckling rat. Pediatr. Res. 6: 119-125 (1972).
38. Sever, L.E. and Emanuel, I.: Is there a connection between maternal zinc deficiency and congenital malformations of the central nervous system in man? Teratology 7: 117-118 (1973).
39. Skulachev, V.P., Chistyakov, V.V., Jasaitis, A.A. and Smirnova, E.G.: Inhibition of the respiratory chain by zinc ions. Biochem. Biophys. Res Commun. 26: 1-6 (1967).
40. Slater, J.P., Mildvan, A.S. and Loeb, L.A.: Zinc in DNA polymerase. Biochem. Biophys. Res. Commun. 44: 37-45 (1971).
41. Smith, J.C., Jr., Zeller, J.A. and Brown, E.D.: Elevated plasma zinc: A heritable anomaly. Science 193: 496-498 (1967).
42. Springgate, C.F., Mildvan, A.S., Abramson, R., Engle, J.L. and Loeb, L.A.: Escherichia coli deoxyribonucleic acid polymerase: I. A zinc metalloenzyme. J. Biol. Chem. 248: 5987-5993 (1973).
43. Strain, W.H., Lascari, A. and Pories, W.J.: Zinc deficiency in babies. Proc. VII Int. Congr. Nutr. 5: 759-765 (1966).
44. Swenerton, H., Shrader, R. and Hurley, L.S.: Zinc-deficient embryos: Reduced thymidine incorporation. Science 166: 1014-1015 (1969).
45. Terhune, M.W. and Sandstead, H.H.: Decreased RNA polymerase activity in mammalian zinc deficiency. Science 177: 68-69 (1972).
46. Van Reen, R.: Zinc toxicity in man and experimental species; in Prasad, A., Zinc Metabolism, pp. 411-426 (Charles C. Thomas, Springfield 1966).
47. Warkany, J. and Petering, H.G.: Congenital malformations of the brain produced by short zinc deficiencies in rats. Am. J. Mental Def. 77: 645-653 (1973).
48. Witham, I.J.: The depression of cytochrome oxidase activities in the livers of Zn intoxicated rats. Biochim. Biophys. Acta 73: 509-511 (1963).

CHAPTER TEN

MORPHOLOGIC AND BIOCHEMICAL EFFECTS OF HORMONES ON THE DEVELOPING NERVOUS SYSTEM IN MAMMALS*

JACQUES LEGRAND, Dc-es-Sc.

This paper will deal with those hormones which may exert an action on the organization of the whole nervous system, i.e. thyroid hormones, pituitary growth hormone (GH), and corticosteroids. The action of testicular androgens, which seems to be localized to well delineated parts of the CNS, e.g. some hypothalamic areas, will not be studied (for a recent review of the effects of testicular androgens on the differentiation of the hypothalamus, see ref. 55).

EFFECTS OF THYROID STATE

Structural, functional and behavioral consequences of thyroid deficiency and treatment with thyroid hormones during infancy. In the human baby suffering from severe congenital hypothyroidism, neurologic and mental troubles appear within the first six months of postnatal life. They can be reversed only if thyroid treatment is begun within the first three months (86). This suggests that the brain of hypothyroid babies presents structural abnormalities appearing at an early and "critical" period of its development, during which thyroid hormones have an "organizing" action on the CNS. In the human species, this period begins long before birth, and continues for several postnatal months. In the rat, the equivalent period is almost entirely postnatal. The first experimental and systematic study of the influence of thyroid hormones upon the developing rat nervous system was done by Eayrs and his colleagues. The behavioral deficits observed by him and others in animals

* The recent work performed in our laboratory (Drs. J. Clos, C. Favre, C. Legrand, A. Rabie, M. Selme-Matrat and myself) or in collaboration with our group (Drs. Crépel and Vigouroux, Laboratory of Comparative Physiology, University of Paris VI) and reported in this paper was supported in part by grants from the DGRST (No 72.7.0102 and 75.7.1325), the CNRS (ATP No 1344) and INSERM (No 75.1.204.6).

made hypothyroid at birth were more or less severe depending on the difficulty of the problems the animals were offered. The severity of impairment of the capacity for adaptive behaviour was also found to be inversely proportional to the interval between birth and the time when the thyroidectomy was performed: thyroid deficiency effectively impaired behaviour only when it was induced between birth and day 10 of postnatal age. Replacement therapy in animals made hypothyroid at birth was effective only when started during the first 10 days of postnatal life (for a review, see ref. 52, 76; see also ref. 47). Changes in the EEG are observed in hypothyroid children (99) and in rats thyroidectomized at birth (18) but not in animals thyroidectomized when they are adult. In the rat, neonatal thyroid deficiency is also accompanied with changes in the evoked potential responses to the stimulation of thalamic structures which may be related to disturbances in axone and dendrite development (16, 17). Thyroid deficiency effectively leads to an increase in the density of cell bodies and a decrease in their mean size in the cerebral cortex, as well as a decrease in the density of axon terminals and a reduction in growth and branching of the dendrites of pyramidal neurones, with a maximum effect at the level of layer IV which receives the specific thalamic afferents. All these structural abnormalities may result in a marked reduction in the number of axodendritic contacts which, in turn, may account for the observed behavioral deficits. However, when rats are thyroidectomized at birth and given thyroxine (T4) in late life, they recover an apparently normal cortical structure but still display behavior troubles (for a review, see ref. 52 and 76). It is possible that irreversible functional deficiencies due to some permanent biochemical disturbances, whose detection would need more specific methods of investigation, add their effects to those of structural abnormalities. Cragg (33) found an increase in the neuronal density of the visual cortex from rats thyroidectomized at birth. Moreover, he demonstrated, by electron microscopy, a 22% decrease in the number of nerve terminals associated with each neurone. According to him, however, the significance of such an observation as an explanation of behaviour disturbances remains questionable.

Conversely, neonatal hyperthyroidism accelerates the morphologic development of cortical neurones, and particularly the ontogenesis of their dendritic spines (134; see however ref. 80). It has adverse effects on the ultimate biochemical composition of the brain (116). Initially it also accelerates the appearance of several automatic and innate responses and improves locomotor coordination and (or) learning abilities (139). Despite this initial precocity, adult rats made hyperthyroid at birth perform less well than littermate

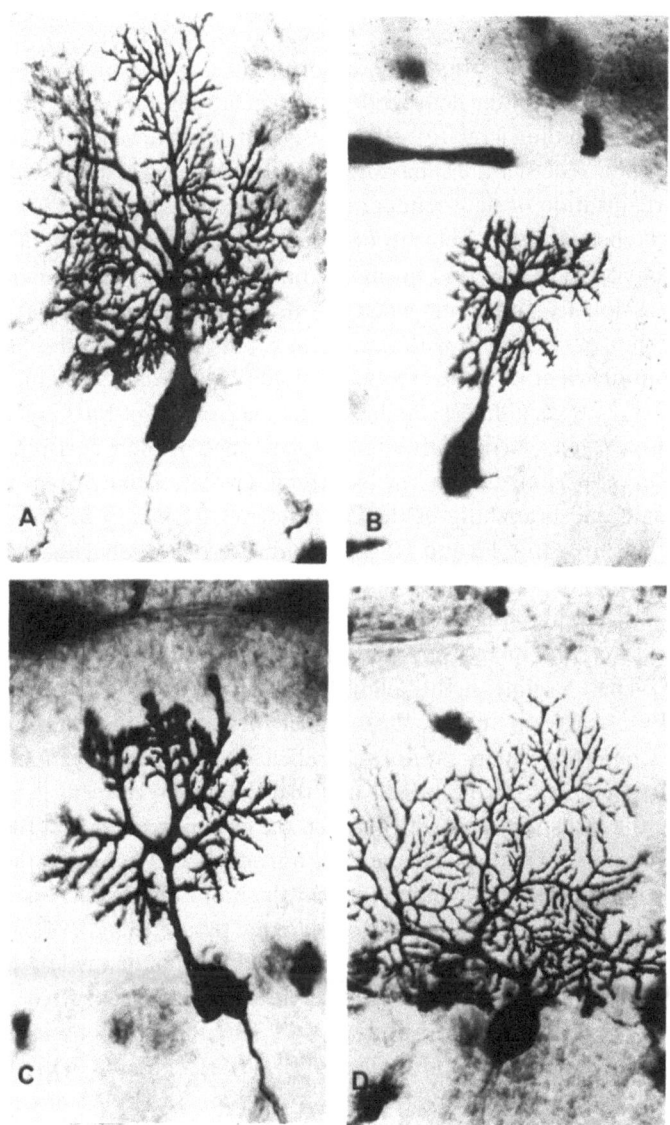

Fig. 1. Development of the dendritic arborization of Purkinje cells in 14-day-old animals. Mid sagittal sections of the cerebellar vermis around the *fissura prima*.
A – Control.
B – Animal surgically thyroidectomized at birth.
C [- PRU-treated animal.
D – PTU-treated animal given excess T4 (2µg DL-T4 per day) during the first postnatal week.
(Golgi Cox preparations, X 330).

controls (for a review, see ref. 52, 76; see also ref. 46). It must be noted that early exposure to excess thyroid hormone induces many profound and permanent alterations of the neuroendocrine regulations, known as the "neo-T4 syndrome" including a relative state of thyroid deficiency (113, 117; see however ref. 136). Whether the behavioral troubles observed result from impaired brain organization or neuroendocrine disturbances is unclear.

The cerebellar cortex seems to be a good model for the study of the morphogenetic action of thyroid hormones on the nervous system, owing to the relative simplicity and homogeneity of its structure, and its relatively late development as compared with that of the cerebral cortex. In the cerebellum of thyroid-deficient rats, the external granular layer persists beyond the normal age of 21 days (78, 97); the histologic maturation of Purkinje cell (PC) bodies is delayed, particularly the disappearance of their perisomatic processes and the organization of the Nissl bodies within their cytoplasm (see fig. 4). Growth and branching of the PC dendritic arborizations are also dramatically delayed (fig. 1B and 1C compared with 1A; see also fig. 4, 5) as well as ontogenesis of the dendritic spines. Finally, the dendritic tree of PC remains markedly hypoplastic long after the disappearance of the external germinative layer, and its basis presents irreversible morphologic abnormalities (95, 125). These troubles in the laying down of granule cells and in the maturation of PC are accompanied by disturbances in the development of cerebellar afferents, climbing and mossy fibres: cerebellar glomeruli differentiate several days later than in control animals and their later development is abnormal (93, 95); the transitory contacts between the climbing fibres and the perisomatic processes of PC persist beyond the normal time (74). Lastly, the density and the final number of synapses within the molecular layer is irreversibly reduced (110, 126), probably as the result of the reduced length of the parallel fibres (37, 125). These structural abnormalities are accompanied by an important delay in the acquisition by PC of their adult spontaneous activity and by a deficit in their activation responses after stimulation of their afferents. Contrariwise, the inhibition responses of PC are accentuated between 20 and 30 days after birth (35, 36, 39). The electrophysiologic studies of Crépel (38) lead to the conclusion that in the CNS of the adult animal rendered hypothyroid during the neonatal period, the bioelectric properties of neuronal membranes are normal, and that the only irreversible functional deficits concern the neuronal connectivity. Finally, besides transient troubles such as the retarded laying down of granule cells and the delayed maturation of PC bodies, more persistent abnormalities in the growth and branching of cell processes are found, resulting in the disappearance of the normal synchronism between the laying down of the interneurones and the morphogenesis of PC: the normal cell interactions during development disappear; in this way the lack of thyroid hormone may result in abnormal connectivity.

Less severe disturbances in the structural maturation of the cerebellar cortex are still observable when thyroidectomy is performed during the second postnatal week; but maturational changes are no longer observed when thyroid deficiency occurs later. Moreover, in animals made hypothyroid at the end of gestation or at birth, the condition is corrected (fig. 1D) only if hormone therapy is begun before the end of the second postnatal week (92, 94).

Neonatal hyperthyroidism accelerates all the processes of the structural maturation of the cerebellar cortex (125, 142, 143) including synaptogenesis in the molecular layer (110), without causing the disappearance of the normal synchronism between the laying down of granule cells and the development of PC, and consequently without leading to morphologic abnormalities of the latter (125; see however ref. 110). Lastly, as with neonatal thyroid deficiency, but for different reasons, early hyperthyroidism causes a terminal deficit in the total number of synapses in the rat cerebellar cortex (110).

Thyroid hormones and cell formation in the CNS. Following the report that excess thyroid hormone in the neonatal period accelerates the disappearance of the external germinative layer of the rat cerebellum (142) and conversely that thyroid deficiency leads to the persistence of this layer and of numerous mitoses within it beyond the normal age of 21 days (95, 97), it was shown that neonatal thyroid deficiency results in a smaller increase in cerebellar DNA content during the second postnatal week, and to a lengthening of the proliferative period in the organ, for subsequently the total number of cerebellar cells is normal in 35-day-old hypothyroid animals (10, 72, 79, 98, 109, 115, and fig. 2C). In contrast, thyroid deficiency never affects the cerebral DNA content (10, 115). Administration of excess thyroid hormone during the first few days following birth results in a slight, transient acceleration in the increase of the cerebellar DNA during the first postnatal week (72, 98, 151), followed after the age of 10 days by a smaller increase than that observed in euthyroid animals (9, 72, 98). In the forebrain, the total DNA content is also lowered after the age of 5 days (9). Lastly, in both forebrain and cerebellum, the proliferative period stops precociously in the hyperthyroid animals. Consequently, the total number of cells in these two structures is irreversibly reduced (fig. 2F).

The lack of effect of thyroid deficiency on cell formation in the forebrain, indicates that thyroid hormones especially affect the neuronal proliferation,

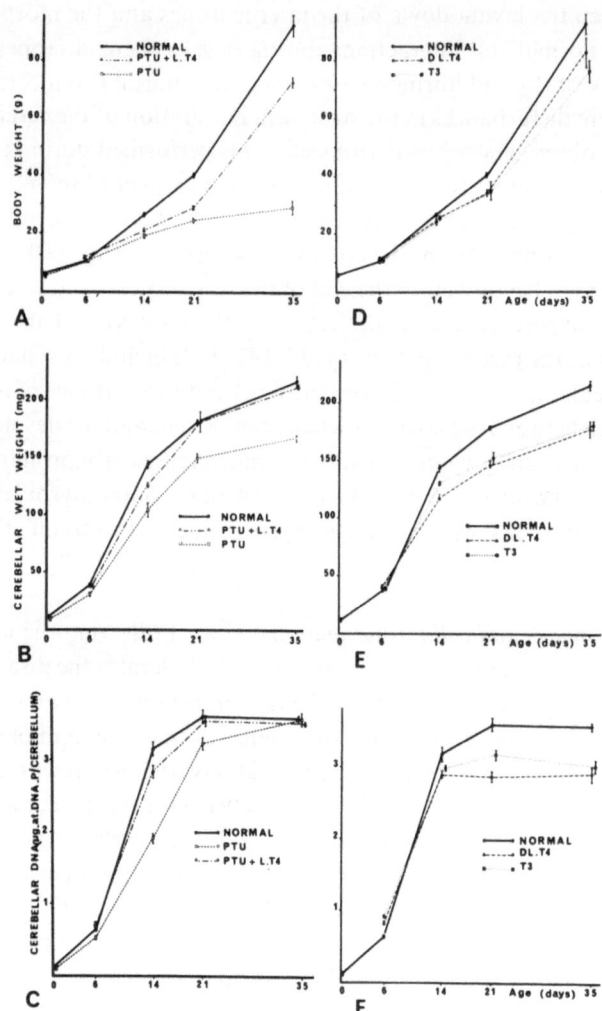

Fig. 2. Developmental curves for body weight (*A* and *D*), cerebellar wet weight (*B* and *E*) and total cerebellar DNA (*C* and *F*) of rats pups treated from the end of gestation with propyl-thiouracil (PTU), with PTU from the end of gestation plus physiologic doses of L-T4 from birth (PTU + L-T4), with 5 µg DL-T4 per day from birth (DL-T4), with 25 µg T3 on day 0 plus 1 µg T3 on alternate days until the end of the experiment (T3), and body weight, cerebellar wet weight and total cerebellar DNA in normal animals (normal).

The low daily doses of L-T4 injected to the PTU-treated rat pups in order to return to normal their plasma T4 concentration just after the injection were the following: 0.05 µg from day 0 to day 2; 0.10 µg from day 3 to day 5; 0.15 µg from day 6 to day 9; 0.20 µg from day 10 to day 14; 0.25 µg from day 15 to day 20, and 0.30 µg thereafter.

Means ± S.E.M. (Data from Legrand et al. 98).

since, in the forebrain, there is no significant formation of neurones after birth, whereas in the cerebellum most of the interneurones are formed post-natally. This point was investigated by counting the cells belonging to the different cell classes of the cerebellum: in thyroid-deficient animals the number of basket cells, which are the first inhibitory interneurones to be formed (1) is lowered, and that of stellate cells, which appear later, is increased, as well as the number of glial cells. The number of granule cells (which constitute the main cell population of the cerebellum) increases more slowly but is finally normal (26) or slightly elevated (109). The increase in the number of glial cells is accompanied by a glial hypertrophy in the molecular layer of the cerebellar cortex (27) and in the internal granular layer around the cerebellar glomeruli (74). According to Nicholson and Altman (109), hyperthyroidism compared to hypothyroidism leads to opposite effects upon the granule cell and the astrocyte cell numbers, but the number of basket cells is decreased in hyperthyroid as in hypothyroid animals when compared with controls. Recently, Rabié et al. (119) reported a reduction in the number of granule cells per PC in the cerebellum of PTU-treated rats given too high doses of T4 (fig. 3C).

Another finding which characterizes cerebellar development in the hypo-thyroid animal, which contributes to the decrease in the number of cerebel-lar cells during the second postnatal week, is an increase in the number of dying cells within the internal granular layer (102). This problem was rein-vestigated in our laboratory. It was found that in the normal cerebellum, as in that of the hypothyroid animal, the number of pyknoses within the internal granular layer was maximal at 10 days (the ages of 10, 14 and 21 days were studied) and that the increase in the number of dying cells in the thyroid-deficient animal was maximal at 14 days, when it amounted to a 20 fold rise (119). We also confirmed, that the increase in thickness of the molecular layer, which reflects the development of the PC dendritic arborizations, was more retarded that the formation and laying down of granule cells in the hypo-thyroid animals. Moreover, the study of the corrective effects of very low doses of T4 demonstrated that the increase in cell death occurs only when the normal synchronism between these two main processes of the cerebellar de-velopment is absent, as is the case in hypothyroidism but not in hyperthy-roidism (fig. 3). In the thyroid-deficient animal, it is conceivable that a greater proportion of granule cells cannot establish a synaptic contact with PC and subsequently die, because PC maturation is particularly retarded.

Thyroid hormones and biochemical maturation of the CNS. The study of the developmental pattern of the DNA concentration and ratios of the wet

144   JACQUES LEGRAND

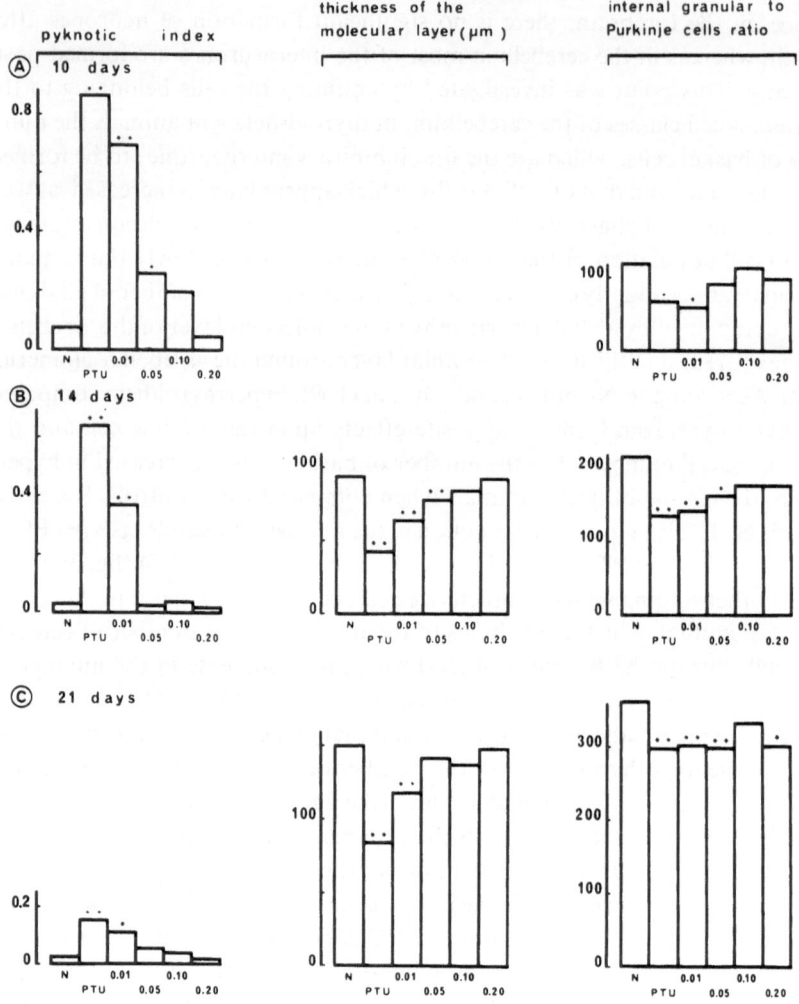

Fig. 3. Effects of PTU-treatment and supplementation with T4 on the pyknotic index in the internal granular layer (left), the thickness of the molecular layer (middle) and the number of internal granule cells per PC (right) in the cerebellar cortex of developing rats, on days 10 (*A*), 14 (*B*) and 21 (*C*). The pyknotic index is the ratio, expressed in percent, of the number of pyknotic nuclei to the total number of internal granule cells. The constant daily dose of T4 administered to the PTU-treated animals (0.01, 0.05, 0.10 and 0.20 μg) is indicated in abscissa. Significance of the difference with the normal animals after Duncan's test: ** P < 0.01; * P < 0.05. (Data from Rabié et al. 119).

weight, RNA and protein contents of the organ to DNA shows that in hypothyroid animals the lower weight of the cerebrum and of the cerebellum is due to smaller mean cell size. Contrariwise, in hyperthyroid animals, the cell size remains normal and the lower weights of the cerebrum and cerebellum result from the deficit in the number of cells (9, 10, 60, 72, 98, 114). It is seen that cell maturation is permanently affected by thyroid deficiency, and only accelerated by hyperthyroidism. Numerous biochemical and histochemical findings argue in favour of selective effects of thyroid dysfunction on the growth of cell processes and on the ontogenesis of nerve endings (see the first section). The developmental pattern of several enzyme activities associated with neuronal membranes and synaptic transmission has been shown to be altered by neonatal hypo- or hyperthyroidism. In hyperthyroid and thyroid-deficient animals, respectively the advance and the delay in the acquisition of the adult normal distribution of acetyl cholinesterase (21, 90) and succinate dehydrogenase (93, 95) activities within the internal granular layer of the cerebellar cortex mainly reflect the acceleration or the retardation in the growth of mossy fibres and in the development of cerebellar glomeruli. Similarly the changes in succinate dehydrogenase activity in the molecular layer (93, 95) reflect the accelerated or delayed appearance of contacts between the parallel fibres and the PC spines. The increase of acetylcholinesterase activity during development is also retarded in the cerebral cortex and hypothalamus of hypothyroid animals (61, 77, 144) and accelerated in the cerebral cortex of hyperthyroid animals (131). In the cerebral cortex, the activity of succinate dehydrogenase (57, 77), which is mainly associated with the mitochondria of synaptosomes and that of glutamate decarboxylase (a constituent of nerve endings) is decreased, whereas the activities of lactate dehydrogenase, a cytoplasmic enzyme, and glutamate dehydrogenase, an enzyme in the mitochondria of neuronal bodies and glia, are less affected (7, 10). The increased conversion of glucose into aminoacids during development, which reflects the increase in the metabolic compartments associated with cell processes during maturation (11), is accelerated or retarded as a consequence of hyper- or hypothyroidism (30, 69). The increase in the amount of synaptosomal proteins is slowed or accelerated to a greater extent than that of total proteins in the cerebellum of hypothyroid or hyperthyroid animals (120).

There exists considerable evidence that thyroid hormone stimulates in vivo protein biosynthesis, at least in the tissues whose metabolism is increased by the hormone (107) and particularly in the immature brain. In vivo incorporation of leucine or phenylalanine into the cerebral or cerebellar proteins is transiently reduced in the hypothyroid young rat (10, 41, 43, 66 141) and transiently increased in the hyperthyroid animal (42, 103). The effects of thyroid dysfunction on the in vivo leucine incorporation in neuronal bodies isolated in bulk from the cerebral cortex and the cerebellum of young rats have been compared with those on the incorporation in the total protein from these two structures (54). It has been demonstrated that the changes in cell proliferation (especially in neurones), due to hypo- or hyperthyroidism, could account for some of the changes in leucine incorporation, and also that T4 stimulated the incorporation of leucine in neurones that do not multiply further, e.g. in PC of the cerebellum. Neonatal thyroidectomy does not affect cerebral RNA synthesis in 25 to 35-day-old rats (7, 10, 58, 59, 63-65; see however ref. 87). However it leads to a significant decrease in the synthesis of "rapidly labelled" cerebral RNA on day 10, accompanied by an altered transport of RNA out of the nucleus (50, 70). Hamburgh and Flexner (77) demonstrated that cortical succinate dehydrogenase activity which is low in rats thyroidectomized at birth in contrast with controls, can return to normal levels if T4 is given after the age of 10 days; but the hormone has no effect if given only after the age of 15 or 20 days. The inhibition of the development of enzyme activities sensitive to thyroid hormone deprivation coincides with the period of the major increase of these enzyme activities and total protein in the brains of controls. Hamburgh and Flexner believed that, in the developing brain, thyroid deficiency leads to reduced synthesis of specific proteins. Nevertheless, the spectrum of proteins affected by T4 must be sufficiently broad to be detectable when total protein synthesis is measured.

Exogenous labelled thyroid hormones injected in vivo (56), or added to organotypic cultures of developing spinal cord and cerebellum (104, 105) can be recognized at the level of neurones. Vigouroux (148) showed in collaboration with our group that the incorporation and utilization (deiodination) of T4 by the brain were greater at 10 than at 30 days. Furthermore, in experiments using isotopic equilibrium with ^{125}I, we demonstrated that the content in endogenous hormonal iodine was 4 to 6 times higher in immature neurones isolated in bulk from the brain of 10-day-old rats than in neurones isolated from 30-day-old animals (29). In the same experiments, it was also shown that at 10 days, T4 and T3 represent respectively 30 and 60% of the total iodine in the neuronal fraction, suggesting that the metabolism of thyroid hormone is intense at the level of the immature neurones. Whether the action of thyroid hormone is at the level of transcription or (and) translation is not clear (62, 137). During the past four years, however, the recognition of specific nuclear binding sites for iodothyronines has reawakened interest in the mechanism of induction of thyroid hormone action (112); increasing evidence suggests that thyroid hormone may initiate its effects by augmentation of the transcription of genetic information.

The formation of a myelin sheath around the axon is perhaps the most obvious feature of the morphologic maturation of the neurone. Myelination is retarded in the hypothyroid animal (5, 13, 14, 24, 25, 44, 53, 106, 150, 152). A transient and slight advance in myelination can be observed in rats given thyroid hormone neonatally (6, 44, 150, 152). However, it seems that this action of thyroid hormone on myelination may result in part from its action on axonal growth (24, 25, 127) or its effect on the number of oligodendrocytes (6).

Concluding remarks concerning the action of thyroid hormone. From the results summarized above, especially those concerning the cerebellum, it appears that thyroid hormone primarily stimulates neuronal maturation. The number of neurones formed before the onset of thyroid function in the fetus, e.g., in the rat, neurones of the cerebral cortex and the cerebellar PC, cannot be modified by a lack of hormone, but the maturation of the same neurones is, on the contrary, affected by thyroid deprivation. The hormone also exerts its maturational effect on the stem cells at the origin of the later-forming neurones. Excess of hormone directs the neuronal elements from the proliferative phase to the maturational phase. Conversely, hormonal deprivation leads to an increased duration of the proliferative period. According to Lewis et al. (102), the cell generation time and the length of the various phases of the cell cycle in both the subependymal layer of the forebrain and the external germinative layer of the cerebellum are not modified by hypothyroidism at 6, 12 and 21 days of age. Thus the hormone may act at earlier stages of development on some maturational events in the stem cells preparing them for mitosis. Some evidence has been found by us which shows a precocious action of T4 (present at birth or even in the fetus) on the number of synapses (91) and on cell formation in the cerebellum. In later stages of development, the delay in the formation of the cerebellar interneurones in thyroid-deficient animals undoubtedly has repercussions on the PC morphogenesis, since the latter was seen to depend on some inductive influences exerted by the interneurones. This last conclusion, as well as the existence of a precocious effect of thyroid hormone upon cell formation, is based on a comparison of the effects of thyroid deficiency on cerebellar morphogenesis with those obtained after administration of an antimitotic drug, methylazoxymethanol (121) or X-irradiation (2-4) at different stages of development. Lastly, the delay in the PC maturation in turn suppresses some inductive influence on the granule cell maturation leading to an increased death of these cells.

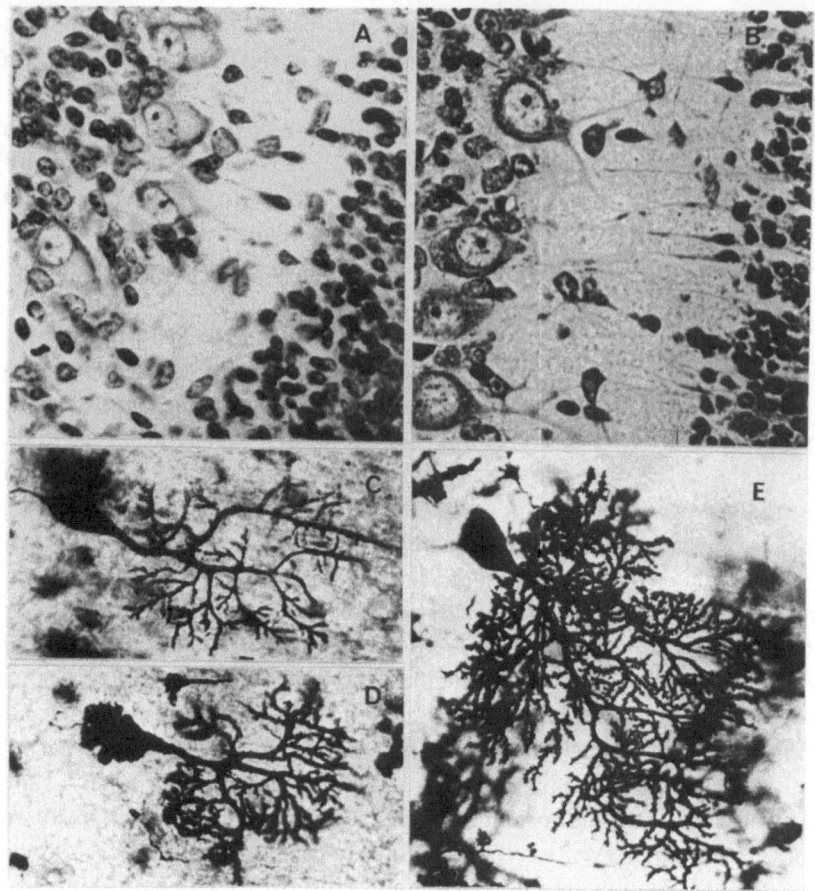

Fig. 4. Morphologic development of Purkinje cells, as seen in preparations stained with cresyl-violet (*A* and *B* ×540) or impregnated according to the Golgi-Cox method (*C, D* and *E,* ×330). Mid sagittal sections of the cerebellar vermis around the *fissura prima.*

A – 10 day-old PTU-treated rat given 0.05 μg L-T4 per day from birth.

B – 10 day-old PTU-treated rat given 0.10 μg L-T4 per day from birth. Note the mature appearance of the PC bodies, which is absent in A.

C and *D* – 14 day-old PTU-treated rat given 0.01 μg L-T4 per day from birth. Same picture as in fig. 1B and 1C.

E – 14 day-old PTU-treated rat given 0.20 μg L-T4 per day from birth. Same picture as in fig. 1D.

Is the action of thyroid hormone "physiologic"? During the past two years, one of our main objectives was to demonstrate the "physiologic" character of thyroid hormone action on the developing cerebellum. It was shown that in

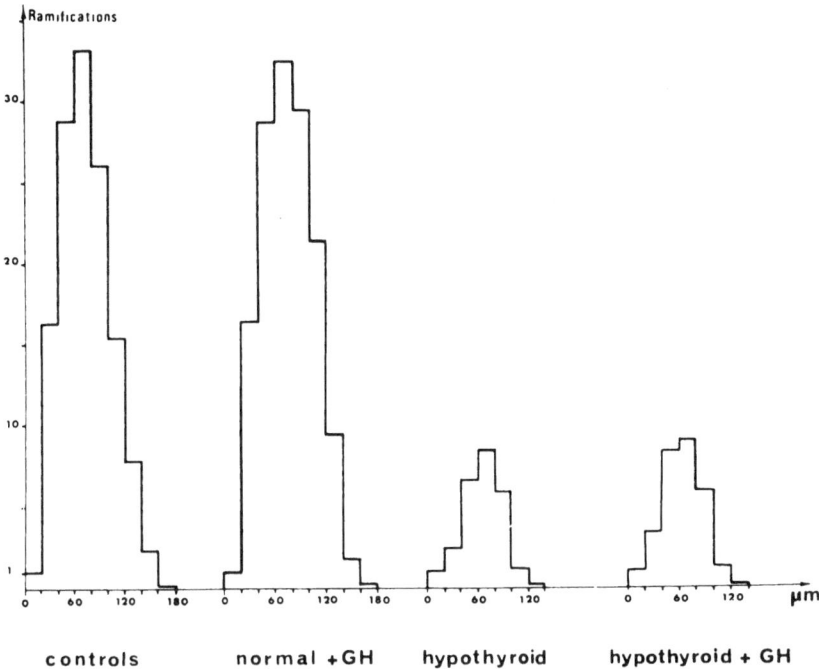

controls   normal +GH   hypothyroid   hypothyroid + GH

Fig. 5. Quantitative estimation of the degree of development of the dendritic arborization of
Purkinje cells in midsagittal sections of the cerebellar vermis and around the *fissura prima* in 14-
day-old rats. In ordinate, the number of dendritic ramifications; in abscissa, the distance from the
basis of the primary dendrite in μm. The influence of PTU-treatment and the lack of marked
effects of GH (20 or 100 μg per day from birth) in normal as well as in PTU-treated animals are
shown (Data from Rebière and Legrand 124).

the young rat treated with propylthiouracil (PTU), the minimum and constant daily dose of L-T4 (0.10 μg) able to correct the disturbances in PC
morphogenesis (fig. 4) also results in normal functional maturation of these
cells (22), normal increase in thickness of the molecular layer, normal
numbers of dying cells within the internal granular layer, and normal numbers
of granule cells per PC (119) (fig. 3). It was shown that this minimum daily
dose of hormone prevents enlargement of the thyroid due to PTU-treatment
during the first ten days of postnatal life, and it was demonstrated that the
sensitivity of cerebellar structures appears earlier and is consequently higher
than that of other structures such as the ossification centers which appear
after birth (22). When the complete study of the development of thyroid
function in the young rat was done by Vigouroux (146, 147), the demonstra-

tion of the "physiologic" character of thyroid hormone action was completed and extended to the effects of the hormone on cell formation: daily administration to PTU-treated animals of progressively increasing doses of T4 nearly equal to the amounts of hormone secreted by the thyroid of the normally developing rat, not only corrects all the disturbances in the structural maturation of the cerebellar cortex, but also results in normal postnatal evolution of the cerebellar wet weight (fig. 2B) and the cerebellar DNA (fig. 2C), even if it does not entirely correct the retardation in body growth (fig. 2A) (98). It is likely that such low doses of T4 which result in normal plasma concentrations of hormone within a short time after the injection cannot completely correct all the biochemical troubles due to PTU-treatment, indicating that the action of the hormone on the structural maturation of the CNS is not only physiologic but also "permissive." During these studies, it was noted that daily administration to PTU-treated rats of a dose of T4 as low as 0.20 µg (i.e. a dose much lower than all those which were previously employed to create hyperthyroidism in the newborn rat) is sufficient to accelerate the histologic maturation of the cerebellum (but not the skeletal maturation) (22), and leads to a deficit in the final number of granule cells (119). Since it is well established that neonatal hyperthyroidism conducts to a reduction in learning ability in the adult rat, such observations must be taken into consideration for the comprehension of the consequences of human perinatal hyperthyroidism on intellectual development, as well as for the treatment of congenital thyroid deficiency.

Is the action of thyroid hormone "specific"? Another objective was to demonstrate the specificity of thyroid hormone action on CNS development. The effects of excess or lack of thyroid hormone were compared to those of other treatments, such as undernourishment and administration of glucocorticoids which are also known to affect CNS cell proliferation. Lastly, they were compared to those of GH administration.

Undernutrition usually accompanies neonatal thyroid deficiency. However, even in the first report of the effects of hypothyroidism on cerebellar development (97), it was noted that the severity of the disturbances did not correspond with those of the body growth of the animals; this was confirmed in subsequent studies in our laboratory. The effects of undernutrition on the morphologic maturation of PC and the ontogenesis of their dendritic spines (125), as well as on synaptogenesis within the molecular layer (123) and on ontogenesis of nerve terminals, determined biochemically from measurements of the protein content of the synaptosomal fraction (120), and also its effects on the final cellular composition of the cerebellum (23), differ considerably. Undernutrition during the proliferative phase of cerebellar development reduces the final number of neurones, especially the number of granule cells, without markedly affecting either the overall histologic maturation of the cerebellar cortex nor the increase in the synaptic density within the molecular layer, and without disrupting the normal synchronism between the laying down of granule cells and the growth and the branching of PC dendritic arborizations. It affects the number of glial cells more than the number of neurones (23) and results in glial hypotrophy within the molecular layer (27). It should be pointed out that the number of synapses per PC is probably reduced in undernourished animals as in hypothyroid animals (but for different reasons). In the cerebral cortex, such a reduction in the number of synapses per neurone was demonstrated by Cragg (34) as a result of undernutrition. Lastly, the effects of undernutrition on growth and myelination of the fibres of the sciatic nerve also differ from those of thyroid deficiency (24, 25, 96).

EFFECTS OF PITUITARY GROWTH HORMONE (GH) ON CNS DEVELOPMENT

Studies on early hypophysectomized rats seem to indicate that GH probably does not exert a direct and essential effect on normal brain growth and development (49, 149). Sufficient circulating thyroid hormones persist in this situation to allow an apparently normal morphologic development of the brain (73). However, since thyroid deficiency is accompanied by degranulation of pituitary acidophil cells (118, 138) which synthetize GH (31, 45, 63, 135), attempts have been made to correct the abnormal development of the hypothyroid brain by treatment with GH. The results are controversial. Even though observations indicate that the action of GH on some aspects of the biochemical development of the CNS (71, 88; see however ref. 63, 64) or of the development of the thermoregulatory mechanisms (75, 78) might be comparable to that of thyroid hormone, all developmental abnormalities of the CNS due to hypothyroidism cannot be corrected by postnatal administration of GH. For example, GH given to the young normal or hypothyroid rat may lead to an increase in the mean size of cortical neurones and in the mean length of their dendrites, and to an increase in the mean length of PC dendrites in the cerebellar cortex; but GH administration does not augment the branching capacity of dendrites in the cerebral cortex (51) or in the cerebellar cortex

(124)(fig. 5). Moreover. in the cerebellum. it does not allow the disappearance of the external germinative layer at the normal time. and does not prevent the delay in the structural maturation of PC bodies (124) and in the ontogenesis of their dendritic spines characteristic of thyroid deficiency (125).

Even more controversial remain the effects of GH administered to pregnant rats on the CNS of its offspring. Zamenhof (153) first reported an increase in cerebral DNA content in the newborn. This was later confirmed by using a highly purified GH preparation (129. 130. 154) but more recently was not found again (40. 155). Some results tending to demonstrate a stimulation of cell maturation in the cerebral cortex and an improvement in the learning abilities in the offspring of mother rats treated with GH during gestation were also obtained (15. 20. 129; see however ref. 67. 122). An action of GH prenatally administered to the mother is not completely excluded. but since it is unlikely that GH can cross the placenta (68. 89). such action is probably mediated via effects upon the mother or the placenta (see ref. 156).

EFFECTS OF CORTICOSTEROIDS

Consequences of treatment with corticosteroids during infancy on cell formation in the CNS. In mice and rats. the administration of corticosterone or cortisol at birth or during the first few days of postnatal life rapidly (i.e. in a shorter time than neonatal administration of excess T4) inhibits the increase in the cerebral and cerebellar DNA content. In these two structures, the proliferative phase stops at nearly the normal age, and the final DNA content is decreased to the same extent as iin hyperthyroid animals (for reviews of the effects or corticosteroids on the cerebral and cerebellar DNA see ref. 12. 82). The final deficit in cell number is not due to an increase in cell death (83). but rather to an inhibition of DNA synthesis. and the decrease in mitotic activity is more marked in the cerebrum than in the cerebellum (8. 32. 140). A final deficit of cells was recently found again in the cerebrum and the cerebellum after administration of cortisol to young rats during the second postnatal week. Moreover it was shown that such a delayed treatment had more marked effects on the DNA gain than the earlier treatments (28). In the cerebellum. the greater inhibition of the DNA increase when the hormone is given later may signify that corticosteroids have the most profound effect on the cells that are arriving near the end of the proliferative phase. In the cerebrum. it is more than likely that the great reduction in the DNA gain. in particular after the administration of hormone during the second postnatal week. especially concerns the glial cells, i.e. the only class of cells whose multiplication is still very active (28, 83). In the cerebellum the cell deficit may bear equally on glial cells and also on interneurones and particularly on granule cells, which represent more than 90% of the total cell population and which still multiply actively during the 2^d and 3^d postnatal weeks. Thus, the effects of corticosteroids on the final cellular composition of the cerebellum (and of the cerebrum) may be compared with those of neonatal hyperthyroidism (and also with those of undernutrition).

Effects of neonatal corticosteroid treatment on cell maturation within the CNS. Corticosteroids can induce, in the CNS, as in other organs, the synthesis of numerous enzymes and they can accelerate certain maturational processes (for review, see ref. 12). They do not affect the development of the mechanisms of conversion of glucose into aminoacids in the brain (32). If one considers the changes in cellular composition of the brain resulting from their administration, they do not seem to affect markedly mean cell size or overall cell maturation as estimated from the ratios of RNA and protein to DNA (8, 12, 28, 32, 83). Moreover, reduction in the lengthening of the dendrites and a delay in the ontogenesis of the dendritic spines of the pyramidal neurones in the visual cortex were reported after injection of cortisol to the newborn rat (131, 134). A reduction in the density of dendritic spines of cerebellar PC was also observed in our laboratory at 35 days of age, after subcutaneous implantation of a corticosterone pellet or injection of cortisol at birth (unpublished results). In the same animals, and at least when their body weight was not greatly affected, the external germinative layer disappeared at the normal age of about 21 days; the time span for the histologic maturation of the PC bodies appeared normal; the dendritic arborization of PC apparently was slightly hypoplastic but did not display obvious morphologic abnormalities. Recent biochemical data suggest a reduction in growth of neuronal processes after corticosterone administration to 3-day-old rats (83). On the other hand, cortisol administration was recently associated with precocious development of cell processes in lamina VI of the neocortex, whereas T4 was ineffective (80). A slight increase in cerebroside concentration in the spinal cord of 16-day-old rats treated with cortisol from the 6th to the 10th days after birth was also reported (19) suggesting accelerated myelination, but in rats given corticosterone on days 2 and 3, a reduction in cerebral sulfatides was observed (83). We do not yet know to what extent corticosteroids may influence myelination. From these often controversial and very fragmentary data it appears that corticosteroids have no marked and unequivocal effect upon cell maturation within the CNS.

Functional and behavioral consequences of neonatal corticosteroid adminis-tration. Studies dealing with the effects of corticosteroid administration in the newborn on subsequent behaviour of the mouse or the rat, and on the functional maturation of the brain are numerous. Cortisol delays the normal ontogenesis of swimming capability and the appearance of the startle reflex in response to a loud noise (128, 133). Mice or rats given corticosteroids at birth display impaired locomotor activity (84) and learning ability (81, 111) in adulthood. In agreement with behavioral effects on the startle response and the swimming development, the ontogenesis of the evoked potential responses to light, sound and sciatic nerve stimulation are delayed by cortisol (128, 133). Cortisol also influences brain excitability as measured by the electroshock seizure threshold technique: it decreases it when given between the 1st and 7th days after birth and, in constrast, increases it when given after the 8th postnatal day (145).

Concluding remarks concerning the effects of administration of an excess of corticosteroids during the neonatal period. Even if corticosteroids have a stimulating effect on certain maturational processes in the nervous system, excess of these hormones during the neonatal period does not enhance nerve cell maturation as does thyroid hormone. As in cases of neonatal hyperthyroidism, corticosteroid administration in the newborn has adverse effects on the organization of the brain and adult behaviour. Although the deficit in final cell number and changes in the cellular composition of the cerebrum and the cerebellum seem similar, the behavioral consequences of treating newborns with excess thyroid hormone or corticosteroid are apparently different (8). One of the factors which seems to be involved in the different effects of the two treatments is the timing of interference with cell formation (see ref. 12); but the effect on the final cell number and on cellular composition is probably only one of the factors contributing to the behavioral disorder which develops after disturbing the hormonal balance in early life. It is clear that the analysis of the effects of corticosteroids must be extended in particular by the application of histologic and electron microscopic methods, in order to better understand how these hormones may modify the neuronal circuitry.

Stimulation of the production of endogenous corticosteroids during early life. All the results dealing with the action of corticosteroids on the developing nervous system which are summarized above are concerned with the effects of excessive dosage of the hormone during early postnatal life. They do not throw light on the physiologic role of the adrenal cortex in brain development. The wide variety of stress situations that elicit adrenal cortical activation in the adult are generally ineffective in infant rats (85, 100, 132). The small stimulation of pituitary-adrenal activity that may occur in response to certain stresses during the neonatal period of relative quiescence of the adrenal glands is a priori of dubious physiologic or behavioral importance to the organism. However, it was shown that adult rats exposed to stresses during this period responded more appropriately to various environmental situations than the controls. This suggests that an increase in endogenous corticosteroids during the critical period of CNS development may influence its organization and (or) permanently modify certain aspects of neuroendocrine regulations and adult behaviour (48, 100, 101). At present there are no data concerning the possible influence of an increase in pituitary-adrenal activity during the neonatal period on cell formation or maturation throughout the brain (see ref. 12). The rat fetus is more sensitive to stressful stimuli than the newborn (85, 108), but to our knowledge, there are no data concerning the possible effects of increased or decreased concentration of corticosteroids in the fetus on subsequent brain development. Thus, contrary to what could be demonstrated with thyroid hormones, one has no clue in favour of a physiologic and organizing action of corticosteroids on the developing nervous system. As with testicular androgens, the action of endogenous corticosteroids may be limited to localized parts of the CNS and not extend to the whole nervous system as is undoubtedly the case for thyroid hormones.

SUMMARY

The morphogenetic action of thyroid hormone upon the mammalian CNS has been studied especially in the rat. Recent research work in this field has been concerned with the developing cerebellum. Thyroid deficiency and neonatal hyperthyroidism respectively retard and accelerate neuronal maturation, and the growth and branching of neuronal processes, the ontogenesis of nerve endings and synaptogenesis. Some of the morphologic abnormalities and disturbances of the biochemical maturation demonstrated in both the cerebrum and the cerebellum of the hypothyroid animal become irreversible after the end of the 2^d week of postnatal life. Thyroid deficiency does not affect the normal increase of cerebral DNA content. In the cerebellum, it leads to a reduction in the rate of cell proliferation. However, since it also leads to a prolongation of the proliferative phase, the final number of cerebellar cells is normal, but the cellular composition of the organ is modified. By acting more deeply on the development of the Purkinje cell dendritic arborization than on the formation and laying down of the granule cells, thyroid deficiency, but not neonatal hyperthyroidism, disrupts the synchronism which normally exists between these two main processes of the cerebellar histogenesis. The increase in the number of dying granule cells observed in the thyroid-deficient animal is probably related to this disturbance in normal cell interactions. In the animal made hyperthyroid during the neonatal period a transient increase in the total number of cells is observed in the cerebellum, but the final number of cells is decreased both the cerebellum and the cerebral cortex. Neonatal hyperthyroidism as thyroid deficiency is associated with behavioral deficits in adulthood. It has been demonstrated that the morphogenetic action of thyroid hormone on both neuronal maturation and cell formation within the cerebellum is physiologic. Moreover it seems to be "permissive." From the comparison of the effects of thyroid dysfunction with those of undernourishment and neonatal administration of excess corticosteroids, two disturbances of the nutritional or hormonal state in which cell proliferation is also disturbed in both cerebrum and cerebellum, it appears that to some extent the action of thyroid hormone is specific. The same conclusion is reached from studying the effects of postnatal administration of GH in normal and hypothyroid animals. The action of GH, without being entirely excluded, remains controversial. As to corticosteroids, even if they are a marked influence on the developing CNS, it is not demonstrated that they exert such an influence in physiologic conditions.

REFERENCES

1. Altman, J.: Autoradiographic and histological studies of postnatal neurogenesis. III. Dating the time of production and onset of differentiation of cerebellar microneurons in rats. J. Comp. Neurol. 136: 269-294 (1969).
2. Altman, J. Experimental reorganization of the cerebellar cortex. V. Effects of early X-irradiation schedules that allow or prevent the acquisition of basket cells. J. comp. Neurol., 165: 31-48 (1976).
3. Altman, j.: Experimental reorganization of the cerebellar cortex. VI. Effects of X-irradiation schedules that allow or prevent cell acquisition after basked cells are formed. J. comp. Neurol., 165: 49-64 (1976).
4. Altman, J.: Experimental reorganization of the cerebellar cortex. VII. Effects of late X-irradiation schedules that interfere with cell acquisition after stellate cells are formed. J. comp. Neurol., 165: 65-76 (1976).
5. Balázs, R., Brooksbank, B.W.L., Davison, A.N., Eayrs, J.T., and Wilson, D.A.: The effect of neonatal thyroidectomy on myelination in the rat brain. Brain Research, 15: 219-232 (1969).
6. Balázs, R., Brooksbank, B.W.L., Patel, A.J. Johnson, A.L. and Wilson, D.A.:

Incorporation of [^{35}S]-sulfate into brain constituents during development and the effects of thyroid hormone on myelination. Brain Research, 30: 273-293 (1971).

7. Balázs, R., Cocks, W.A., Eayrs, J.T. and Kovács, S.: Biochemical effects of thyroid hormones on the developing brain; in Hamburgh and Barrington, Hormones in development, pp. 357-379 (Appleton Century Crofts, New York, 1971).

8. Balázs, R. and Cotterrell, M.: Effect of hormonal state on cell number and functional maturation of the brain. Nature, 236: 348-350 (1972).

9. Balázs, R., Kovács, S., Cocks, W.A., Johnson, A.L. and Eayrs, J.T.: Effect of thyroid hormone on the biochemical maturation of rat brain. Postnatal cell formation. Brain Research, 25: 555-570 (1971).

10. Balázs, R., Kovács, S., Teichgräber, P., Cocks, W.A. and Eayrs, J.T.: Biochemical effects of thyroid deficiency on the developing brain. J. Neurochem. 15: 1335-1349 (1968).

11. Balázs, Patel, A.J. and Richter, D.: Metabolic compartments in the brain: their properties and relation to morphological structures; in Balázs and Cremer, Metabolic compartmentation in the brain, pp. 167-184 (Macmillan, London, 1973).

12. Balázs, R. and Richter, D.: Effects of hormones on the biochemical maturation of the brain; in Himwich, Biochemistry of the developing brain, vol. 1, pp. 253-299 (Dekker, New York, 1973).

13. Barrnett, R.J.: Some aspects of the physiology of the experimental cretin-like animal. Thesis, Yale Univ. School of Medicine (1948).

14. Bass, N.H. and Young, E.: Effects of hypothyroidism on the differentiation of neurons and glia in developing rat cerebrum. J. neurol. Sci., 18: 155-173 (1973).

15. Block, J.B. and Essman, W.B.: Growth hormone administration during pregnancy – a behavioural difference in offspring rats. Nature, 205: 1136-1137 (1965).

16. Bradley, P.B., Eayrs, J.T., Glass, A. and Heath, R.W.: The maturational and metabolic consequences of neonatal thyroidectomy upon the recruiting response in the rat. Electroenceph. Clin. Neurophysiol. 13: 577-586 (1961).

17. Bradley, P.B., Eayrs, J.T. and Richards, N.M.: Factors influencing potentials in normal and cretinous rats. Electroenceph. Clin. Neurophysiol. 17: 308-313 (1964).

18. Bradley, P.B., Eayrs, J.T. and Schmalbach, K.: The EEG of normal and hypothyroid rats. Electroenceph. Clin. Neurophysiol. 12: 467-477 (1960).

19. Casper, R., Vernadakis, A. and Timiras, P.S.: Influence of oestradiol and cortisol on lipids and cerebrosides in the developing brain and spinal cord of the rat. Brain Research, 5: 524-526 (1967).

20. Clendinnen, B.G. and Eayrs, J.T.: The anatomical and physiological effects of prenatally administered somatotrophin on cerebral development in rats. J. Endocr., 22: 183-193 (1961).

21. Clos, J.: Etude histochimique et biochimique des effets de la déficience thyroïdienne de la thyroxine sur le développement de l'activité cholinestérasique dans le cervelet du jeune rat. C.R. Acad. Sc. (Paris), série D, 275: 2917-2920 (1972).

22. Clos, J., Crépel, F., Legrand, C., Legrand, J., Rabié, A. and Vigouroux, E.: Thyroid physiology during the postnatal period in the rat: a study of the development of thyroid function and of the morphogenetic effects of thyroxine with special reference to cerebellar maturation. Gen. comp. Endocr., 23: 178-192 (1974).

23. Clos, J., Favre, C., Selme-Matrat, M. and Legrand, J.: Effects of undernutrition on cell formation in the rat brain and specially on cellular composition of the cerebellum. Brain Research, 123: 13-26 (1977).

24. Clos, J. et Legrand, J.: Influence de la déficience thyroïdienne et de la sous-alimentation sur la croissance et la myélinisation des fibres nerveuses de la moelle cervicale et du nerf sciatique chez le jeune rat blanc. Arch. Anat. micr. Morphol. exp., 58: 339-354 (1969).

25. Clos, J. et Legrand, J.: Influence de la déficience thyroïdienne et de la sous-alimentation sur la croissance et la myélinisation des fibres nerveuses du nerf sciatique chez le jeune rat blanc.

Etude au microscope électronique. Brain Research, 22: 285-297 (1970).

26. Clos, J. and Legrand, J.: Effects of thyroid deficiency on the different cell populations of the cerebellum in the young rat. Brain Research. 63: 450-455 (1973).

27. Clos, J., Rebière, A. and Legrand, J.: Differential effects of hypothyroidism and undernutrition on the development of glia in the rat cerebellum. Brain Research. 63: 445-449 (1973).

28. Clos, J., Selme-Matrat, M., Rabié, A. et Legrand, J.: Effects du cortisol sur la prolifération et la maturation cellulaires dans le cerveau et le cervelet du rat: importance de l'âge des animaux au début du traitement. J. Physiol. (Paris) 70: 207-218 (1975).

29. Clos, J., Vigouroux, E. et Legrand, J.: Recherche des hormones thyroïdiennes endogènes dans les neurones isolés du cerveau de jeunes rats. J. Physiol. (Paris) 72: 37A (1976).

30. Cocks, J.A., Balázs, R., Johnson, A.L. and Eayrs, J.T.: Effect of thyroid hormone on the biochemical maturation of rat brain: conversion of glucose carbon into aminoacids. J. Neurochem. 17: 1275-1285 (1970).

31. Contopoulos, A.N., Simpson, M.E. and Koneff, A.A.: Pituitary function in the thyroidectomized rat. Endocrinology, 63: 642-653 (1958).

32. Cotterrell, M., Balázs, R. and Johnson, A.L.: Effects of corticosteroids on the biochemical maturation of rat brain: postnatal cell formation. J. Neurochem., 19: 2151-2167 (1972).

33. Cragg, B.G.: Synapses and membranous bodies in experimental hypothyroidism. Brain Research, 18: 297-307 (1970).

34. Cragg, B.G.: The development of cortical synapses during starvation in the rat. Brain, 95: 143-150 (1972).

35. Crépel, F.: Maturation of the cerebellar Purkinje cells. II. Hypothyroidism and ontogenesis of cerebellar Purkinje cells spontaneous firing. Exp. Brain Res. 14: 472-479 (1972).

36. Crépel, F.: Excitatory and inhibitory processes acting upon cerebellar Purkinje cells during maturation in the rat; influence of hypothyroidism. Exp. Brain Res. 20: 403-420 (1974).

37. Crépel, F.: Consequences of hypothyroidism during infancy on the function of cerebellar neurons in the adult rat. Brain Research. 85: 157-160 (1975).

38. Crépel, F.: Développement du cervelet chez le Rat; influence des hormones thyroïdiennes et rôle des interactions cellulaires. Thèse de Doctorat-es-Sciences, Université Paris VI, 1976. No CNRS: AO 12327.

39. Crépel, F., and Legrand, J.: Electrophysiological and structural correlates of the effects of thyroid deficiency on the cerebellum of the young rat; in Jilek and Trojan, Ontogenesis of the brain, vol. 2, pp. 259-270 (Charles University, Prague, 1974).

40. Croskerry, P.G., Smith, G.K., Shepard, B.J. and Freeman, K.B.: Perinatal brain DNA in the normal and growth hormone-treated rat. Brain Research, 52: 413-418 (1973).

41. Dainat, J., Gourdon, J., et Legrand, J.: Variations avec l'âge de l'incorporation de la leucine dans les protéines du cervelet après traitement par le propylthiouracile chez le Rat. C.R. Soc. Biol., 164: 1550-1554.

42. Dainat, J. et Legrand, J.: influence de l'hyperthyroïdisme néonatal sur l'incorporation in vivo de la L-^{3}H-leucine dans les protéines du cervelet chez le jeune rat. C.R. Soc. Biol. 169: 1377-1381 (1971).

43. Dainat, J. Rebière, A. et Legrand, J.: Influence de la déficience thyroïdienne sur l'incorporation in vivo de la L-^{3}H-leucine dans les proteines du cervelet chez le jeune rat. J. Neurochem. 17: 581-586 (1970).

44. Dalal, K.B., Valcana, T., Timiras, P.S. and Einstein, E.R.: Regulatory role of thyroxine on myelinogenesis in the developing rat. Neurobiology, 1: 211-224 (1971).

45. Daughaday, W.H., Peake, G.T., Birge, C.A. and Mariz, I.K.: The influence of endocrine factors on the concentration of growth hormone in rat pituitary, in Pecile and Müller, Proc. Int. Symp. Growth Hormone, pp. 238-252 (Excepta Medica, Amsterdam, 1968).

46. Davenport, J.W., Hagquist, W.W. and Hennies, R.S.: Neonatal hyperthyroidism: maturational acceleration and learning deficit in triodothyronine-stimulated rats. Physiol.

Psychology. 3: 231-236 (1975).

47. Davenport, J.W. and Hennies, R.S.: Perinatal hypothyroidism in rats: persistent motivational and metabolic effects. Development Psychobiology. 9: 67-82 (1976).

48. Denenberg, V.H.: A consideration of the usefulness of the critical period hypothesis as applied to the stimulation of rodents in infancy; in Newton and Levine, Early experience and behaviour: psychobiology of development, pp. 142-167 (Charles C. Thomas, Springfield, 1968).

49. Diamond, M.C.: The effects of early hypophysectomy and hormone therapy on brain development. Brain Research, 7: 407-418 (1968).

50. Duvilanski, B.H., Soto, A.M., De Guglielmone, A.E.R. and Gómez, C.J.: Age dependentchanges of uridine nucleotide and RNA metabolism in the brain of normal and hypothyroid rats. Acta Physiol. Lat. Amer., 25: 165-171 (1975).

51. Eayrs, J.T.: Protein anabolism as a factor ameliorating the effect of early thyroid deficiency. Growth, 25: 175-189 (1961).

52. Eayrs, J.T.: Thyroid and developing brain: anatomical and behavioural effects; in Hamburgh and Barrington, Hormones in development, pp. 345-355 (Appleton Century Crofts, New York, 1971).

53. Faryna de Raveglia, I., Gómez, C.J. and Ghittoni, N.E.: Hormonal regulation of brain development. V. Effect of neonatal thyroidectomy on lipid changes in cerebral cortex and cerebellum of developing rats. Brain Research, 43: 181-195 (1972).

54. Favre, C.: Influence des hormones thyroïdiennes et de la sous-alimentation sur l'incorporation de leucine tritiée dans les neurones isolés du cortex cérébral et du cervelet chez le jeune rat. Thèse de Spécialité, Physiologie animale, Université Montpellier II, 1976.

55. Flerkó, B.: Perinatal androgen action and the differentiation of the hypothalamus; in Brazier, Growth and development of brain: nutritional, genetic, and environment factors, pp. 117-137 (Raven Press, New York, 1975).

56. Ford, D.H., Rhines, R.K. and Stieg, C.: Hormone localization in the nervous system; in Ford, Influence of hormones on the nervous system, pp. 2-16 (Karger, Basel, 1971).

57. Garciá Argiz, C.A., Pasquini, J.M., Kaplún, B. and Gómez, C.J.: Hormonal regulation of brain development. II. Effect of neonatal thyroidectomy on succinate deshydrogenase and other enzymes in developing cerebral cortex and cerebellum of the rat. Brain Research, 6: 635-646 (1967).

58. Geel, S.E.: Neonatal hypothyroidism: enhanced incorporation of precursors into cerebral RNA in vivo and normalizing effect of a semi-acute injection of thyroxine. Life Sc. 17: 539-544 (1975).

59. Geel, S.E. and Gonzales, L.K.: In vitro studies of cerebral cortical RNA and nucleotide metabolism in hypothyroidism. J. Neurochem., 25: 377-385 (1975).

60. Geel, S.E. and Timiras, P.S.: The influence of neonatal hypothyroidism and of thyroxine on the ribonucleic acid and deoxyribonucleic acid concentrations of rat cerebral cortex. Brain Research, 4: 135-142 (1967).

61. Geel, S.E. and Timiras, P.S.: Influence of neonatal hypothyroidism and of thyroxine on the acetylcholinesterase and cholinesterase activities in the developing central nervous system of the rat. Endocrinology, 80: 1069-1074 (1967).

62. Geel, S.E. and Timiras, P.S.: The role of hormones in cerebral protein metabolism; in Lajtha, Protein metabolism of the nervous system, pp. 335-354 (Plenum Press, New York, 1970).

63. Geel, S.E. and Timiras, P.S.: Influence of growth hormone on cerebral cortical RNA metabolism in immature hypothyroid rats. Brain Research, 22, 63-72 (1970).

64. Geel, S.E. and Timiras, P.S.: The role of thyroid and growth hormones on RNA metabolism in the immature brain; in Hamburgh and Barrington, Hormones in development, pp. 391-401 (Appleton Century Crofts, New York, 1971).

65. Geel, S.E. and Valcana, T.: Cerebral RNA metabolism and thyroid function in early life; in

Ford, Influence of hormones on the nervous system, pp. 165-173 (Karger, Basel, 1971).

66. Gill, J.H., Reid, L.D., McClellan, R. and Porter, P.B.: Somatotrophin effects on fetal growth: failure to replicate. Psychon. Sci., 9: 289-290 (1967).

68. Gitlin, D., Kumate, J. and Morales, C.: Metabolism and maternofoetal transfer of human growth hormone in the pregnant woman at term. J. Clin. Endocrinol. Metab., 25: 1599-1608 (1965).

69. Gómez, C.J. and De Guglielmone, A.E.R.: Influence of neonatal thyroidectomy on glucose-aminoacids interrelations in developing rat cerebral cortex. J. Neurochem. 14: 1119-1128 (1967).

70. Gómez, C.J., De Guglielmone, A.E.R. and Duvilanski, B.: Effect of neonatal thyroidectomy on ribonucleic acid synthesis in developing rat brain. Acta Physiol. Lat. Amer., 21: 152-155 (1971).

71. Gómez, C.J., Ghittoni, N.E. and Dellacha, J.M.: Effect of L-thyroxine or somatotrophin on body growth and cerebral development in neonatally thyroidectomized rats. Life Sc. 5: 243-246 (1966).

72. Gourdon, J., Clos, J., Coste, C., Dainat, J. and Legrand, J.: Comparative effects of hypothyroidism, hyperthyroidism and undernutrition on the protein and nucleic acid contents of the cerebellum in the young at. J. Neurochem. 21: 861-871 (1973).

73. Gregory, K.M. and Diamond, M.C.: Effects of early hypophysectomy on brain morphogenesis in the rat. Exp. Neurol., 20: 394-414 (1968).

74. Hajós, F., Patel, A.J. and Balázs, R.: Effects of thyroid deficiency on the synaptic organization of the rat cerebellar cortex. Brain Research, 50: 387-401 (1973).

75. Hamburgh, M.: An analysis of the action of thyroid hormone on development based on in vivo and in vitro studies. Gen. comp. Endocrinol., 10: 198-213 (1968).

76. Hamburgh, M.: The role of thyroid and growth hormones in neurogenesis; in Moscona and Monroy, Current topics in developmental biology, vol. 4, pp. 109-148 (Academic Press, New York 1969).

77. Hamburgh, M. and Flexner, L.B.: Biochemical and physiological differentiation during morphogenesis. XXI. Effect of hypothyroidism and hormone therapy on enzyme activities of the developing cerebral cortex of the rat. J. Neurochem. 1: 279-288 (1957).
J. Physiol. (Paris) 72: 102 A (1976).

78. Hamburgh, M., Lynn, E. and Weis, E.P.: Analysis of the influence of thyroid hormone on prenatal and postnatal maturation of the rat. Anat. Rec. 150: 147-162 (1964).

79. Hamburgh, M. Mendoza, L.A., Burkart, J.J. and Weil, F.: Thyroid-dependent processes in the developing nervous system; in Hamburgh and Barrington, Hormones in development, pp. 403-415 (Appleton Century Crofts, New York, 1971).

80. Hodge, G.K., Butcher, L.L. and Geller, E.: Hormonal effects on the morphologic differentiation of layer VI cortical cells in the rat. Brain Research, 104: 137-141 (1976).

81. Howard, E.: Increased reactivity and impaired adaptability in operant behaviour of adult mice given corticosterone in infancy. J. comp. Physiol. Psychol., 85: 211-220 (1973).

82. Howard, E.: Hormonal effects on the growth and DNA content of the developing brain; in Himwich, Biochemistry of the developing brain, vol. 2, pp. 1-68 (Dekker, New York, 1974).

83. Howard, E. and Benjamins, J.A.: DNA, ganglioside and sulfatide in brains of rats given corticosterone in infancy, with an estimate of cell loss during development. Brain Research, 92: 73-87 (1975).

84. Howard, E. and Granoff, D.M.: Increased voluntary running and decreased motor coordination in mice after neonatal corticosterone implantation. Exp. Neurol., 22: 661-673 (1968).

85. Jost, A.: Problems of fetal endocrinology: the adrenal glands; in Pincus, Recent progress in hormone research, vol. 22, pp. 541-574 (Academic Press, New York, 1966).

86. Klein, A.H., Meltzer, S. and Kenny, F.M.: Improved prognosis in congenital hypothyroidism treated before age three months. J. Pediatrics. 81: 912-915 (1972).

87. Kohl, H.H.: Depressed RNA synthesis in the brains and livers of thyroidectomized, normal and hormone injected rats. Brain Research, 40: 445-458 (1972).

88. Krawiec, L., García-Argiz, C.A., Gómez, C.J. and Pasquini, J.M.: Hormonal regulation of brain development. III Effects of triiodothyronine and growth hormone on the biochemical changes in the cerebral cortex and cerebellum of neonatally thyroidectomized rats. Brain Research, 15: 209-218 (1969).

89. Laron, Z., Pertzelan, A., Mannheimer, S., Goldman, J. and Guttman, S.: Lack of placental transfer of human growth hormone. Acta Endocrinol., 53: 687-692 (1966).

90. Lefranc, G., George, Y. et Tusques, J.: Etude de l'activité acétylcholinestérasique du cortex cérébelleux du rat nouveau-né au cours de sa maturation sous l'action de la thyroxine. C.R. Soc. Biol. 162: 219-221 (1968).

91. Legrand, C., Clos, J. et Selme, Matrat, M.: Effects précoces des hormones thyroïdiennes sur la synaptogenèse dans le cervelet du Rat. J. Physiol. (Paris) 72: 102 A (1976).

92. Legrand, J.: Maturation du cervelet et déficience thyroïdienne: données chronologiques. Arch. Anat. micr. Morphol. exp. 52: 205-214 (1963).

93. Legrand, J.: La maturation du cervelet chez le rat blanc hypothyroïdien, in Minkowski, Regional development of the brain in early life, pp. 485-493 (Blackwell, Oxford, 1967).

94. Legrand, J.: Variations, en fonction de l'âge, de la réponse du cervelet à l'action morphogénétique de la thyroïde chez le Rat. Arch. Anat. micr. Morphol. exp. 56: 291-307 (1967).

95. Legrand, J.: Analyse de l'action morphogénétique des hormones thyroïdiennes sur le cervelet du jeune rat. Arch. Anat. micr. Morphol. exp. 56: 205-244 (1967).

96. Legrand, J.: Comparative effects of thyroid deficiency and undernutrition on maturation of the nervous system and particularly on myelination in the young rat; in Hamburgh and Barrington, Hormones in development, pp. 381-390 (Appleton Century Crofts, New York, 1971).

97. Legrand, J., Kriegel, A. et Jost, A.: Déficience thyroïdienne et maturation du cervelet chez le Rat blanc. Arch. Anat. micr. Morphol. exp. 50: 507-519 (1961).

98. Legrand, J., Selme-Matrat, M., Rabié, A., Clos, J. and Legrand, C.: Thyroid hormone and cell formation in the developing rat cerebellum. Biol. Neonate, 29: 368-380 (1976).

99. Lenard, H.G. and Bell, E.F.: Bioelectric brain development in hypothyroidism. A quantitative analysis with EEG power spectra. Electroenceph. Clin. Neurophysiol. 35: 545-549 (1973).

100. Levine, S.: The pituitary-adrenal system and the developing brain; in De Wied and Weijnen, Progress in Brain Research, vol. 32, pp. 79-85 (Elsevier, Amsterdam, 1970).

101. Levine, S. and Mullins, R.F. Jr.: Hormones in infancy; in Newton and Levine, Early Experience and behaviour: psychobiology of development, pp. 168-197 (Charles C. Thomas, Springfield, 1968).

102. Lewis, P.D., Patel, A.J., Johnson, A.L. and Balázs, R.: Effect of thyroid deficiency on cell acquisition in the postnatal rat brain. A quantitative histological study. Brain Research, 104: 49-62 (1976).

103. Macho, L., Štrbák, V. and Hromadová, M.: The effect of thyroxine on amino acid incorporation into protein during ontogenesis in rats. Hormones, 3: 354-360 (1972).

104. Manuelidis, L.: Studies with electron microscopic autoradiography of thyroxine ^{125}I in organotypic cultures of the CNS. II. Sites of cellular localization of thyroxine [125]I. Yale J. Biol. Med., 45: 501-518 (1972).

105. Manuelidis, L. and Bornstein, M.: I[125] labelled thyroid hormones in cultured mammalian nerve tissue. Z. Zellforsch., 106: 189-199 (1970).

106. Matthieu, J.M., Reier, P.J. and Sawchak, J.A.: Proteins of rat brain myelin in neonatal hypothyroidism. Brain Research, 84: 443-451 (1975).

107. Michels, R., Cason, J. and Sokoloff, L.: Thyroxine: effects on amino acid incorporation into protein in vivo. Science, 140: 1417-1418 (1963).

108. Negellen-Perchellet, E. and Cohen, A.: Effect of ether inhalation by adrenalectomized

pregnant rats on the adrenal corticosterone concentration in normal, decapitated and encephalectomized fetuses. Neuroendrocrinology, 17: 225-235 (1975).

109. Nicholson, J.L. and Altman, J.: The effects of early hypo- and hyperthyroidism on the development of rat cerebellar cortex. I. Cell proliferation and differentiation. Brain Research, 44: 13-23 (1972).

110. Nicholson, J.L. and Altman, J.: The effects of early hypo- and hyperthyroidism on the development of the rat cerebellar cortex. II. Synaptogenesis in the molecular layer. Brain Research, 44: 25-36 (1972).

111. Olton, D.S., Johnson, C.T. and Howard, E.: Impairment of conditioned active avoidance in adult rats given corticosterone in infancy. Develop. Psychobiol. 8: 55-61 (1974).

112. Oppenheimer, J.H., Schwartz, H.L. and Surks, M.I.: Tissue differences in the concentration of triiodothyronine-nuclear binding sites in the rat: liver, kidney, pituitary, heart, brain, spleen and testis. Endocrinology, 95: 897-903 (1974).

113. Pascual-Leone, A.M., Garcia, M.D., Hervás, F. and Morreale De Escobar, G.: Decreased pituitary growth hormone content in rats treated neonatally with high doses of L-thyroxine. Horm. Metab. Res., 8: 215-217 (1976).

114. Pasquini, J.M., Kaplún, B., Garciá Argiz, C.A. and Gómez, C.J.: Hormonal regulation of brain development. I. The effect of neonatal thyroidectomy upon nucleic acids, protein and two enzymes in developing cerebral cortex and cerebellum of the rat. Brain Research, 6: 621-634 (1967).

115. Patel, A.J., Rabié, A., Lewis, P.D. and Balázs, R.: Effects of thyroid deficiency on postnatal cell formation in the rat brain. A biochemical investigation. Brain Research, 104: 33-48 (1976).

116. Pelton, E.W. and Bass, N.H.: Adverse effects of excess thyroid hormone on the maturation of rat cerebrum. Arch. Neurol., 29: 145-150 (1973).

117. Phelps, C.P. and Leathem, J.H.: Effects of postnatal thyroxine administration on brain development, response to postnatal androgen and thyroid regulation in female rats. J. Endocr. 69: 175-182 (1976).

118. Purves, H.D. and Griesbach, W.E.: Observations on the acidophil changes in the pituitary in thyroxine deficiency states. I. Acidophil degranulation in relation to goitrogenic agents and extrathyroidal synthesis. Brit. J. exp. Path., 27: 170-179 (1946).

119. Rabié, A., Favre, C., Clavel, M.C. and Legrand, J.: Effects of thyroid dysfunction on the development of the rat cerebellum, with special reference to cell death within the internal granular layer. Brain Research, 120: 521-531 (1977).

120. Rabie, A. and Legrand, J.: Effects of thyroid hormone and undernourishment on the amount of synaptosomal fraction in the cerebellum of the young rat. Brain Research, 61: 267-278 (1973).

121. Rabié, A., Selme-Matrat, M., Clavel, M.C., Clos, J. and Legrand, J.: Effects of methylazoxymethanol given at different stages of postnatal life on development of the rat brain. Comparison with those of thyroid deficiency: J. Neurobiol. 8: 337-354 (1977).

122. Ray, O.S. and Hochauser, S.: Growth hormone and environmental complexity effects on behaviour in the rat. Develop. Psychol., 1: 311-317 (1969).

123. Rebière, A.: Aspects quantitatifs de la synaptogenèse dans le cervelet du rat sous-alimenté dès la naissance. Comparaison avec l'animal rendu hypothyroïdien, C.R. Acad. Sc. (Paris), série D, 276: 2317-2320 (1973).

124. Rebière, A. et Legrand, J.: Absence d'effets marqués de l'hormone hypophysaire de croissance sur la maturation histologique du cortex cérébelleux chez le jeune rat normal ou hypothyroïdien. Brain Research, 22: 299-312 (1970).

125. Rebière, A. et Legrand, J.: Effets comparés de la sous-alimentation, de l'hypothyroïdisme et de l'hyperthyroïdisme sur la maturation histologique de la zone moléculaire du cortex cérébelleux chez le jeune rat. Arch. Anat. micr. Morphol. exp. 61: 105-126 (1972).

126. Rebière, A. et Legrand, J.: Données quantitatives sur la synaptogenèse dans le cervelet du

rat normal et rendu hypothyroïdien par le propylthiouracile. C.R. Acad. Sc. (Paris), serie D. 274: 3581-3584 (1972).

127. Reier, P.J. and Hughes, A.F.: An effect of neonatal radiothyroidectomy upon nonmyelinated axons and associated Schwann cells during maturation of the mouse sciatic nerve. Brain Research, 41: 263-282 (1972).

128. Salas, M. and Schapiro, S.: Hormonal influences upon the maturation of the rat brain's responsiveness to sensory stimuli. Physiol. Behav., 5: 7-11 (1970).

129. Sara, V.R. and Lazarus, L.: Prenatal action of growth hormone on brain and behaviour. Nature, 250: 257-258 (1974).

130. Sara, V.R., Lazarus, L., Stuart, M.C. and King, T.: Fetal brain growth: selective action by growth hormone. Science, 186: 446-447 (1974).

131. Schapiro, S.: Some physiological, biochemical, and behavioural consequences of neonatal hormone administration: cortisol and thyroxine. Gen. comp. Endocr. 214-228 (1968).

132. Schapiro, S.: Maturation of the neuroendocrine S.: Maturation of the neuroendocrine response to stress in the rat; in Newton and Levine, Early experience and behaviour: psychobiology of development, pp. 198-257 (Charles C. Thomas, Springfield, 1968).

133. Schapiro, S., Salas, M. and Vukovich, K.: Hormonal effects on ontogeny of swimming ability in the rat: assessment of central nervous system development. Science, 168: 147-151 (1970).

134. Schapiro, S., Vukovich, K. and Globus, A.: Effects of neonatal thyroxine and hydrocortisone administration on the development of dendritic spines in Exp. Neurol. visual cortex of rats. Neurol. 40: 286-296 (1973).

135. Schooley, R.A., Friedkin, S. and Evans, E.S.: Re-examination of the discrepancy between acidophil numbers and growth hormone concentration in the anterior pituitary following thyroidectomy. Endocrinology, 79: 1053-1057 (1966).

136. Sjöden, P.O. and Söderberg, U.: Effects of neonatal thyroxine stimulation on adult open-field behaviour and thyroid activity in rats. Physiol. Psychol. 4: 50-56 (1976).

137. Sokoloff, L.: The mechanism of action of thyroid hormones on protein synthesis and its relationship to the differences in sensitivities of mature and immature brain; in Lajtha, Protein metabolism of the nervous system, pp. 367-382 (Plenum Press, New York, 1970).

138. Solomon, J. and Greep, R.O.: The effect of alterations in thyroid function on the pituitary growth hormone content and acidophil cytology. Endocrinology, 65: 158-164 (1959).

139. Stone, J.M. and Greenough, W.T.: Excess neonatal thyroxine effects on learning in infant and adolescent rats. Develop. Psychobiol., 8: 479-488 (1975).

140. Szijan, I. and Burdman, J.A.: The relationship between DNA synthesis and the synthesis of nuclear proteins in rat brain. Effect of hydrocortisone acetate. Biochem. Biophys. Acta,

141. Szijan, I., Kalbermann, L.E. and Gómez, C.J.: Hormonal regulation of brain development. IV. Effect of neonatal thyroidectomy upon incorporation in vivo of L^3H phenylanaline into proteins of developing rat cerebral tissues and pituitary gland. Brain Research, 27: 309-318 (1971).

142. Tusques, J.: Recherches expérimentales sur le rôle de la thyroïde dans le développement du système nerveux. Biologie médicale, 45: 395-413 (1956).

143. Tusques, J., Lefranc, G. et George, Y.: Analyse par la technique de Golgi-Cox de la maturation du cortex cérébelleux sous l'influence de la thyroxine chez le Rat nouveau-né. C.R. Soc. Biol. 161: 2256-2260 (1967).

144. Valcana, T.: Effect of neonatal hypothyroidism on the development of acetylcholinesterase and choline acetyltransferase activities in the rat brain; in Ford, Influence of hormones on the nervous system, pp. 174-184 (Karger, Basel, 1971).

145. Vernadakis, A. and Woodbury, D.M.: Effects of cortisol on maturation of the central nervous system, in Ford, Influence of hormones on the nervous system, pp. 85-97 (Karger, Basel, 1971).

146. Vigouroux, E.: Dynamic study of postnatal thyroid function in the rat. Acta

Endocrinologica. 83: 752-762 (1976).

147. Vigouroux. E.: Développement de la fonction thyroïdienne chez le jeune rat. Thèse de Doctorat-es-Sciences. Université Paris VI. 1974. No CNRS: AO 10057.

148. Vigouroux. E.: Etude in vivo de quelques aspects du métabolisme de la thyroxine et des iodures dans le cerveau du jeune rat. J. Physiol. (Paris). 71: 151-152A (1975).

149. Walker. D.G.. Simpson. M.E.. Asling. C.W. and Evans. H.M.: Growth and differentiation in the rat following hypophysectomy at 6 days of age. Anat. Record. 106: 539-554 (1950).

150. Walravens. P. and Chase. H.P.: Influence of thyroid on formation of myelin lipids. J. Neurochem.. 16: 1477-1484 (1969).

151. Weichsel. M.E.. jr.: Effect of thyroxine on DNA synthesis and thymidine kinase activity during cerebellar development. Brain Research. 78: 455-465 (1974).

152. Wysocki. S.J. and Segal. W.: Influence of thyroid hormones on enzyme activities of myelinating rat central nervous tissues. Eur. J. Biochem.. 28: 183-189 (1972).

153. Zamenhof. S.: Stimulation of cortical cell proliferation by the growth hormone. III. Experiments on albino rats. Physiol. Zool. 15: 281-292 (1942).

154. Zamenhof. S.. Mosley. J. and Schuller. E.: Stimulation of the proliferation of cortical neurons by prenatal treatment with growth hormone. Science. 152: 1296-1397 (1966).

155. Zamenhof. S.. Van Marthens. E. and Grauel. L.: Prenatal cerebral development; effect of restricted diet. reversal by growth hormone. Science. 174: 954-955 (1971).

156. Zamenhof. S. and Van Marthens. E.: Study of factors influencing prenatal brain development. Molec. Cell. Biochem.. 4: 157-168 (1974).

CHAPTER ELEVEN

EFFECTS OF DRUGS, NARCOTICS AND TOXINS ON THE CHEMICAL MATURATION OF THE INFANT BRAIN*

ELLSWORTH C. ALVORD, JR. AND S. MARK SUMI

The remark by Norman (1958) that "the act of birth marks no particular milestone in the development of the human brain" has made it essential that embryologists, neonatologists, pediatricians, pathologists and others work together to bridge gaps created by professional but artificial time-based periods of study. Lemire et al. (1975) have argued that "malformations of the nervous system" can much more easily be considered as part of the spectrum of "diseases of the developing nervous system."

It seems likely that such a generalist's approach provoked the present invitation to neuropathologists to discuss chemistry, which must be our weakest subject! Just to anticipate potential problems, let us assure you that the most complicated chemistry that we shall present will consist of sudanophilia, the staining of materials (predominantly cholesterol esters) by sudan dyes such as oil-red-0.

First, though, let us review briefly some of the known effects of drugs, narcotics and toxins on the immature nervous system. Shepard (1973) has catalogued the numerous agents which have teratogenic potential in animals and man. In spite of the fact that many drugs are teratogenic to the developing nervous system in lower species, they present only minor problems in humans (Shepard 1973; Wilson 1973).

Infants born after an unsuccessful attempt at abortion with aminopterin have been found to have several anomalies, including delayed ossification of the neurocranium, hydrocephalus, encephalocele, anencephaly, cerebral aplasia and partial craniosynostosis (Warkany 1971).

Methotrexate has been associated with absence of the frontal bone and craniosynostosis of other sutures.

Even though thalidomide was responsible for numerous malformations of other parts of the body, it has only occasionally affected the CNS and produced hydrocephalus, microcephaly, meningomyelocele, microphthalmia

* Supported in part by research grants number HD-02274-11 and HD-08633-03 from the National Institutes of Health, U.S. Public Health Service.

and cobolomata of the iris and retina (Mellin and Katzenstein 1962; Warkany 1971).

Methyl mercury is known to have adverse effects on the developing nervous system in the human. Many people living near Minamata Bay in Japan during the 1950s acquired a neurologic illness from eating fish and shellfish which had a high content of methyl mercury. Several offspring of mothers residing in this area developed cerebral palsy. Snyder (1971) reported a newborn infant with grossly abnormal neurologic findings and elevated urinary mercury levels, whose mother had ingested methyl mercury-contaminated pork between the third and sixth months of pregnancy. Matsumoto et al. (1965) provided neuropathologic descriptions of two infants with fetal onset of Minamata disease. Both brains were very small with a reduction in size of deep nuclei and grey and white matter. The corpus callosum was hypoplastic. There was a reduction in size and number of nerve cells in the cerebral cortex and a glial proliferation in areas where neurons were deficient. The cerebellum was small and the folia atrophic. There were no granule cells in one case and a decreased number in the second. Purkinje cells were also decreased.

Many problems remain, especially in defining the minimal toxic dose of organic mercury and the relationships of gestational ages, doses and sites affected in the nervous system. Marked differences in sites and types of reactions in adult monkeys subjected to acute or chronic methylmercury poisoning have been reported by Shaw et al. (1975), with differential effects on deep nuclei and cortex, and we may expect that perhaps even more complicated relationships will become recognized when the fetus is examined experimentally.

Whether cyanate, a proposed therapy for sickle cell anemia, can damage the fetus is not yet known, but it is toxic to adult monkeys, affecting both central and peripheral nervous structures (Shaw et al. 1974), and we should be on the lookout for effects on the fetus.

Hexachlorophene (HCP) is a myelinotoxic agent that can be absorbed through normal mucosa or abnormal skin (e.g., burns, desquamating dermatitis) at any age or through normal skin in small premature infants (under 1400 g birth weight, about 30 weeks gestation). Since most myelin is formed after birth, the potential damage to the developing nervous system would seem to be relatively limited. However, since some myelin is being formed in utero, it can be damaged when small premature infants are bathed in 3% HCP, as described by Shuman et al. (1973a, 1974, 1975a). As few as two such whole-body baths can produce a vacuolar encephalopathy of the brain stem reticular formation with fatal apnea and no other lesions detectable at autopsy. Remarkably similar age-dose relationships have been found experimentally in baby rats. Two-week-old baby rats are extremely sensitive to HCP, as few as two baths one day apart being sufficient to kill them (Shuman et al. 1973b, 1975b; Alvord et al. 1975).

In older children with normal skin the vacuolar encephalopathy of HCP does not occur. However, burned skin or desquamating dermatitis allows HCP to be absorbed in toxic amounts. Typically, the lesion is much more diffuse, death resulting from massive cerebral edema due to the vacuolization of myelin sheaths throughout the CNS. Had such lesions been named "hexachlorophene encephalopathy" rather than "burn syndrome" (Larson 1968), it is likely that cases would no longer be occurring.

Since an objection was raised as to how HCP produces the vacuoles before myelin is known to be present in the reticular formation, we considered two approaches to this problem: 1) by analysing a second problem which had appeared, Wallerian degeneration in the immature pyramidal tracts, also supposedly unmyelinated (Leech and Alvord 1975), and 2) by examining by electron microscopy the two areas involved, looking for early myelin forms (Alvord et al. 1976).

In some cases, the process of Wallerian degeneration in premature humans is grossly visible in the pyramidal tract (Leech and Alvord 1975). In most cases, however, it requires special stains, such as sudan (oil-red-0) and hematoxylin on frozen sections. The degree of sudanophilia of the pyramids in most premature infants is so slight that these extreme accumulations stand out clearly, and occasionally the rostral encephaloclastic lesions are so asymmetric as to produce such an asymmetry in the degree of sudanophilia of the pyramidal tracts that one can practically eliminate the possibility of normal variation. In the adult CNS, as every neuropathologist knows, tissue reactions generally and Wallerian degeneration in particular are remarkably slow. In the infant it is not so well known that reactions generally are much more rapid (Sumi and Hager 1968; Sumi 1970). Even we, however, were startled at the rapidity of the process of Wallerian degeneration in the premature infant: Wallerian degeneration is practically over in the infant before it begins in the adult. Of course, the timings are by no means very accurate in the infant, since there is nothing to go on clinically which resembles the adult's massive stroke, but the estimates cannot be very far off as judged by the degree of development of gyral patterns which can be recognized in many cases even through the area of necrosis.

Thus, there are at least two areas where presumably unmyelinated structures can be damaged by a myelinotoxin (HCP) or by a myelin-specific process (Wallerian degeneration). This forced us to question the basis for the assumption that the reticular formation and pyramidal tract are not myelinated in the premature infant. Almost a century of light microscopic studies of myelinogenesis has culminated in the recent report by Yakovlev and LeCours (1967), who reported that myelinization in these areas does not begin before full-term birth. Light microscopy, however, cannot resolve the early stages of wrapping of membranes into myelin. Electron microscopy of myelin in the human CNS is complicated by the post mortem instability of myelin, but Alvord et al. (1976) have obtained many specimens which quite adequately show 4 or more lamellae of myelin in both the pyramidal tract and reticular formation in small premature humans of considerably less than 30 weeks gestation, even down to 20 weeks gestation. The lack of earlier myelin forms with fewer lamellae could easily be due to the relatively late and inadequate fixation of these human brains as compared to the perfusion-fixation possible in experimental animals.

Let us now consider the question whether hypoxia and acidosis are capable of producing malformations of the nervous system in human embryos and fetuses as well as cerebral damage in the premature and newborn. The embryo and young fetus are generally thought to be quite resistant to anoxia in utero unless it is profound or focal (e.g., encephaloclastic forms of porencephaly-hydranencephaly as discussed in chapter fifteen). Villee (1967) believes that this resistance is due to a combination of metabolic factors. Warkany (1971) cites several examples of cases where carbon monoxide poisoning has been associated with malformations, such as hydrocephalus, microcephaly, cortical atrophy, microgyria and softening and cavitation of the basal ganglia. Whether this is the effect of the profound anoxia or a direct effect of the carbon monoxide is not known.

Leech and Alvord (1974a, b), Sumi (1974) and Sumi et al. (1972, 1973) have been investigating the accumulations of lipids in immature glial cells that occur in practically all human newborns coming to autopsy and comparing them with similar accumulations observed in certain immature monkeys. These lipid accumulations have been recognized classically to occur in two forms, focal accumulations in macrophages (usually seen in foci of necrosis known as periventricular leucomalacia, or PVL) and more diffuse accumulations in glial cells (glial fatty metamorphosis, or GFM). These glial cells are generally too immature to be classified morphologically as committed to one or another form of differentiation, and are probably small and large glioblasts, as described by Vaughn (1969) in the development of normal rat optic nerves. Although Vaughn made no quantitative estimates of the lipids he saw, we doubt that all of the lipids which we see can be normal. A third form, accumulation in astrocytes, has been recently found by Sumi (1976, in prep.) electron microscopically in infant monkeys and also recognized by us in oil-red-0-stained sections of acutely necrotic gray matter in premature humans. This necrosis of gray matter, which is common in full-term infants but very unusual in premature infants, has been attributed by Grunnet (1976) to wide variations in blood pH with hyperoxygenation accompanying treatment in the Bird respirator.

Since a certain amount of "necrobiosis" occurs normally in development, not all macrophages can be considered abnormal, but most investigators agree that the focal necroses characteristic of PVL are abnormal (DeRueck et al. 1972). However, considerable disagreement persists concerning the significance of GFM. Since no truly normal human infants die and come to autopsy, and since subprimates do not show large amounts of these sudanophilic lipids, the question whether the diffuse type of lipid accumulation (GFM) is merely a stage in normal pre-myelin lipidogenesis (Mickel and Gilles 1970), as described by Vaughn (1969) electron microscopically, or whether it is an acute reaction to metabolic stress (Leech and Alvord 1974a), cannot be resolved without an experimental model.

Recently Sumi (1974) and Sumi et al. (1972, 1973) have found monkeys to be remarkably similar to humans in showing both necrotic lesions of PVL and subnecrotic lesions of GFM under obviously abnormal situations (e.g., stillbirth, death during infancy, transplacentally induced fetal hypoxia and acidosis due to drug-induced maternal hypertension or hypotension). Most infant monkeys of comparable ages do not show these changes, especially when obtained under strictly normal situations (e.g., Caesarean section followed by immediate fixation by perfusion with formaldehyde or glutaraldehyde). However, some lipids were noted in a large minority, 30-40%, of supposedly normal infant monkeys which we received from other investigators. Since these infants were obtained from pregnant monkeys without regard for their past breeding habits, and since about the same incidence of neonatal wastage is characteristic of this breeding colony, we could not determine whether this large minority should be considered really normal. Since foci of necrosis were present in some of these infants, we obviously suspected that at least some of the 30-40% of infants with sudanophilic lipids were coming from those pregnancies which would have proven to be "bad" if they had been allowed to proceed naturally.

During the past 2 years these preliminary observations have been subjected to a much more strict experimental design, including light and electron microscopic "blind" examination of brains from naturally delivered baby monkeys from parents of known past breeding habits. Some of these parents are known to be of high or low risk for abortion, stillbirth and/or neonatal death (and half of each group has been stressed daily during pregnancy to make 4 experimental stress-risk study groups). These baby monkeys (group 1) are being sacrificed at either 1 or 36 hours after birth. Babies in group 2 are animals who were stillborn, who died neonatally, or who were sacrificed by other investigators because of progressive respiratory failure.

Histologic studies are being made on paraffin-embedded sections stained with hematoxylin and eosin, and on frozen sections stained with oil-red-0 and hematoxylin. In those brains fixed by perfusion with paraformaldehyde-glutaraldehyde solution, tissues for both light and electron microscopy are obtained. Those brains fixed by immersion in formalin are examined by light microscopy alone. All histologic evaluations have been carried out by a "blind" procedure, without knowing from which experimental group the infant monkey was derived. The differences between perfused and non-perfused specimens are so great, however, that we cannot claim to have been completely "blind" as to whether the baby belongs to group 1 or group 2. Since the experiment is still in progress, and since new animals are still being received for study, the stress risk code for the live-born (group 1) animals has not yet been broken.

Thus far we have had 9 live-births, of which 7 have been fully evaluated histologically. As shown in table 1, there are no more than traces of sudanophilic lipids in the corpus callosum, except for 2 animals (#5142 and 5329),

Table 1. Histologic findings in live-born monkeys. Data are given as a ratio: Number of lipid-containing cells/degree of myelination.*

Case No.	Myelin stage	corpus collosum			Occipital tapetum	Germinal matrix
		ant.	mid	post		
4910	2	tr/0	0/0		0/2	
5142	3a	2/1	0/1	1/1	tr/0	
5092	3b	tr/1	0/1			
5091	3b	tr/1	tr/1	4/1		
5099	3b	0/1	tr/1	tr/1		
5368	3b	tr/1	0/tr	3/tr	1/0	
5329	3b	1/tr	0/1			1/0

* Degree of change or development: 0 = None. Tr = Trace. 1 = Slight. 2 = Moderate. 3 = Marked. 4 = Very Marked.

where slight-to-moderate amounts are present anteriorly. The lipid-containing cells in all animals have been small, spindle-shaped or irregularly shaped, with a few lipid-filled processes. Electron-microscopically in 3 animals we have confirmed our earlier observation that these lipid droplets are in thin or oval cells with dark cell bodies, which we have identified as young undifferentiated "pre-myelin" glial cells. Large macrophage-like cells have been found in the germinal matrix in only one animal (#5329).

Group 2 consists of 10 stillborn and neonatally dead or sacrificed animals, of which 10 have been examined by light microscopy and 2 by electron microscopy. Table 2 lists the histologic findings (observed blindly) and the clinical

Table 2. Histologic findings in nonviable newborn monkeys. Data as in table 1, with N = necrosis, SB = stillborn.

Case No.	Age Gest.	Postnatal	Myelin stage	Corpus callosum ant.	mid.	post.	Occipital tapetum	Germinal matrix
5291	132	3 d.*	1b	4/0	N, 3/0	4/0		
5037	169	8 d.*	2	2/0		2/0	3	
5030	166	SB	3a	4/0	1/1	3/tr	1/0	
5029	170	SB	3b	1/1	2/1	4/0	4/tr	
5284	134	10 d.*	3b	1/2	3/2	4/0	3/0	
5293	152	½ hr.**	3b		tr/1		4/0	
5329		SB	3b	3/tr	0/1			1/0
5292	186	**	4	3/tr	1/1	3/1		
5294		26 d.	4	0/2	tr/1	tr/1	0/0	
5298		28 d.	4	0/1	tr/1			

* respiratory failure.
** bite wounds.
*** meningitis, abscess.

features (subsequently obtained, this information not being capable of biasing any future observations since these represent extra animals over and beyond those eventually to be provided in the live-born experimental group). The striking histologic finding in 8 of these 10 animals is the presence of large numbers of the small lipid-containing cells. Just 2 animals had only trace amounts of lipids comparable to all but 2 of the infants in group 1. Many of these cells were present in the tapetum in the occipital periventricular white matter as well as in the corpus callosum. In the 3 animals (#5291, 5037 and 5284) suffering progressive respiratory failure after survival for 3 to 10 days, there were also many large, globular, lipid-filled, macrophage-like cells in the cerebral white matter.

Electron-microscopically, these animals had 3 different types of lipid-containing cells: 1) "pre-myelin" glial cells, without other distinguishing cyto-logic features, similar to those seen in group 1; 2) astrocytes containing bundles of cytoplasmic fibrils, corresponding to irregularly shaped cells with 2 or more processes noted by light microscopy as resembling the "hypertrophic" astrocytes of Gilles and Murphy (1969), these cells being most numerous in the 2 animals who survived 8 and 10 days; and 3) macrophages containing crystalline inclusions and cellular debris as well as lipid droplets. In addition, there were several large degenerating axons filled with dense bodies and vesicles.

Tables 3 and 4 summarize these results according to our estimates of the

Table 3. Degree of lipid-accumulation related to degree of myelination of the same region of the corpus callosum (CC) in 7 live-born monkeys from pregnancies of known risk* and compared to 10 non-viable monkeys from pregnancies of unknown risk (in parentheses).

Lipids in CC	Myelination of corpus callosum				
	0	tr	+	+ +	Total
0	1(0)	1(0)	4(2)	0(1)	6(3)
tr	2(0)		3(4)	1(0)	6(4)
+	1(1)	1(2)	0(2)	0(1)	2(6)
+ +	0(2)		1(1)		1(3)
+ + +	0(3)	0(1)	0(2)	0(1)	0(7)
+ + + +	0(6)	0(1)			0(7)
Totaal	4(12)	2(4)	8(11)	1(3)	15(30)

* code not yet broken.

degrees of development of the brains. These tables indicate that there are marked qualitative and quantitative histologic differences in the brains of sacrificed viable baby monkeys (from table 1) as compared to nonviable baby monkeys (from table 2). We conclude, therefore, that prenatal and neonatal distress is accompanied by GFM-type lipid accumulations in glial cells, some of which are "pre-myelin" cells and others astrocytes. In addition, areas of frank PVL-like necrosis of the cerebral white matter develop in animals suffering prolonged hypoxia postnatally due to respiratory failure. These changes correspond closely to what we see in the human in comparable clinical circumstances. Although the possible deleterious effect of the birth process itself has not been ruled out, abnormalities found in the stillborn animals indicate that prenatal factors can be sufficient and that natal and postnatal factors are not required to induce a moderate-to-marked GFM-like accumulation of lipids.

Finally, but tentatively, if we include the 3 stillborn animals (# 5030, 5029

Table 4. Degree of lipid-accumulation in the corpus callosum (CC) related to degree of maturation (myelinogenesis) of the rest of the brain in 7 live-born monkeys from pregnancies of known risk* and compared to 10 non-viable monkeys from pregnancies of unknown risk (in parentheses).

Lipids in CC	Myelinogenetic stage 1b.	2	3a	3b	4	Total
0		1(0)	1(0)	4(1)	0(2)	6(3)
tr		1(0)	1(0)	4(1)	0(3)	6(4)
+			0(2)	2(2)	0(2)	2(6)
++		0(2)	1(0)	0(1)		1(3)
+++	0(1)	0(1)	0(1)	0(3)	0(1)	0(7)
++++	0(2)		0(1)	0(4)		0(7)
Total	0(3)	2(3)	3(4)	10(12)	0(8)	15(30)

* code not yet broken.

and 5329 in table 2) with 2 of the experimental study animals (#5142 and 5329 in table 1), we can see a pattern of results which strongly suggests that more than trace amounts of lipid accumulations in supposedly normal infants may only be occurring in infants at high risk for subsequent neonatal death. Within the next 6 months or so we hope to finish this first part of our study and to break the stress-risk code. We hope to find the explanation for the previously reported 30-40% incidence of lipid accumulations in the probability that these are not the result of really normal pregnancies.

Thus, we are at an exciting moment in the course of our experiments. Regardless of how imperfect the model proves to be, we believe that the immature glial cell is highly sensitive to metabolic stress, most likely hypoxia-acidosis. Such stress can be induced in the fetus by maternal hypertension or hypotension, so far experimentally drug-induced; probably these changes in maternal blood pressure can also be induced by such psychogenic factors as environmental stress. These stresses are so commonly experienced by the mother as to be considered "normal" and, as such, are generally overlooked by both patient and physician, and may not be recorded in the obstetrical records.

The possible importance of these lesions for the production of some of the syndromes of mental retardation and "minimal brain damage" (or "minimal cerebral dysfunction") probably lies in the damage to the developing white matter (multifocally in PVL and either diffusely or focally, especially in the corpus callosum, in GFM). In many cases of mental retardation coming to autopsy years to decades after birth, anatomic changes are slight. The difficulty in quantitating the amount of cerebral white matter is well known, and the difficulty in detecting small glial scars is almost as well known. Since the degree of reversibility of GFM is as yet completely unknown, we can only speculate on the possibility that GFM may play a role in these cases.

Patients with "minimal brain damage" simply do not die while they demonstrate the syndrome; they tend to recover from it, so that the syndrome has long since been forgotten when the patient finally dies of some other disease. Thus, it seems unlikely that the pathoanatomic substrate will ever be defined in actual clinicopathologic studies. Gazzaniga (1973) has suggested that at least some of the elements of the syndrome may relate to delayed maturation of the corpus callosum, which is the site where GFM most strikingly occurs (Leech and Alvord 1974c). In many cases this GFM-like accumulation of lipids may actually represent Wallerian degeneration secondary to multifocal necrotic lesions, PVL, in the adjacent white matter. In some cases, however, such PVL lesions do not appear to be present. We can only suggest that it may be Wallerian degeneration secondary to compression of the corpus callosum against the falx cerebri as the head is molded during delivery, rather than GFM itself. Such a lead is tempting to follow both experimentally and clinically.

Although the bipolar distribution of lipids in both GFM and Wallerian degeneration suggests that the two processes may be similar, we doubt that they can be identical. We would not be surprised to discover that GFM-like changes in premyelin glial cells can be induced by the degeneration of adjacent unmyelinated axons and that the Wallerian degeneration of early myelin surrounding axons induces GFM-like changes which may be quite different from the same process occurring in well-formed adult myelin.

SUMMARY

Of the many drugs, narcotics and toxins that are known to be teratogenic to the developing nervous system of animals, only a few are known to have affected the human in utero: aminopterin, methotrexate, thalidomide and methyl mercury. Others may have age-related effects:

Hexachlorophene (HCP) can penetrate the skin of small premature infants as easily as the severely burned or inflamed skin of older individuals. This drug appears to have a predilection for actively myelinating fibers, as in the brainstem reticular formation of premature infants who die of apnea probably due to paralysis of "vital centers." When more mature myelin is affected, death usually results from cerebral edema.

Hypoxia-acidosis, a frequent concomitant of stillbirth and infantile respiratory distress syndrome (the most common cause of death in premature infants), can produce glial fatty metamorphosis (GFM), characterized by accumulation of sudanophilic lipids (probably predominantly cholesterol esters) in immature glia and in astrocytes. Whether GFM is completely reversible is not yet known, but from studies in humans and monkeys it seems likely that periventricular leucomalacia (PVL), characterized by similar lipid accumulations in macrophages at sites of focal necrosis, may also occur in areas of severe GFM.

The problem of myelin-specific toxins (such as HCP) and degenerations (such as Wallerian) occurring in immature tracts not known by light microscopy to be myelinated before full term has been at least partially resolved by discovering electron-microscopically that myelin is present in the reticular formation and pyramidal tracts many weeks earlier, even as early as 20 weeks gestation. Whether Wallerian degeneration of unmyelinated axons can induce a GFM-like change in adjacent glioblasts to augment the more classical changes in myelin-forming oligodendroglia remains an interesting possibility still being studied.

REFERENCES

Alvord, E.C., Jr., Kogon, M.B. and Shuman, R.M.: Onset of myelination in the human nervous system. J. Neuropathol. Exp. Neurol. 35: 115 (1976).
Alvord, E.C., Jr., Shuman, R.M. and Leech, R.W.: Reply to letter to the editor. Pediatrics 55: 743-744 (1975).
De Reuck, J., Chattha, A.S. and Richardson, E.P., Jr.: Pathogenesis and evolution of periventricular leukomalacia in infancy. Arch. Neurol. 27: 229-236 (1972).
Gazzaniga, M.S.: Brain theory and minimal brain dysfunction. Ann. N.Y. Acad. Sci. 205: 89-92 (1973).
Gilles, F.H. and Murphy, S.F.: Perinatal telencephalic leucoencephalopathy. J. Neurol. Neurosurg. Psychiat. 32: 404-413 (1969).
Grunnet, M.L.: Atypical periventricular leukomalacia in a neonatal intensive care unit. J. Neuropath. Exp. Neurol. 35:308 (1976).
Larson, D.L.: Studies show hexachlorophene causes burn syndrome. J. Am. Hosp. Assoc. 42: 63 (1968).
Leech, R.W. and Alvord, E.C., Jr.: Glial fatty metamorphosis, an abnormal response of pre-myelin glia frequently accompanying periventricular leukomalacia. Am. J. Pathol. 74: 603-613 (1974a).
Leech, R.W. and Alvord, E.C., Jr.: Morphologic variations in periventricular leukomalacia. Am. J. Pathol. 74: 591-602 (1974b).
Leech, R.W. and Alvord, E.C., Jr.: Perinatal leucoencephalopathy: An expanded concept. J. Neuropathol. Exp. Neurol. 33: 568-569 (1974c).

Leech, R.W. and Alvord, E.C., Jr.: Wallerian degeneration in the premature human. J. Neuropathol. Exp. Neurol. 34: 92 (1975).

Lemire, R.J., Loeser, J.D., Leech, R.W. and Alvord, E.C., Jr.: Normal and Abnormal Development of the Human Nervous System. Harper & Row, Publishers, Inc., Hagerstown, Md. (1975).

Matsumoto, H., Koyo, G. and Takeuchi, T.: Fetal Minamata disease. A neuropathological study of two cases of intrauterine intoxication by a methyl mercury compound. J. Neuropathol. Exp. Neurol. 24: 563-574 (1965).

Mellin, G.W. and Katzenstein, M.: The saga of thalidomide: Neuropathy to embryopathy, with case reports of congenital anomalies. N. Engl. J. Med. 267: 1184-1193, 1238-1244 (1962).

Mickel, H.S. and Gilles, F.H.: Changes in glial cells during human telencephalic myelinogenesis. Brain 93: 337-346 (1970).

Norman, R.M.: Malformations of the nervous system, birth injury and diseases of early life, in Greenfield, Blackwood, McMenemy, Meyer and Norman, Neuropathology, pp. 300-407 (Edward Arnold Ltd., London 1958).

Shaw, C.M., Mottet, K.M., Body, R.L. and Luschei, E.S.: Variability of neuropathologic lesions in experimental methylmercury encephalopathy in primates. Am. J. Pathol. 80: 451-470 (1975).

Shaw, C.M., Papayannopoulou, T. and Stamatoyannopoulos, G.: Neuropathology of cyanate toxicity in rhesus monkeys, preliminary report. Pharmacology 12: 166-176 (1974).

Shepard, T.H.: Catalog of Teratogenic Agents. (Johns Hopkins Press, Baltimore 1973).

Shuman, R.M., Leech, R.W. and Alvord, E.C., Jr.: Neurotoxicity of hexachlorophene in human infants. Am. J. Pathol. 70:19a (1973a).

Shuman, R.M., Leech, R.W. and Alvord, E.C., Jr.: Neurotoxicity of hexachlorophene in the human: I. A clinico-pathologic study of 248 children. Pediatrics 54: 689-695 (1974).

Shuman, R.M., Leech, R.W. and Alvord, E.C., Jr.: Neurotoxicity of hexachlorophene in humans. II. A clinico-pathologic study of 46 premature infants. Arch. Neurol. 32: 320-325 (1975a).

Shuman, R.M., Leech, R.W. and Alvord, E.C., Jr.: Neurotoxicity of topically applied hexachlorophene in the young rat. Arch. Neurol. 32: 315-319 (1975b).

Shuman, R.M., Leech, R.W., Alvord, E.C., Jr. and Sumi, S.M.: Experimental neurotoxicity of pHisoHex. J. Neuropathol. Exp. Neurol. 33: 195 (1973b).

Snyder, R.D.: Congenital mercury poisoning. N. Engl. J. Med. 284: 1014-1016 (1971).

Sumi, S.M.: Reaction of the immature brain to injury, in Angle and Bering, Physical Trauma as an Etiologic Agent in Mental Retardation, pp. 177-189 (U. S. Gov't. Printing Office, Washington, D.C. 1970).

Sumi, S.M.: Periventricular leukoencephalopathy in the monkey, a search for the "normal control" and the "early lesion." Arch. Neurol. 31: 38-44 (1974).

Sumi, S.M., Alvord, E.C., Jr., Parer, J., Eng, M. and Ueland, K.: Accumulation of sudanophilic lipids in the cerebral white matter of premature primates: An experimental inquiry into the pathogenesis of the Virchow-Schwartz-Banker-Larroche lesion. J. Neuropathol. Exp. Neurol. 31: 183 (1972).

Sumi, S.M. and Hager, H.: Electron microscopic study of the reaction of the newborn rat brain to injury. Acta Neuropathol. 10: 324-335 (1968).

Sumi, S.M., Leech, R. W., Alvord, E.C., Jr., Eng, M. and Ueland, K.: Sudanophilic lipids in unmyelinated primate cerebral white matter after intrauterine hypoxia and acidosis. Res. Publ. Assoc. Res. Nerv. Ment. Dis. 51: 176-197 (1973).

Vaughn, J.E.: An electron microscopic analysis of gliogenesis in rat optic nerves. Zts. Zellforsch. 94: 293-324 (1969).

Villee, C.A.: Bioenergetic consideration in fetal and mature tissues. in James, Myers and Gaull, Brain Damage in the Fetus and Newborn from Hypoxia or Asphyxia, pp. 47-56 (57th Ross Conference on Pediatric Research, Columbus 1967).

Warkany, J.: Congenital Malformations. (Year Book Med. Publ., Chicago 1971).
Wilson, J.G.: Present status of drugs as teratogens in man. Teratology 7: 3-16 (1973).
Yakovlev, P.I. and LeCours, A.R.: The myelogenetic cycles of regional maturation of the brain. in Minkowski, Regional Development of the Brain in Early Life, pp. 3-70 (Blackwell Publ., Oxford 1967).

CHAPTER TWELVE

TOXIC EFFECTS OF ELEVATED OXYGEN TENSION ON BRAIN MATURATION

LOUIS SOKOLOFF

Because mammalian organisms are uniformly aerobic and dependent for survival on oxidative metabolic processes, the potential toxic effects of oxygen are often neglected. Nevertheless, as with most substances, natural and otherwise, excessive oxygen has deleterious effects in a variety of tissues including those of the nervous system. It has long been known that high concentrations of oxygen at elevated atmospheric pressure produce seizures (Bean, 1954) and neuronal (Balentine, 1968) and retinal ganglion cell necrosis (Margolis & Brown, 1966). Hyperoxia at normal atmospheric pressure is not known to produce pathologic effects in the mature nervous system, but there is evidence that during periods of rapid maturation and development in perinatal life nervous tissues may be particularly susceptible to toxic effects of oxygen.

The elucidation of the role of oxygen in the pathogenesis of retrolental fibroplasia (Ashton, Ward & Serpell, 1954; Patz, 1957) demonstrated that hyperoxia at normal atmospheric pressure may alter the normal development of vascular components of neural-like tissue with subsequent disruption of its structure and function. The question naturally arose whether effects like those in the retina could also occur, perhaps, to a lesser degree, in the brain of the newborn. Various clinical observations on the incidence of mental retardation in children with retrolental fibroplasia have been consistent with this possibility but have been inconclusive because of difficulties in defining and obtaining suitable control subjects for comparison (Potter, 1954; Krause, 1955; Williams, 1958; Genn & Silverman, 1964). Some experimental support for this possibility has been obtained by Gyllensten (1959a, 1959b), who observed that newborn mice exposed to hyperoxia at normal atmospheric pressure for the first ten days of life exhibit deficiencies in cerebral capillary beds and abnormal neuronal morphology. In the present studies newborn rats were exposed continuously to 70-80% O_2 at one atmosphere during their first nine days of life, and the effects on some aspects of the chemical development of their brains were examined. The results indicate that a high con-

Table 1. Effects of hyperoxia on brain and body growth in early postnatal life.*

Animal	No. of Animals	Age (days)	Total Body Weight Birth (g)	Final (g)	Brain Weight at 9 days (mg)	Brain Weight/Body Weight (%)
Control	4	0	6.7 ± 0.2		248 ± 17	3.7 ± 0.2
Control	32	9	6.8 ± 0.5	21.5 ± 2.0	853 ± 49	4.0 ± 0.2
O₂-Exposed	32	9	6.9 ± 0.7	21.6 ± 2.3	818 ± 52**	3.8 ± 0.2**

* Data from Grave, Kennedy & Sokoloff (1972). The values presented are the means ± S.D. of measurements in the number of animals indicated.
** Indicates statistically significant difference from values obtained in control animals of same age ($p < 0.01$).

Table 2. Effects of hyperoxia on early postnatal changes in cerebral nucleic acids and protein contents.*

Animal	No. of Animals	Age (days)	DNA-P (µg/brain)	RNA-P (µg/brain)	Total Protein (mg/brain)	Proteolipid Protein (mg/brain)
Control	4	0	53.9 ± 2.8	45.6 ± 4.2	16.8 ± 2.4	0.34 ± 0.02
Control	32	9	120.4 ± 8.2	146.0 ± 11.0	65.8 ± 8.9	0.98 ± 0.29
O₂-Exposed	32	9	115.6 ± 11.1	140.1 ± 15.4	64.0 ± 8.5	0.92 ± 0.26
p value**			$p \sim 0.05$	$p \sim 0.05$	n.s.	$0.05 < p < 0.1$

* Data from Grave, Kennedy & Sokoloff (1972). The values presented are the means ± S.D. of measurements in the number of animals indicated.
** The p values refer to the differences between the 9-day old control and experimental animals.

centration of oxygen at ambient barometric pressure significantly impedes the growth of the brain and inhibits the normal accumulation of nuclei acids that occur during this early postnatal period of cerebral development.

Sixty-four newborn Sprague-Dawley rats, taken from eight mothers on the day of birth, were randomly mixed and arbitrarily assigned to eight litters of eight rats each. The rats in each litter were individually weighed and tattooed for subsequent identification. Four of the litters were placed in four separate cages and kept in an Isolette which was maintained at ambient temperature and pressure but ventilated with oxygen at a rate sufficient to maintain the atmosphere at a mean oxygen tension of 557 mm Hg, equivalent to 73^{o}_{o} oxygen at one atmosphere. The other four litters were kept in four other cages in normal room air. The environmental temperatures of all groups of animals were kept the same at normal room temperature. One lactating mother was assigned to each cage, but the mothers were rotated daily among the cages, alternately between cages in air and cages in oxygen, to prevent oxygen-induced pulmonary damage in the mothers and to minimize the possibility of artifactual effects arising from maternal differences. At the end of the nine-day interval, the animals were weighed and killed by freezing in Freon XII chilled to -60° C with dry ice. The brains were dissected away from the cranium, weighed, and analyzed for DNA, RNA, protein, and proteolipid protein as described previously (Grave, Kennedy & Sokoloff, 1972). Four newborn rats were weighed and killed and their brains analyzed chemically on the day of their birth to obtain the base-line values at the onset of the nine-day experimental period.

There was substantial body and brain growth in both the control and experimental animals during the nine-day interval (table 1). The chronic exposure to 73^{o}_{o} oxygen had no significant effect on body growth, indicating that nutritional status of the experimental animals was comparable to that of the control group. Brain growth, however, was significantly retarded so that at nine days the brain weight and the brain/body ratio were both significantly lower in the O_2-exposed animals than in the normal controls (table 1).

Along with increased mass, there were substantial increases in the contents of DNA, RNA, total protein, and proteolipid protein in the brains of the rats during the first nine days of life. O_2 exposure diminished the accumulation of DNA and RNA during this period (table 2); there appeared to be similar effects on total protein and proteolipid protein also, but these were not statistically significant.

The absolute differences in brain weights and chemical composition of the normal and O_2-exposed animals at nine days of age may not appear to be impressive, but the potential influence of the elevated oxygen on brain development may, perhaps, best be appreciated by its effects on the increments in these components of the brain. Although O_2 had no effect on the total body growth, brain growth was retarded by about 6% and the increase in DNA content by 7% (table 3). The accumulation of RNA and proteolipid protein were diminished to about the same degree (table 3).

Table 3. Effects of hyperoxia on the increments in brain and body weights and cerebral chemical components during first nine postnatal days of life.*

Exposure	△ Body Weight	△ Brain Weight	△ DNA-P	△ RNA-P	△ Total Protein	△ Proteolipid Brain
	(g)	(mg)	(μg brain)	μg brain)	(mg brain)	(mg brain)
Room Air (32)	+ 14.7	+ 605	+ 66.5**	+ 100.4	+ 49.0	+ 0.64
70-80% O₂ (32)	+ 14.7	+ 570	+ 61.7**	+ 94.5	+ 47.2	+ 0.58
% O₂ Effect	0	− 5.8	− 7.2	− 5.9	− 3.7	− 9.3
p value	−	< 0.01	< 0.05	∼ 0.05	n.s.	0.05 < p < 0.1

* Data from Grave. Kennedy & Sokoloff (1972). The values presented are the means ± S.D. of measurements in the number of animals indicated in parentheses.
** The difference of 4.8 μg of DNA-P is equivalent to a difference of approximately 65 μg of total DNA.

Although a 7% reduction in postnatal DNA synthesis is small, the deficit of 65 μg in the rat brain is equivalent to a diminution of cell number of 10 million cells if one assumes the value of 6.2×10^{-12}g of DNA per diploid rat cell (Enesco & Leblond, 1962). The functional significance of such a cellular deficit to the young rat may be difficult to asses, but there may be implications of these results for the care of human infants of young gestational age. A human seven-month fetus has about $1/3$ its final total brain cell population and, if delivered at that stage, must synthesize ex utero twice as much DNA as a term baby or about 110 billion cells (Winick, 1968). A 7% decrement in this number would represent more than seven billion cells.

The types of cells affected in this study have not been established. By terminating the experiments on the ninth day of life, however, we have exempted from consideration many cerebellar neurons and cerebral granular cells which proliferate most rapidly during the second week of life (Altman & Das, 1965; Altman, 1967; Fish & Winick, 1969). If the cellular effect of oxygen is a function of the local tissue pO_2, one might expect inhibition of cell division to be greatest in regions of highest pO_2, namely in the walls and perivascular zones of arterioles and arterial ends of capillaries. That such a location plays a role is suggested by the known O_2-sensitivity of vascular endothelium and glial cells during maturation (Ashton et al., 1954; Gyllensten, 1959a; Brand & Bignami, 1969; Shivers & Roofe, 1966), but it is possible that there are differences in susceptibility according to cell type and that rapidly proliferating neuronal elements of the germinal matrix which later migrate to the cerebral cortex may also be affected.

Although it is still uncertain whether the effects observed in these studies persist through the remainder of the life of the animals and alter brain function at maturity, it appears that the developing brain is not only threatened by deficient oxygen supply but may be adversely affected by hyperoxia at even normal atmospheric pressure. Ambient hyperoxia can, therefore, be added to the list of chemical affronts, such as malnutrition (Winick & Noble, 1966) and hypoxia (Cheek, Graystone & Rowe, 1969), which jeopardize the normal development of the brain.

SUMMARY

The continuous exposure of newborn rats to 70-80% oxygen at atmospheric pressure throughout the first nine days of life significantly inhibited the growth of the brain which normally occurs during this period of life. The accumulations of DNA, RNA, total protein, and proteolipid protein which accompany brain growth during this period were all approximately proportionately depressed by the oxygen-enriched atmosphere. The increase in brain mass in the first week of life reflects mainly cell proliferation, and since the decreased DNA accumulation occurred with no change in RNA/DNA ratios, we conclude that the effect of oxygen was to inhibit cellular division. We estimate that the oxygen exposure caused an approximately 7% deficit in the cell population of the brain. These results indicate that the use of elevated concentrations of oxygen may have deleterious effects on the growth and development of the brain.

REFERENCES

Altman, J.: Postnatal growth and differentiation of the mammalian brain, with implications for a morphological theory of memory; in Quarton, Melnechuk and Schmitt, The Neurosciences, pp. 723-743 (Rockefeller University Press, New York 1967).
Altman, J. and Das, G.D.: Autoradiographic and histological evidence of postnatal hippocampal neurogenesis in rats. J. comp. Neurol. 124: 319-336 (1965).
Ashton, N., Ward, B. and Serpell, G.: Effect of oxygen on developing retinal vessels with particular reference to the problem of retrolental fibroplasia. Br. J. Ophthal. 38: 397-432 (1954).

Balentine, J. D.: Pathogenesis of central nervous system lesions induced by exposure to hyperbaric oxygen. Am. J. Path. 53: 1097-1109 (1968).

Bean, J. W.: Effects of oxygen at increased pressure. Physiol. Rev. 25: 1-147 (1945).

Brand, M.M. and Bignami, A.: The effects of chronic hypoxia on the neonatal and infantile brain. A neuropathological study of five premature infants with the respiratory distress syndrome treated by prolonged artificial ventilation. Brain 92:233-254 (1969).

Cheek, D.B., Graystone, J.E. and Rowe, R.D.: Hypoxia and malnutrition in newborn rats: effects on RNA, DNA, and protein in tissues. Am. J. Physiol. 217: 642-651 (1969).

Enesco, M. and Leblond, C.P.: Increase in cell number as a factor in the growth of the organs and tissues of the young male rat. J. Embryol. exp. Morph. 10:530-562 (1962).

Fish, I. and Winick, M.: Cellular growth in various regions of the developing rat brain. Pediat. Res. 3:407-412 (1969).

Genn, M.M. and Silverman, W.A.: The mental development of ex-premature children with retrolental fibroplasia. J. nerv. ment. Dis. 138: 79-86 (1964).

Grave, G.D., Kennedy, C. and Sokoloff, L.: Impairment of growth and development of the rat brain by hyperoxia at atmospheric pressure. J. Neurochem. 19: 187-194 (1972).

Gyllensten, L.: Influence of oxygen exposure on the postnatal vascularization of the cerebral cortex in mice. Acta morph. neerl.-scand. 2: 289-310 (1959a).

Gyllensten, L.: Influence of oxygen exposure on the differentiation of the cerebral cortex in growing mice. Acta morph. neerl.-scand. 2: 311-330 (1959b).

Krause, A.C.: Effect of retrolental fibroplasia in children. A.M.A. Arch. Ophthalmology 53: 522-529 (1955).

Margolis, G. and Brown, I.W. Jr.: Hyperbaric oxygenation: The eye as a limiting factor. Science 151: 466-468 (1969).

Patz, A.: The role of oxygen in retrolental fibroplasia. Pediatrics, 19: 504-524 (1957).

Potter, C.T.: The problem of blind children and the responsibilities of the paediatrician. Proc. Royal Soc. Med. 47: 715-720 (1954).

Shivers, R.R. and Roofe, P.G.: Cerebral cell population under hypoxia. Anat. Rec. 154: 841-846 (1966).

Williams, C.E.: Retrolental fibroplasia in association with mental defect. Br. J. Ophthal. 42: 549-557 (1958).

Winick, M.: Changes in nucleic acid and protein content of the human brain during growth. Pediat. Res. 2: 352-355 (1968).

Winick, M. and Noble, A.: Cellular response in rats during malnutrition at various ages. J. Nutr. 89: 300-306 (1966).

CHAPTER THIRTEEN

NERVE GROWTH FACTOR*

Barry G.W. Arnason and Michael Young

Growth and differentiation of both the sensory and motor components of the peripheral nervous system (PNS) is favored by factors released from peripheral tissues and heterotransplant experiments have established that these effects are not species restricted. Similarly an intact innervation and the release of "supporting factors" from nerve endings are essential for normal development of the periphery and, at least in lower forms, for regeneration following injuries such as limb amputation. The chemical basis for neuron-target and target-neuron interactions within the central nervous system (CNS) has only begun to be explored, but evidence for factors which specifically and selectively influence growth and differentiation of both neural and glial elements has begun to accumulate.

Nerve growth factor (NGF) is a potent hormone released by target tissues which is essential for the growth and differentiation of peripheral sympathetic neurons and for their maintenance throughout life. NGF also stimulates growth of exteroceptive sensory ganglion neurons during a restricted period of life. In the context of this colloquium it is appropriate to consider NGF as a peripheral nervous system prototype for factors, yet undiscovered, which control cell growth, cell differentiation and possibly cell orientation within the CNS. Evidence which argues for a role for NGF itself in CNS development and in the response to CNS injury will also be reviewed.

Historical Background: In 1948, Bueker (25) reported that mouse sarcomas 37 and 180, when transplanted into chick embryos, induced a pathologic hypertrophy of chick embryonic sensory and sympathetic ganglia. Levi-Montalcini followed through on this provocative observation and demonstrated that neuronal hypertrophy as observed by Bueker depended upon a humoral factor released by the transplanted tumors (64). She named this substance nerve growth factor (NGF). Shortly thereafter Levi-Montalcini developed an in

* This work was supported by grant #CA-21043-01 from the National Cancer Institute, National Institutes of Health.

vitro bioassay for NGF (67). The assay relies on the fact that explanted chicken embryo sensory or sympathetic ganglia will, when exposed to appropriate concentrations of NGF in vitro, exhibit a visibly detectable outgrowth of nerve fibers. The bioassay is sensitive but only roughly quantitative.

Levi-Montalcini and Cohen next set out to purify NGF from tumor extracts using their bioassay to monitor fractions. In the course of this work snake venom, which is rich in phophodiesterase, was-added to the tumor extracts in order to release nuclease resistant DNA from protein. This led to the startling finding that snake venom contained 3000-6000 times as much NGF as the tumor extracts (35). A survey of the oral cavity glands of other species was then undertaken; this led to the further discovery that the submandibular gland of the adult male mouse is particularly rich in NGF (33). In recent years, most work on NGF has been done with mouse submandibular gland material.

Why adult male mouse submandibular glands should contain such large quantities of NGF remains unknown. The tubular cells of mouse submandibular glands exhibit sexual dimorphism, a property shared to some extent by the salivary glands of other rodents (rats, hamsters) though not by those of other taxons, man included. Tubular cells develop fully only in male mice after puberty. NGF is synthesized in submandibular gland tubular cells of adult male mice and is secreted in large amounts into the saliva (80, 81). When testosterone is given to female mice, hypertrophy and hyperplasia of tubular cells occurs and salivary gland NGF content rises markedly; when male mice are castrated, the tubular portion of the gland atrophies and salivary gland content of NGF falls precipitously. Salivary gland NGF levels in female mice rise (by bioassay) during pregnancy and particularly lactation; at these times androgen levels rise and salivary gland tubule cells enlarge (28). Salivary gland NGF synthesis, then, is androgen inducible. This finding is of potential interest in view of recent evidence for androgen dependent sexual dimorphism in the CNS, as for example, in the neuropil of the rat preoptic area (96).

Biologically active mouse submandibular gland NGF has how been purified. Availability of pure NGF has permitted production of specific anti-NGF antibodies and the development of sensitive and quantitatively accurate immunoassays to measure NGF levels in animals and man. Surprisingly, blood levels of NGF as measured by radioimmunoassay do not differ appreciably between male and female mice (80). Further, salivary gland extirpation (which has no obvious deleterious effects upon the animal) does not alter circulating NGF levels and, except for reduction in cell size and norepinephrine content of those neurons in the superior cervical ganglion which innervate the salivary glands, is without effect on the sympathetic nervous system (129, 130 but for a contrary opinion see 53). In addition, little if any NGF is present in the submandibular glands of newborn mice, yet NGF can be found in fetal and newborn sympathetic ganglia, humans included (27, 44, 130). Taken together, these data indicate that NGF in blood and tissues does not derive from the salivary glands and argue against any significant androgen dependence of NGF synthesis by those tissues (apart from the submandibular glands) which produce it.

When antiserum to NGF is injected into newborn mice or chick embryos, the sympathetic nervous system is ablated. The procedure has come to be known as an immunosympathectomy (see 66, 113, 134 for reviews). When antiserum to NGF is administered to adult animals, deleterious effects on sympathetic neurons are seen to occur but most neurons survive and when antiserum treatment is stopped the neurons recover (5, 19). These dramatic findings point up two facts: 1) NGF is absolutely critical for sympathetic nervous system development. 2) NGF has a major role in the maintenance of sympathetic neurons throughout life.

When large amounts of NGF are given to embryos, a generalized overgrowth of sympathetic ganglia and axons occurs. It follows that tissue levels of the hormone must be finely modulated during normal development.

The NGF Molecule: NGF from mouse salivary glands exists in at least two molecular weight classes depending upon the method of its purification. One of these forms is called 2.5S (referring to its sedimentation coefficient). Ideally this protein is composed of two noncovalently joined identical A chains, each of molecular weight 13,259 (3) but, during extraction, limited proteolysis of 2.5S NGF commonly occurs and some A chains are cleaved between residues 8 and 9 to form so-called B chains. For this reason, most 2.5S NGF preparations contain a mixture of A and B chains. The NH_2 terminal octapeptide cleaved from some A chains of 2.5S NGF is not required for biological activity, nor does it exhibit nerve growth promoting activity. Interestingly, the NGF octapeptide has four sequence homologies with the octapeptide angiotensin, although it appears not to possess angiotensin-like activity (8).

Another molecular weight class of salivary gland NGF has been called 7S NGF (124). This molecule is a complex of three different proteins known as the α submit, β NGF and the γ subunit. NGF dissociates at acid pH to yield a mixture of these three components. β NGF differs from 2.5S NGF only in

that both of its structurally identical constituent polypeptide chains are protected from proteolysis during the isolation procedure so that it contains only
A chains.

The biologic roles of the α and γ subunits of NGF have yet to be determined;
neither subunit promotes nerve outgrowth. The γ subunit, which has a molecular weight of 26,000, has recently been shown by isoelectric focussing to be
subdividable into 5 subcomponents which differ slightly in amino acid composition from one to another (110). Isolated γ subunits display arginine este-
ropeptidase activity while 7S NGF itself does not. The carboxy-terminal
amino acid of β NGF is arginine; when this terminal arginine is cleaved the
capacity of β NGF to bind the α and γ subunits is lost although biologic
activity of β NGF is unchanged. This observation has led to the suggestion
that the γ subunit may normally function to cleave active NGF from a pro-
NGF precursor molecule. The γ subunit has also been found to release con-
fluent cultures of chick embryo fibroblasts from contact inhibition (45); the
relationship of this effect to the action of NGF remains obscure.

The NGF protein monomer A chain (whether derived from 2.5S NGF or
from β NGF) contains 118 amino acid residues (mw 13,259) and its primary
structure has been solved (3). The isolectric point is 9.3 consistent with the
content of 8 lysines, 7 arginines and 8 (of 19) amidated acidic residues. The
monomer contains 6 half-cystinyl residues, all of which must participate in
disulfide linkages since there are no free sulfhydryl groups in the native
protein.

Cobra venom NGF has also been purified and in large part sequenced. The
molecule has a mw of 28,000 and, like 2.5S NGF, is made up of 2 identical
subunits (59). Approximately 60 percent of the primary structure of cobra
NGF is identical to that of mouse NGF (106). Thus, the NGF molecule has
been highly conserved throughout evolution. The extensive immunologic
crossreactivity of NGF from various mammalian species, from mouse to
man, also attests to the conservation of the molecule throughout evolution.
The isoelectric point of cobra venom NGF is 6.75, a reflection of the fact that
it has 7 more acidic and 3 fewer basic residues than the mouse molecule. The
carboxyterminal of cobra venom NGF is not arginine and, as would be expected, cobra NGF will not combine with α and γ subunits from the mouse.

Cobra venom NGF shows a maximum in its biologic activity (as measured
by bioassay on chick ganglia) at the same concentration (10 ng/ml) as mouse
NGF although the magnitude of neurite outgrowth is less than with mouse
NGF. NGF inhibits outward migration of supporting cells from the ganglia
for reasons unknown. Mouse salivary gland NGF is more potent in this

regard than is snake venom NGF. Whether these differences in potency of the two hormones would hold for target tissues from other phyla (e.g. reptiles, amphibia) is not known.

Recently, attention has been drawn to structural homologies between NGF, guinea pig insulin and human proinsulin (3, 41). The sequence of mouse NGF can be aligned with human proinsulin so that, with only 5 deletions, a maximum similarity of 21 per cent identical residues can be achieved. The majority of identical residues cluster in those segments of NGF which align with the functionally significant A and B chain sections of proinsulin. Further, these regions are separated by exactly the 35 residues required to accommodate the C-peptide of proinsulin. Some regions of NGF can also be envisaged as having a three-dimensional structure similar to analogous regions of the insulin molecule. For these reasons the suggestion has been advanced that both NGF and insulin derive from a common evolutionary precursor. Snake venom NGF is no closer structurally to insulin than is mouse NGF; if the two proteins diverged from a common ancestor protein, they must have done so very early in evolution.

The putative homology between NGF and insulin has been challenged on the grounds that residues shared between NGF and human and guinea pig insulins are variable in other insulins, that many residues conserved in the insulins do not have NGF counterparts, and that the predicted helical regions of NGF do not correlate with those of the insulins (10).

The notion of homology, however distant, remains attractive nonetheless, since striking similarities in the biologic properties of the two molecules exist. Both have similar cell surface receptor binding properties and both appear to act, at least initially, upon the cell surface. Both insulin and NGF stimulate many anabolic processes in their respective target cells in what has come to be called a pleiotypic response. Insulin stimulates uridine uptake and RNA synthesis, polysome formation, protein synthesis, glucose utilization and decreases protein degradation. NGF likewise increases uridine uptake, RNA synthesis, polysome formation, protein and lipid synthesis, and glucose utilization (65). All the above suggests that both hormones could possess similar biologic mechanisms of action. Interestingly, in this connection, proinsulin is cleaved at one site by an arginine esterase; it will be recalled that γ subunit of 7S NGF is an arginine esterase. Insulin will not substitute for NGF although it may potentiate the effects of NGF to some extent. The binding sites for the two hormones on the surfaces of target cells are also distinct, although NGF receptors and insulin receptors may both be absent from, or expressed on, the cell surface concurrently (60).

During studies on the chemical properties of 2.5S NGF in serum, one of us (MY) observed that the gel filtration properties of dilute solutions of 2.5S NGF were not those of a dimer of mw 26,000. Since it is established that 2.5S NGF exists as a dimer at concentrations of 1 mg/ml and above the existence of a monomer $\rightleftharpoons$ dimer equilibrium at lower concentrations seemed possible. Analysis of the sedimentation equilibrium, sedimentation velocity and gel filtration properties of NGF demonstrated this to be the case (132). We calculate that NGF at a concentration of 1.4 μg/ml contains equal quantities of monomer and dimer. At 1 ng/ml the mixture contains greater than 99 per cent monomer. Similar studies with 7S NGF have demonstrated that this molecule is somewhat unstable, spontaneous dissociation of the complex begins at concentrations as high as 10 μg/ml (90). It follows that, at the NGF concentrations habitually employed in bioassay (1-10 ng/ml), the monomeric A or B chain is the biologically active molecule.

A maximum in the biologic dose-response curve of NGF exists as assessed both by morphologic and biochemical criteria. Neurite growth promoting activity increases as NGF concentration is increased but then decreases sharply as concentrations of NGF exceed 1 μg/ml. It is at just these concentrations that dimer begins to predominate. The intriguing possibility is thus raised that NGF may have different sets of actions on target cells depending on its concentration. Some support for this notion may be provided by observations that certain effects of NGF are seen only with concentrations of the hormone too high to support neurite outgrowth from ganglionic explants.

Evidence that dimer will fix to NGF binding sites has been presented (39, 93). The question whether NGF dimer is biologically active has been addressed by Stack and Shooter (111), who prepared NGF covalently cross-linked with dimethyl suberimidate. Their preparation proved indistinguishable from native NGF in the sensory ganglion assay. This result would seem to argue against the idea that NGF dimer differs in biologic properties from NGF monomer but since it is not known how extensively (in a 3-dimensional sense) the two chains were cross-linked, unfolding of cross-linked dimer to expose biologically active monomeric segments might have occurred at those dilutions where biologic activity was maximal. Stack and Shooter did observe a maximum in the dose-response curve for cross-linked NGF dimer closely similar to that of native NGF.

Recent studies from a number of laboratories have established that NGF is synthesized by a wide range of normal and tumor tissues both in vitro and in vivo, as discussed more extensively in a subsequent section. The NGF produced by L cells (transformed mouse fibroblasts) and by rat muscle cells in vitro has a molecular weight similar to that of 7S NGF as judged by its elution profile (82, 89). Unlike 7S NGF, however, it is stable in dilute solutions (in the range of its maximal biological activity) yet when treated with denaturing solvents it dissociates to yield a molecule electrophoretically and chromatographically identical to purified mouse gland NGF A chains. The NGF secreted in mouse saliva has been calculated to have a mw of 114,000, and has been shown to be stable in dilute solution (80). The precise relationship of these forms of NGF to 7S NGF is unclear at present.

2.5S and chemically unstable 7S NGF are cleared rapidly after injection into the circulation (9). Yet, it is known that circulating NGF levels in the mouse are of the order of 10 ng ml. Possibly the form of NGF which circulates normally in the mouse is chemically stable, of high molecular weight and akin to that secreted into mouse saliva and by tumor cells into their culture media. At least in man circulating NGF is of high mw (as judged by elution profiles from gels of material which cross-reacts immunologically with mouse 2.5S NGF) and remains stable at dilutions at which mouse salivary gland 7S NGF dissociates.

SYNTHESIS OF NGF BY TARGET TISSUES

NGF activity was first demonstrated in mouse sarcomas 180 and 37 (25). Although NGF from these two tumors has never been purified chemically, the biologic effect of tumor extracts can be blocked by antiserum to mouse submandibular gland NGF, so that a close similarity between the molecules doubtless exists.

In 1974, the highly malignant mouse L cell line (which produces fibrosarcomas in vitro) was shown to secrete a protein which elicited intense neurite outgrowth in the sensory ganglion assay and which by two independent immunoassays proved indistinguishable immunochemically from submandibular gland NGF (85). Since L cells are a transformed and malignant cell line, as are sarcomas 180 and 37, secretion of NGF could merely reflect this property. For this reason other cells in culture have been examined for NGF synthesizing capacity both by us and by others (table 1). From inspection of this table, several points emerge:

1. In addition to transformed fibroblast lines (L, 3T3, SV403T3), primary explant fibroblasts synthesize NGF (131), a finding foreshadowed many years ago by the observation of Levi-Montalcini that nerve growth promoting

Table 1. Tissues shown to secrete NGF in vitro.

	Bioassay	Immunoassay		References
L cells	+	+	mouse	85, 89
3T3	+	+	mouse	85
SV40 3T3 cells	+	+	mouse	85
Primary fibroblasts	+	+	chick, man	131, 133
Primary synovial fibroblasts	+	+	man	102, 133
Iris	+	N.D.*	rat	62, 109
Skeletal muscle cells	+	+	rat	82
Heart	+	N.D.	chick, rat	32
Vas deferens	+	N.D.	rat	32
Myosarcoma	+	+	rat	82
Neuroblastoma	+	+	mouse, man	63, 79a
Melanoma	+	+	man	133
Glioma	+	+	rat	11, 71, 79
Glial cells	+	N.D.	rat, man	30, 36
Adrenal medulla	+	+	mouse	51
Schwann and satellite cells	+	N.D.	mouse	30

* Not done.

activity was present in granulation tissue (65). Fibroblasts can be considered an appropriate "target" for sympathetic neurons and production of NGF by them is perhaps to be expected. Similarly, iris, heart and vas deferens are all organs which normally are supplied with an extensive sympathetic innervation and, again, that they should synthesize NGF is hardly surprising (32).

2. The synthesis of NGF by a wide array of tumors is less readily explained.

Table 2. C-1300 Neuroblastoma growth in A/J mice treated with 6-hydroxydopamine or with N.G.F.

Group	Number of Mice	Age at Tumor Cell Inoculation	Tumor Size in mgm***	Significance****
6-hydroxydopamine treated*	16	21 days	47 ± 20	< .001
Saline injected control	17	21 days	328 ± 70	
NGF treated**	10	12 days	450 ± 140	< .02
Saline injected control	13	12 days	160 ± 40	

* 100 μgm/gm body weight for 10 days from birth intraperitoneally.
** 2 μgm/gm body weight for 10 days from birth intraperitoneally.
*** All mice received 10^5 viable dispersed neuroblastoma cells in one flank. All tumors excised 10 days post-inoculation.
**** Student's t-test.

Why tumors should attract a sympathetic innervation to themselves and the advantage which might accrue to a tumor from such an innervation is not immediately apparent.

In an attempt to address this question we have treated mice at birth with 6-hydroxydopamine, a drug which ablates the sympathetic nervous system. When some weeks later C-1300 neuroblastoma is transplanted into chemically sympathectomized mice, tumor growth is markedly slowed as compared to controls injected with saline at birth (32a, b). Data are given in table 2. In contrast, in mice in which sympathetic overgrowth has been induced by pretreatment with NGF, growth of transplanted neuroblastoma is favored (table 2). It must be stressed that comparable differences have not been found with certain other tumor lines. Nonetheless it seems clear that an intact responsive sympathetic innervation can favor tumor growth, although the means by which it does so remains obscure. Additional evidence supports this contention. Immunosympathectomy of newborn mice has been reported to delay the formation of tumors induced by benzopyrene (16) and, in those tumors that did develop, rate of tumor growth was retarded. Treatment at birth with NGF shortens the latent period for appearance of chemically induced neural tumors (though not their incidence) and seems to favor the development of bladder tumors (112).

3. Cells of neural crest derivation are capable of synthesizing NGF. Satellite cells, schwann cells (30), adrenal medullary cells (51), neuroblastomas (79a) and melanomas (24) can all be included in this category. Sympathetic neurons and embryonic dorsal root ganglion neurons are the prime responders to NGF; these cells likewise derive from the neural crest. Certain neuroblastoma lines (42, 63, 72) and one line of pheochromocytoma (46, 123) have also been found to respond to NGF; again both types of tumor derive from neural crest progenitors. Neural crest derived cells then can both make and respond to NGF.

In the case of neuroblastoma, NGF is released from the tumor cells during the late S and g_2 stages of the cell cycle (R. Revoltella, personal communication) but the NGF receptpr (and in consequence the potential to respond) is expressed on the cell surface only during the late G_1 stage of the cell cycle (23, 97, 98). Some form of autoregulatory mechanism seems possible.

Normal chromaffin cells secrete NGF (51) while a tumor derived from chromaffin cells (pheochromocytoma) responds to NGF by process extension (46, 123). The situation may be analogous to that just described for neuroblastoma. While normal chromaffin cells are reported not to respond to NGF (65 but see 133), they possess the potential to extend processes when placed in an appropriate milieu. Adrenal medulla transplants into the anterior chamber of the eye put out fibers similar to normal iris nerves with the exception of their peculiar cellular origin (87), an indication of the ontogenetic relationship between these cells and adrenergic nerves.

4. Both normal glial cells and glioma cells are capable of making NGF in culture (11, 36, 71, 79). The observation has obvious relevance to the putative role of NGF in CNS development. A second growth promoting factor is also produced by glioma cells. This factor promotes outgrowth of neurites from neuroblastoma lines which are refractory to the action of NGF and is not blocked by anti-NGF serum (77, 78). It should be noted that other cell lines which synthesize NGF make additional growth factors as well. L cells, for example, make a macrophage stimulating factor (126), plus NGF.

5. Demonstration of NGF synthesis by muscle cells (82) is of particular interest. Fusion of myoblasts to form myotubes is associated with an increased synthesis of those proteins which are associated with muscle contraction but has no consistent effect on NGF production. Muscle cells are known to somehow enhance the development of anterior horn cell neurons in culture. This effect does not depend on NGF. Nonetheless, the demonstration that muscle cells make NGF perhaps argues for synthesis by them of an additional growth factor with specificity for anterior horn cells. The analogy would be with L cells and glioma cells, both of which make more than one growth factor.

CELLS RESPONDING TO NGF

a. *Peripheral nervous system.* Peripheral sympathetic neurons are the primary target cells for NGF and respond to NGF throughout life (reviewed in 65). Interestingly the cells of the intrinsic ganglia of the stomach and intestine (Meissner and Auerbach plexuses) appear to be refractory to the hormone.

Dorsal root ganglion neurons respond to NGF during a restricted period only. In the chick embryo, response can be demonstrated from day 6 to day 15 in ovo. In mice and rats, sensory ganglion cells remain responsive for a brief period after birth (2, 48, 125), but it should be recalled that these species are very immature at birth. The late differentiating medial dorsal exteroceptive neurons of the sensory ganglia respond to NGF, whereas the earlier differentiating dorsal lateral proprioceptive neurons fail to do so. Dorsal lateral neurons of the trigeminal ganglion are of placodermal origin, whereas medial dorsal neurons derive from the neural crest (49). This basic difference in cell origin may be germane to the difference in response to NGF. In dorsal root ganglia, dorsal lateral neurons are believed to derive from the neural crest although their lateral position in the ganglion suggests that they originate at the most lateral parts of the neural fold, where it continues into the epidermis, and may be akin to epidermal derivatives (50).

As mentioned earlier, some clones of neuroblastoma (42, 63, 72) and one clone of pheochromocytoma (46, 123) have been found to be capable of responding to NGF. Finally, chicken embryo pelvic rudiment cartilage cells have been reported to react to NGF by an inhibition of chondromucoprotein synthesis even though total protein synthesis is not inhibited (37).

All of the cells listed above (except cartilage) derive from the neural crest and a role for NGF in the earliest organization and development of both the neural tube and the neural crest has been proposed. Electron microscopic study of newt embryos exposed to NGF ($32\mu g/ml$ αg/ml in the bathing medium) has revealed an acceleration of the process of neurulation and of differentiation of neuroblasts within the neural tube (73, 95). Morphologic findings included increased polysome formation followed by increased development of endoplasmic reticulum, precocious myotome contraction and accelerated synapse development. These were taken as the morphologic expressions of accelerated protein synthesis.

In vitro studies with 3-day chick embryos have suggested that ventral neural tube promotes a developmental change in young somites which then become competent to promote or permit differentiation of sympathetic neurons from neural crest cells (84). NGF can substitute for ventral tube in this regard. NGF ($1~\mu gm/ml$) has also been reported to induce neural differentiation in vitro of stage 3 chick embryo ectoderm. This holds not only for presumptive neural plate ectoderm but also for presumptive epidermal ectoderm (18a). Many years ago, Bueker and his colleagues reported that protein fractions of combined axial structures (spinal cord, spinal ganglia, notochord and somite derivatives) from 7-day chick embryos possessed potent nerve growth promoting activity (127). There is thus evidence that NGF has a role in the earliest organization of the peripheral nervous system.

CENTRAL NERVOUS SYSTEM

a. *Adults.* Transected axons of central monoaminergic neurons in adult rats possess a considerable capacity for regenerative sprouting. When iris fragments are transplanted into brain (20), sprouts from transected monoaminergic neurons are attracted towards and grow into the fragments. The iris is normally extensively innervated by sympathetic noradrenergic nerves and iris explants in culture stimulate sympathetic neuron growth presumably because they synthesize NGF (62, 109). Possibly they do likewise when transplanted into brain. Consistent with this notion is the observation that intraventricular injection of NGF at the time of axonal damage accelerates the formation and growth of noradrenergic sprouts into iris transplants; the magnitude of the effect depends on the dose of NGF given. Also in keeping with this formulation is the finding that intraventricular anti-NGF blocks regenerative sprouting (18b).

Rats with bilateral hypothalamic lesions develop an anorexic syndrome attributable at least in part to a deficit in noradrenergic function. When rats with this anorexic syndrome are given intraventricular NGF they eat more food, regain body weight more rapidly and feed more vigorously in response to intraventricular administration of norepinephrine than do untreated controls (15). Thus NGF facilitates functional recovery from this syndrome, possibly by stimulating growth of regenerating noradrenergic neurons in the brain.

Material which cross reacts with NGF antibody, material which competes with ^{125}I-NGF for binding to dorsal root ganglia, and NGF receptors can all be detected in brain (116, 117), although the levels are consistently low. All the above suggests that NGF or a structural analogue may have a role in CNS growth.

One report that vestibular neurons from adult rabbits respond to NGF in vitro has appeared (58). Only a small percentage of explanted cells responded and the possibility that the sample was contaminated with central adrenergic neurons from the locus ceruleus cannot be excluded. One report that regeneration of dorsal column neurons in mature cats is favored by NGF has also appeared but the data are not persuasive (105).

b. *Embryonic CNS Development.* A limited amount of work has been reported on the effects of NGF on embryonic brain development. NGF receptors have been reported to appear at 7 to 8 days in chick embryo brain with a second wave of receptors appearing at 13 to 14 days (116, 117). Also, embryonic chick spinal cord extract will substitute for NGF in maintenance of dissociated chick embryo spinal ganglia neurons in culture (27). Thus indirect evidence that NGF production and potentially responsive cells are present in fetal CNS is at hand.

It has been shown that the cell surface adhesive properties of chick embryo tectal cells change between day 7 and 8 of fetal development. This change is inducible in culture with NGF (1.5 µg/ml) and can be blocked by anti-NGF (74). The concentration of NGF required to demonstrate the tectal effect is better than 100 times greater than that required to maximally stimulate outgrowth from dorsal root ganglia, and cobra venom NGF is inactive in the tectal system. Further, NGF in which all three tryptophan residues have been oxidized with N-bromosuccinimide is active in the tectal system yet inactive in the dorsal root ganglion assay. NGF has no effect on protein or RNA synthesis of tectal cells, nor is it required for survival of tectal cells. Surprisingly, the effect of NGF on tectal cells can still be seen when protein synthesis is blocked with cycloheximide, but it should be pointed out that certain effects of epidermal growth factor (EGF)*, a hormone which shares many properties with NGF, will proceed in the presence of cycloheximide (34). In particular, polysome formation and membrane transport are both augmented in responsive EGF treated epithelial cells, even when protein synthesis has been blocked. NGF may serve in the tectal system as an imperfect analog for some other CNS specific factor. Possibly it somehow modifies a pre-existing membrane component. In any case, the finding that an external trophic factor can alter cell surface specificity in the absence of de novo protein synthesis is of potential importance in understanding cell surface specificity changes.

In amphibia, exposure to high concentrations of NGF (10-50 µgm/ml) in vitro has been reported to favor cellular maturation of gastrula explants and to induce precocious CNS neuronal differentiation (95). Disaggregated fetal mouse brain cultures exhibit neurite bridge formation between adjacent brain fragments in culture and bridge formation between fragments has been reported to be encouraged by NGF (47).

Morphologic Effects of NGF. The in vivo findings will be described; morphologic effects observed in vitro are essentially comparable.

When NGF is given to embryos or newborns, a dramatic overgrowth of the sympathetic nervous system ensues (13, 14, 65, 104). An increase in the number of sympathetic neurons, in the size of the individual neurons and in the tempo of their differentiation are all observed. A striking overgrowth of nerve fibers is also seen so that target organs become swamped with nerve fibers.

* Epidermal growth factor (EGF) is a polypeptide hormone found in large quantity in the salivary gland of the adult male mouse. It stimulates growth of epidermal tissues both in vivo and in culture. The hormone enhances membrane transport of amino acids, favors ribosomal polymerization and augments protein and RNA synthesis. It is structurally distinct from NGF but shares many biologic properties with NGF. In salivary gland preparations, EGF, like NGF, is associated with an arginine esterase (34).

These nerve fibers may end abruptly and fail to establish contact with target cells. In extreme instances, the lumina of veins may be invaded by nerve endings and at times even occluded by them. This dramatic neurite overgrowth has no obvious deleterious effect on the animal.

At the fine structural level, equally dramatic changes are observed. An increase within the soma of neurofilaments and neurotubules, which tend to form bundles, is an early change and is accompanied by an increase in polysomes. Later there is an enrichment of membrane constituents, particularly of rough endoplasmic reticulum and of Golgi vesicles. These changes bespeak intense metabolic activity and point to an orchestrated increase in protein synthesis. Curious blebbings of the nuclear membrane with distortions, foldings, and at times disruptions of the nuclear membrane, have been noted as a transient early alteration after NGF treatment (22, 65). Similar alterations have been seen in other cells undergoing transformation. While it has been speculated that these nuclear changes are indicative of a conversion of undifferentiated stem cells into neuroblasts under the influence of NGF, the precise significance of the observation remains obscure. Increase in cytoplasmic volume is a relatively late change after NGF treatment. Mitochondrial number remains constant (104).

In older animals treated with NGF, neuronal cell size increases but cell number does not (18, 65), a situation which stands in contrast to that seen in younger animals. When NGF treatment of newborns is stopped, sympathetic neurons begin to die and the hyperplastic sympathetic ganglia revert over several weeks to their normal cellular complement (13, 14, 18). A critical period exists in post-natal development during which neurons must succeed in making contact with target cells or be eliminated. A substantial loss of sympathetic neurons normally occurs during the post-natal period as unsuccessful neurons die according to "program." This phenomenon can be manipulated experimentally. For example, crush of the post-ganglionic nerve trunk of the rat superior cervical ganglion before 12 days, a maneuver which precludes synaptic contact, causes a permanent ablation of the ganglion. Crush after 12 days leads merely to chromatolysis followed by regeneration. Post-crush nerve cell death in newborns can be prevented by NGF (52). It is possible that much of the increased cell population seen in ganglia after NGF treatment in the neonatal period reflects the capacity of the hormone to extend the life of effect neurons. If this be so, the falloff in neuron number once NGF treatment is stopped and the failure of NGF treatment to alter neuron number in older animals with populations of exclusively "successful" neurons is perhaps more readily understood. In 9-month-old mice treated with NGF at birth, catecholamine content of target organs has been found to be subnormal (31). This paradoxical observation is unexplained at present.

Substantial evidence exists to indicate that trans-synaptic factors have a regulatory role in the development of sympathetic neurons. Increased activity of preganglionic nerves produces characteristic changes in the enzyme pattern of post-ganglionic adrenergic neurons, notably an increase in tyrosine hydroxylase, whereas deafferentation of sympathetic ganglia in newborns prevents normal development of target cell innervation (121). While NGF treatment of deafferented rats raises tyrosine hydroxylase levels in sympathetic perikarya, nerve terminals fail to develop (21). It follows that the ortho-grade action of cholinergic neurons on adrenergic neurons cannot be mediated by NGF alone. At the same time is should be noted that NGF stimulates terminal aborization of sympathetic ganglion neurons in tissue culture in circumstances where cholinergic innervation is apparently lacking. An explanation for the seeming discrepancy between the in vivo and in vitro findings is not at hand.

Choline acetyltransferase activity is increased in the sympathetic ganglia of animals treated with NGF (120). This enzyme is considered to be a marker for pre-ganglionic cholinergic nerve endings and the data have been interpreted as indicating a positive retrograde trans-synaptic effect for NGF. The finding that NGF antiserum treatment results in impaired development of choline acetyltransferase activity in sympathetic ganglia has been interpreted as a negative retrograde trans-synaptic effect and in keeping with this formulation. It should be recalled, however, that cholinergic neurons may exist in sympathetic ganglia (54a) and that dissociated sympathetic neurons can express cholinergic properties when exposed in culture to factors released by fibroblastic, muscle or glial cells (86). The cholinergic inducing factor, although uncharacterized, appears to be distinct from NGF.

NGF can prevent the synaptic depression seen in adult sympathetic ganglion cells after interruption of their axons (94). NGF also prevents the development of degenerative responses in the dendrites of axotomized sympathetic neurons. Synaptic depression reflects a loss of synaptic contacts. Application of colchicine to postganglionic sympathetic nerves causes synapse loss in the absence of mechanical axon interruption. For this reason a trophic signal from the periphery is thought responsible for normal maintenance of preganglionic synaptic contacts by sympathetic neurons.

These findings provide the first firm evidence for an electrophysiologic effect of NGF on mammalian sympathetic neurons. They further suggest that NGF may be the trophic signal which regulates synaptic contact in mature sympathetic ganglia. The loss of synaptic contact would be expected to affect preganglionic nerve endings and provides a reasonable mechanistic explanation for the postulated trans-synaptic action of NGF discussed above.

BIOCHEMICAL EFFECTS OF NGF AND ITS MECHANISM OF ACTION

a. *NGF Receptors*. Several lines of evidence indicate the existence of an NGF receptor on the cell surface of NGF sensitive nervous tissues. Binding of radioactive NGF to sympathetic ganglia or ganglionic microsomal fractions is saturable and the avidity of binding appears comparable to that calculated for other hormones such as insulin (12,55). NGF treated with N-bromosuccinimide which destroys biologic activity will not bind to microsomal fractions. In chicken sensory ganglia, saturable binding has been found between 8 and 14 days in ovo with a drastic decline in binding capacity thereafter. This decline coincides temporally with loss of the biologic response to NGF (55). There is also good evidence for saturable NGF binding to neuroblastoma cells as discussed in an antecedent section (23, 97, 98). NGF binding to transformed lymphocytes but not to untransformed lymphocytes has also been reported (60).

Reports of non-saturable NGF binding to sensory and sympathetic neurons, as well as to brain and a variety of non-neuronal tissues, have recently been published (40, 93, 116, 117). Binding in peripheral tissues was still demonstrable after denervation, indicating that binding was not simply to sympathetic nerve terminals in the tissues. A reasonable interpretation for these surprising data is not available at present.

Evidence that tubulin may be involved in surface binding of NGF has been advanced (68). NGF in solution will bind to tubulin in solution in a ratio of 2 moles per mole, but the specificity of the effect is not established. Mouse NGF is basic, tubulin is acidic and a charge affinity between the two molecules would be expected. Whether snake venom NGF, a less basic molecule, also binds tubulin is not known.

The NGF receptor is a protein as judged by the fact that treatment of receptor bearing cells with proteolytic enzymes abolishes binding, whereas neuraminidase and phospholipase treatment are without effect. Binding is not significantly influenced by the presence or absence of divalent cations. After trypsin treatment binding capacity reappears after one to two hours – a rapid turnover of NGF receptor protein is suggested.

Hormones which bind to cell surface receptors frequently induce a "second messenger" which mediates the intracellular actions of the hormone. For example, adenylcyclase is induced when glucagon binds to the glucagon receptor on the surface of glucagon responsive cells. Adenylcyclase in turn acts to raise intracellular levels of 3′ 5′, adenosine monophosphate (c AMP) and c AMP acts as the intracellular second messenger for glucagon.

Search for a second messenger for NGF has proven unrewarding to date. c AMP was at one time considered a potential second messenger for NGF since c AMP treatment of sensory ganglia leads to neurite extension reminiscent of that seen with NGF (48, 99-101). However, the increase in protein synthesis which is seen after NGF treatment is not seen after c AMP treatment and intracellular c AMP and adenylcyclase levels are normal at the time of maximal response to NGF (57). c AMP is believed to favor polymerization of tubulin which in turn is thought to favor neurite extension. NGF in contrast augments tubulin synthesis and increases the size of the tubulin pool (56). Whether NGF favors polymerization of tubulin as an additional effect is not known.

Recently a transient rise in c AMP levels after NGF treatment has been reported. The response was confined to the first 5 minutes after exposure to the hormone (83). The response was not seen with oxidized and biologically inactive NGF and was blocked by addition of antiserum to NGF to the culture medium. The quantity of NGF required to demonstrate an increase in c AMP was many times that required to elicit increased tubulin synthesis and neurite extension; for this reason the biologic significance of the observation must remain in doubt.

There is evidence that catecholamine uptake at nerve endings is augmented by NGF treatment (38). This effect could depend on an NGF induced local increase in c AMP, since c AMP is known to favor exocytosis and endocytosis. The chemistry of local effects of NGF on nerve endings has not been studied systematically. The dose sensitivity of nerve endings could be quite different from that of the cell as a whole.

b. *Retrograde Transport.* Retrograde axonal transport of NGF from adrenergic nerve terminals and pre-terminal regions to the cell body was first proposed by Bueker in 1957 (26), but has only recently been demonstrated convincingly (54). Following injection of radioiodinated NGF into the anterior eye chamber, or into the submandibular gland, a preferential accumulation of NGF in the superior cervical ganglion of the injected side occurs after a lag period of some hours. The effect is lessened by destruction of sympathetic endings with 6-hydroxy dopamine (9), is blocked by axotomy and, as is the case with orthograde rapid axonal transport, is colchicine sensitive (54). Retrograde flow of NGF in sympathetic nerves occurs at a rate of 2.5 mm per hour.

Systemically administered NGF accumulates rapidly in sympathetic ganglia as NGF binds to sympathetic neuronal perikarya but a second larger wave of accumulation occurs after some hours (91). This secondary accumulation is impeded by colchicine and blocked by axotomy as is the greater part of the increase in tyrosine hydroxylase synthesis which follows systemic NGF administration (91). These findings suggest that under normal circumstances most NGF reaches the neuron by rapid retrograde axonal transport and that the biologic effects of the hormone within the soma are dependent in large part on the moiety of NGF which reaches the cell body in this way.

Retrograde transport of NGF seems to be relatively specific; other proteins of like size and charge are transported to a much lesser extent. Oxidation of NGF so as to destroy its biologic activity reduces retrograde transport drastically. Continuous pinocytosis occurs at nerve endings as membrane added to the nerve terminal during release of packaged transmitter is reincorporated into the cell so as to prevent expansion of the terminal membrane. Extracellular proteins may be trapped during this process, internalized, and then transported retrogradely. The phenomenon has been taken advantage of experimentally as, for example, in the use of horseradish peroxidase as a sensitive marker for neuronal pathways. A preliminary binding of NGF to specific surface receptor sites best accounts for the amounts of the hormone which are internalized, however. This view is reinforced by the observation that motor nerves which internalize and transport tetanus toxin fail to do so with NGF (114). Similarly cholinergic parasympathetic nerves, which like motor nerves are refractory to the biologic actions of NGF, do not transport NGF. Sensory endings, on the other hand, do pick up NGF and move it proximally at a rapid rate (13 mm/hr) to the large cell bodies of the dorsal root ganglia (115). This is true even in adult animals in which no biologic response to NGF by sensory ganglion cells occurs. The magnitude of the effect is much less than in the case of sympathetic nerves but is clearly detectable after both foot pad or systemic NGF administration. Possibly a limited number of NGF receptor sites persists at sensory endings even after the biologic response has been lost. Alternatively, receptors for a growth factor specific for sensory neurons may exist at sensory nerve endings to which NGF is able to bind.

Neuroblastoma cells lacking axons also bind NGF and, if the NGF has been attached to red blood cells, the red blood cells are interiorized (23, 97, 98). The finding suggests that the potential for internalization and transport of NGF is not confined to the terminal axon. NGF administration to new-born animals simultaneously treated with 6-hydroxydopamine induces a greater enlargement of sympathetic ganglia than is seen after treatment with NGF alone (64a). Under these circumstances, NGF uptake from nerve endings cannot occur and the phenomenon is best explained by an effect of NGF on the soma itself.

NGF transported retrogradely from nerve endings to the perikaryon remains structurally intact as judged by its elution properties from gels (114). How NGF might act once it has gained the cell body remains unclear. Possibly it is re-externalized; more probably it acts intracellularly.

In spite of the impressive evidence for axonal transport of NGF, it seems unlikely that all the biologic effects of NGF can be exerted within the cell soma. In terms of the distal sprouting and orientation of sprouts to appropriate targets (tropism) which follows NGF administration it is difficult to envisage how a molecule acting solely in the soma could favor development of one sprout at the expense of others unless the growth of sprouts stripped of receptor by internalization of receptor-NGF complex is somehow favored. The impetus to growth of the most peripheral parts of a neuron is probably a local one and independent of the nerve cell body, as evidenced by the observation that after severance of the axon of a motorneuron both ends of the cut axon sprout initially. Certainly some effects of NGF occur before axonal transport could have occurred. Glucose uptake and amino acid uptake augment within minutes of NGF treatment, as is the case with amino acid uptake in EGF treated cells, and probably are independent of de novo protein synthesis in the soma.

c. *Metabolic Actions.* NGF exerts multiple effects on cellular metabolism. These effects are not necessarily coupled to each other; this suggests that the hormone may have more than one site of action.

Glucose utilization is rapidly augmented after NGF treatment, both in vivo and in vitro (4, 7). This may reflect a direct membrane effect of the hormone as discussed above. Lipid synthesis, including ganglioside synthesis, is also augmented as might be expected if new cell membrane were being formed (6, 43, 69). ^{14}C acetate incorporation into lipids is increased but ^{14}C mevalonic acid incorporation is decreased particularly into sterols (70). These findings indicate that lipid synthesis is preferentially directed along particular pathways under the influence of NGF.

Protein synthesis rises in the soma of NGF treated neurons within a few hours (2). Much of the delay may reflect time required for axonal transport of

NGF to the soma. Acidic proteins, such as tubulin, are preferentially synthesized in the early hours after NGF treatment (41a, 56, but see 75), and this finds its morphologic counterpart in the increased number of neurotubules and neurofilaments seen microscopically. Intact neurotubules are essential for neurite extension and drugs such as colchicine, which depolymerizes tubulin, or vincristine which cross-links neurotubules, block neurite outgrowth. When neurite outgrowth is blocked with vincristine, augmented tubulin synthesis will still occur in NGF treated ganglia (56). This finding indicates that neurite extension induced by NGF is not the trigger for increased tubulin synthesis. Puromycin, a drug which blocks protein synthesis, also blocks NGF induced neurite extension. It follows that de novo protein synthesis is required before NGF-induced neurite extension can occur.

Tyrosine hydroxylase levels rise markedly in sympathetic ganglion cells within 12 hours of exposure to NGF, although the concentrations of NGF required (300 ngm/ml for a minimal response and 3 μgm/ml for a maximal response) are 10 to 100 times greater than those required to elicit neurite outgrowth (118, 119). Interestingly, tyrosine hydroxylase induction by NGF can be potentiated by glucocorticoids (88). It has been suggested that a preferential action of NGF on enzymes involved in transmitter processing occurs (118). Yet, in a line of pheochromocytoma which responds to NGF by neurite extension (and presumably by tubulin synthesis) tyrosine hydroxylase levels do not rise (46, 123), and in fetal sensory ganglion cells tyrosine hydroxylase is not induced by NGF. Thus, synthesis of tubulin and of tyrosine hydroxylase under the influence of NGF appear not to be coupled (see also 115a).

There is some evidence that intracellular protein turnover is increased after NGF treatment, presumably because of augmented proteolysis (65). Selective synthesis of those proteins such as tubulin required for the growth of responsive cells if accompanied by a generalized increase in intracellular turnover of proteins would provide a means whereby cells could rapidly change their chemical composition, enzyme pattern, distribution of macromolecules, and hence their morphology.

RNA synthesis is increased in NGF responsive cells. Chemical blocking of RNA synthesis, as with actinomycin D, does not block the early increase in protein synthesis which follows NGF treatment (76, 92) and increased protein synthesis can even be demonstrated in enucleated neurons in which de novo RNA synthesis is impossible (Arnason, B.G. and Darzynciewicz, Z., unpublished observation). It follows that protein synthesis, at least during the early hours after NGF administration, does not require newly synthesized RNA. An action by NGF on translation seems likely. In this connection it is of interest to note that NGF favors ribosomal aggregation, a mechanism which might favor "readout" of pretranscribed messenger RNA.

NGF has been reported to accelerate the rate of fast axoplasmic transport of intraganglionically injected radioactive leucine and to slow the rate of slow axoplasmic transport (1). The data have been interpreted as indicating a selective effect of NGF on transport.

In spite of much information, it must be admitted that the mechanisms of action of NGF remain imperfectly understood. What can be stated is that NGF induces a wide array of biochemical effects in responsive neurons, the net result of which is to favor growth and maturation. Which effects are primary, which secondary, and the relationships between the various effects remain areas of active controversy, based on bioassays; recent studies emply radioimmunoassay.

NGF Levels in Man in Health and Disease. Several groups have measured blood NGF levels in man. Earlier studies relied on bioassay; more recent studies have employed radioimmunoassay (RIA) and sometimes both methods. Substantial differences in results reported exist between laboratories. Recent studies using RIA have, in general, found normal levels of NGF to be 50-100 ngm/ml with no difference noted between males and females (61, 102, 108, 133). Systematic studies of various age groups have not been reported but no dramatic differences between infants, children and adults have been observed.

Increased blood NGF levels as measured RIA have been reported in familial dysautonomia (Riley-Day syndrome) (107-108). Bioassays in this disease in contrast have revealed normal levels of response. It is possible that the bioassay of human serum, which uses extension of neurites from chick ganglia in culture as the end-point, is measuring some factor other than NGF which is capable of stimulating neurite outgrowth. The RIA data perhaps suggest a failure of NGF receptor function in familial dysautonomia, or that the NGF secreted in this disease is functionally defective as a consequence of genetic mutation. Either abnormality would be expected to result in the failure of peripheral sympathetic nervous system development which characterizes familial dysautonomia.

It should be recalled that the CNS is not indemnified in familial dysautonomia. Behavioral abnormalities are common accompaniments of this disorder and there is evidence for failure of development of central catecholaminergic neurons. Whether these tie to the abnormalities in NGF is not known, but it should be recalled that there is evidence indicating that central catecholaminergic neurons do respond to NGF (15, 20). General growth failure is also seen in familial dysautonomia; birth weight is usually low and sub-sequent growth is tardy. The point is of interest in view of the effects of chemical sympathectomy on tumor growth discussed earlier.

Increased blood NGF levels have been reported in neurofibromatosis (Von Recklingshausen's disease) on the basis of bioassay (103, 122). This has not been confirmed in other laboratories and measurements of NGF by RIA, both by ourselves and by others, have failed to reveal increased levels of NGF in this condition. Among other phakomatoses we have found increased blood NGF levels by RIA in 2 cases of Sturge-Weber syndrome.

Curiously, blood NGF levels are elevated (by RIA) in Paget's disease of bone (133). In our series, levels tended to return towards normal with success-ful treatment of the disease with diphosphonate. The increased vascularity and intense metabolic activity which characterize this disease may bear on this unexpected response.

There are several reports that blood NGF levels are increased in patients bearing tumors. Most studies have employed bioassay and for this reason must be viewed with some skepticism. At the same time it should be recalled that the NGF effect was first demonstrated in tumor-bearing chick embryos and it was this core discovery which led to the purification of NGF and the elaboration of its function. The notion that tumor-bearing patients should have increased NGF levels seems highly plausible, and merits systematic study. Increased NGF levels have been reported by bioassay in a patient with a liposarcoma (127) and several groups have studied blood NGF levels in patients with neuroblastomas. The situation with regard to neuroblastomas is confused. Levels, as measured by bioassay, have been reported as elevated (17, 29, 44) and normal (128), and a need for further study of patients with this condition by RIA methods is evident. We have examined a limited number of glioma patients by RIA and have found blood NGF levels to be mod-estly elevated in some cases.

Little attention has been directed to NGF production in vitro by non-tumor tissue from patients with disease. We have compared NGF production in vitro by synovial fibroblasts from patients with rheumatoid arthritis (RA) and compared it to that of synovial fibroblasts from controls. Although doubling time of the two types of cells was comparable, NGF production per cell was 4 times greater with the RA synovial fibroblasts than with the con-trols. Interestingly, skin fibroblasts from RA patients and controls produced comparable amounts of NGF (102, 133).

REFERENCES

1. Almon, R.R. and McClure, W.O.: The effect of nerve growth factor (NGF) upon axoplasmic transport in sympathetic neurons of the mouse. Brain Res. 74: 255-267 (1974).
2. Angeletti, P.U., Gandini-Attardi, D., Toschi, G., Salvi, M.L. and Levi-Montalcini, R.: Metabolic aspects of the effect of nerve growth factor on sympathetic and sensory ganglia: protein and ribonucleic acid synthesis. Biochem. Biophys. Acta. 95: 111-120 (1965).
3. Angeletti, P.U., Hogue-Angeletti, R., Frazier, W.A. and Bradshaw, R.A.: Nerve Growth Factor; in Proteins of the Nervous System, pp. 133-153. (Raven Press, New York 1973).
4. Angeletti, P.U., Levi-Montalcini, R. and Calissano, P.: The nerve growth factor (NGF): chemical properties and metabolic effects. Advances Enzymol. 31: 51-75 (1968).
5. Angeletti, P.U., Levi-Montalcini, R. and Caramia, F.: Analysis of the effects of the antiserum to the nerve growth factor in adult mice. Brain Res. 27: 343-355 (1971).
6. Angeletti, P.U., Liuzzi, A. and Levi-Montalcini, R.: Stimulation of lipid biosynthesis in sympathetic and sensory ganglia by a specific nerve growth factor. Biochem. Biophys. Acta. 84: 778-781 (1964).
7. Angeletti, P.U., Luizzi, A., Levi-Montalcini, R. and Gandini-Attardi, D.: Effect of a nerve growth factor on glucose metabolism by sympathetic and sensory nerve cells. Biochem. Biophys. Acta 90: 445-450 (1964).
8. Angeletti, R.A., Bradshaw, R.A. and Marshall, G.R.: The synthesis and characterization of the amino-terminal octapeptide of mouse nerve growth factor. Int. J. Peptide Protein Res. 6: 321-328 (1974).
9. Angeletti, R.H., Angeletti, P.U. and Levi-Montalcini, R.: Selective accumulation of [125I] labelled nerve growth factor in sympathetic ganglia. Brain Res. 46: 421-425 (1972).
10. Argos, P.: Prediction of the secondary structure of mouse nerve growth factor and its comparison with insulin. Biochem. Biophys. Res. Comm. 70: 805-811 (1976).
11. Arnason, B.G.W., Oger, J., Pantazis, N.J. and Young, M.: Secretion of nerve growth factor by cancer cells. J. Clin. Invest. 53: 2a (1974).
12. Banerjee, S.P., Snyder, S.H., Cuatrecasas, P. and Greene, L.A.: Binding of nerve growth factor receptor in sympathetic ganglia. Proc. Nat. Acad. Sci. (U.S.A.) 70: 2519-2523 (1973).
13. Banks, B.E.C., Charlwood, K.A., Edwards, D.C., Vernon, C.A. and Walter, S.J.: Effects of nerve growth factors from mouse salivary glands and snake venom on the sympathetic ganglia of neonatal and developing mice. J. Physiol. 247: 289-298 (1975).
14. Banks, B.E.C. and Walter, S.J.: The effects of axotomy and nerve growth factor on the neuronal population of the superior cervical ganglion of the mouse. J. Physiol. (London) 249: 61P-62P (1975).
15. Berger, B.D., Wise, C.D. and Stein, L.: Nerve growth factor: enhanced recovery of feeding after hypothalamic damage. Science 180: 506-508 (1973).
16. Bhagat, B. and Rana, M.W.: Antitumor activity of antiserum nerve growth factor (anti-NGF). Proc. Soc. Exper. Biol. Med. 138: 983-984 (1971).
17. Bill, A.H., Seibert, E.S., Beckwith, J.B. et al.: Nerve growth factor and nerve growth stimulating activity in sera from normal and neuroblastoma patients. J. Nat. Cancer Inst. 43: 1221-1230 (1969).
18. Bjerre, B., Bjorklund, A., Mobley, W. and Rosengren, E.: Short- and long term effects of nerve growth factor on the sympathetic nervous system in the adult mouse. Brain Res. 94: 263-277 (1975).
18a.Bjerre, B. and Nord, L.: Effects of nerve growth factor on competent chick ectoderm. Experentia 29: 1018-1019 (1973).
18b.Bjerre, B., Bjorklund, A. and Stenevi, V.: Inhibition of the regenerative growth of central noradrenergic neurons by intracerebrally administered anti-NGF serum. Brain Res. 74: 1-18 (1974).
19. Bjerre, B., Wiklund, L. and Edwards, D.C.: A study of the de- and regenerative changes in

the sympathetic nervous system of the adult mouse after treatment with the antiserum to nerve growth factor. Brain Res. 92: 257-278 (1975).

20. Bjorklund, A. and Stenevi, V.: Nerve growth factor: stimulation of regenerative growth of central noradrenergic neurons. Science 175: 1251-1253 (1972).

21. Black, I.B. and Mytilineou, C.: The interaction of nerve growth factor and trans-synaptic regulation in the development of target organ innervation by sympathetic neurons. Brain Res. 108: 199-204 (1976).

22. Blood, L.A.: A note on some peculiar membrane structures in NGF-treated sensory neuroblasts. J. Anat. 112: 309-313 (1972).

23. Bosman, C., Revoltella, R. and Bertolini, L.: Phagocytosis of nerve growth factor-coated erythrocytes in neuroblastoma rosette-forming cells. Cancer Res. 35: 896-905 (1975).

24. Bradshaw, R.A. and Young, M.: Nerve growth factor-recent developments and perspectives. Biochem. Pharmacol. 25: 1445-1449 (1976).

25. Bueker, E.D.: Implantation of tumors in the hind limb field of the embryonic chick and the developmental response of the lumbosacral nervous system. Anat. Rec. 102: 369-385 (1948).

26. Bueker, E.D.: Screening tumors (in vivo) for their effects on the growth of spinal and sympathetic ganglia of the embryonic chick. Cancer Res. 17: 190-199 (1957).

27. Bueker, E.D., Scheinkein, I. and Bane, J.L.: The problem of distribution of a nerve growth factor specific for spinal and sympathetic ganglia. Cancer Res. 20: 1220-1228 (1960).

28. Bueker, E.D., Weis, P. and Scheinkein, I.: Sexual dimorphism of mouse submaxillary glands and its relationship to nerve growth stimulating protein. Proc. Exper. Biol. Med. 118: 204-207 (1965).

29. Burdman, J.A. and Goldstein, M.N.: Long-term tissue culture of neuroblastomas III. In vitro studies of a nerve growth-stimulating factor in sera of children with neuroblastoma. J. Nat. Cancer Inst. 33: 123-133 (1964).

30. Burnham, P., Raiborn, C. and Varon, S.: Replacement of nerve-growth factor by ganglionic non-neuronal cells for the survival in vitro of dissociated ganglionic neurons. Proc. Nat. Acad. Sci. (U.S.A.) 69: 3556-3560 (1972).

31. Campbell, R.J., Wilson, L.G.M., Herschman, H.H., Di Cara, L.V. and Stone, E.A.: Paradoxical decrease in norepinephrine content of adult mouse spleen and heart after neonatal nerve growth factor treatment. Biochem. Pharmacol. 24: 2213-2216 (1975).

32. Chamley, J.H., Goller, I. and Burnstock, G.: Selective growth of sympathetic nerve fibers to explants of normally densely innervated autonomic effect or organs in tissue culture. Develop. Biol. 31: 362-379 (1973).

32a. Chelmicka-Szorc, E. and Arnason, B.G.W.: Effect of 6-hydroxydopamine on tumor growth. Cancer Res. 36: 2382-2384 (1976).

32b. Chelmicka-Schorr, E. and Arnason, B.G.W.: The sympathetic nervous system and neuroblastoma growth in mice. Neurology (in press).

33. Cohen, S.: Purification of a nerve-growth promoting protein from the mouse salivary gland and its neuro-cytotoxic antiserum. Proc. Nat. Acad. Sci. (U.S.A.) 46: 302-311 (1960).

34. Cohen, S.: Epidermal growth factor. J. Investig. Dermatol. 59: 13-16 (1972).

35. Cohen, S. and Levi-Montalcini, R.: A nerve growth-stimulating factor isolated from snake venom. Proc. Nat. Acad. Sci. (U.S.A.) 42: 571-574 (1956).

36. Ebendal, T. and Jacobson, C.-O.: Human glial cells stimulating outgrowth of axons in cultured chick embryo ganglia. Zoon 3: 169-172 (1975).

37. Eisenbarth, G.S., Drezner, M.K. and Lebovits, H.E.: Inhibition of chondromucoprotein synthesis: an extraneuronal effect of nerve growth factor. J. Pharmacol. Exper. Therap. 192: 630-634 (1975).

38. England, J.M. and Goldstein, M.N.: The uptake and localization of catecholamines in chick embryo sympathetic neurons in tissue culture. Cell Sci. 4: 677-691 (1969).

39. Frazier, W.A., Boyd, L.F. and Bradshaw, R.A.: Properties of the specific binding of [125]I-

nerve growth factor to responsive peripheral neurons. J. biol. Chem. 249: 5513-5519 (1974).

40. Frazier, W.A., Boyd, L.F., Pulliam, M.W., Szutowicz, A. and Bradshaw, R.A.: Properties and specificity of binding sites for ^{125}I-nerve growth factor in embryonic heart and brain. J. Biol. Chem. 249: 5918-5923 (1974).

41. Frazier, W.A., Hogue-Angeletti, R. and Bradshaw, R.A.: Nerve growth factor and insulin. Science 176: 482-488 (1972).

41a. Gandini-Attardi, D., Calissano, P. and Angeletti, P.: Protein synthesis in embryonic sensory ganglia: effect of the nerve growth factor on soluble proteins. Brain Res. 6: 367-370 (1967).

42. Goldstein, M.N., Brodeur, G.M. and Ross, D.: The effect of nerve growth factor and dibutyryl cyclic AMP on acetyl-cholinesterase in human and mouse neuroblastoma. Anat. Record. 175: 330 (1973).

43. Graves, M., Varon, S. and McKhann, G.: The effect of nerve growth factor (NGF) on the synthesis of gangliosides. J. Neurochem. 16: 1533-1541 (1969).

44. Greenberg, R.E., Winick, M. and Reilly, T.A.: Studies of a nerve-growth-promoting protein. Demonstration in human fetal tissue. J. Pediatr. 63: 753-754 (1963).

45. Greene, L.A., Tomita, J.T. and Varon, S.: Growth-stimulating activities of mouse submaxillary esteropeptidases on chick embryo fibroblasts in vitro. Exper. Cell Res. 64: 387-395 (1971).

46. Greene, L.A. and Tischler, A.S.: Establishment of a noradrenergic clonal line of rat adrenal pheochromocytoma cells which respond to nerve growth factor. Proc. Nat. Acad. Sci. (U.S.A.) 73: 2424-2428 (1976).

47. Greenham, L.W., Hill, T.J., Moss, C.A. and Peacock, D.B.: Preparation of disaggregated mouse brain cultures and observations on early neurite and axon formation therein. Exper. Cell Res. 84: 287-299 (1974).

48. Haas, D.C., Hier, D.B., Arnason, B.G.W. and Young, M.: On a possible relationship of cyclic AMP to the mechanism of action of nerve growth factor. Proc. Soc. Exper. Biol. Med. 140: 45-47 (1972).

49. Hamburger, V.: Experimental analysis of the dual origin of the trigeminal ganglion in the chick embryo. J. Exper. Zool. 148: 91-123 (1961).

50. Hamburger, V.: Specificity in neurogenesis. J. Cell Comp. Physiol 60 (Suppl.): 81-92 (1962).

51. Harper, G.P., Pearce, F.L. and Vernon, C.A.: Production of nerve growth factor by the mouse adrenal medulla. Nature 261: 251-253 (1976).

52. Hendry, I.A.: The response of adrenergic neurones to axotomy and nerve growth factor. Brain Res. 94: 87-97 (1975).

53. Hendry, I.A. and Iversen, L.L.: Reduction in the concentration of nerve growth factor in mice after sialectomy and castration. Nature 243: 500-504 (1973).

54. Hendry, I.A., Stöckel, K., Thoenen, H. and Iversen, L.L.: The retrograde axonal transport of nerve growth factor. Brain Res. 68: 103-121 (1974).

54a. Hermetet, J.C., Treska, J. and Mandel, P.: Histochemical study of isolated neurons in culture from chick embryo sympathetic ganglia. Histochimie 22: 177-186 (1970).

55. Herrup, K. and Shooter, E.M.: Properties of the β-nerve growth factor receptor in development. J. Cell Biol. 67: 118-125 (1975).

56. Hier, D.B., Arnason, B.G.W. and Young, M.: Studies on the mechanism of action of nerve growth factor. Proc. Nat. Acad. Sci. (U.S.A.) 69: 2268-2272 (1972).

57. Hier, D.B., Arnason, B.G.W. and Young, M.: Nerve growth factor: relationship to the cyclic AMP system of sensory ganglia. Science 182: 78-81 (1973).

58. Hillman, H. and Sheikh, K.: The growth in vitro of new processes from vestibular neurons isolated from adult and young rabbits. Exper. Cell. Res. 50: 315-322 (1968).

59. Hogue-Angeletti, R.A., Frazier, W.A., Jacobs, J.W., Niall, H.D. and Bradshaw, R.A.: Purification, characterization and partial amino acid sequence of nerve growth factor from cobra venom. Biochemistry 15: 26-34 (1976).

60. Hollenberg, M.D. and Cuatrecasas, P.: Hormone receptors and membrane glycoproteins during in vitro transformation of lymphocytes; in Clarkson and Baserga, Control and Proliferation in Animal Cells, pp. 423-433 (Cold Spring Harbor Laboratory 1974).
61. Johnson, D.B., Gorden, P. and Kopen, I.J.: A sensitive radioimmunoassay for 7S nerve growth factor antigens in serum and tissues. J. Neurochem. 18: 2355-2362 (1971).
62. Johnson, D.G., Silberstein, S.D., Hanbauer, I. and Kopin, I.J.: The role of nerve growth factor in the ramification of sympathetic nerve fibres into the rat iris in organ culture. J. Neurochem. 19: 2025-2029 (1972).
63. Kolber, A.R., Goldstein, M.N. and Moore, B.W.: Effect of nerve growth factor on the expression of colchicine-binding activity and 14-3-2 protein in an established line of human neuroblastoma. Proc. Nat. Acad. Sci. (U.S.A.) 71: 4203-4207 (1974).
64. Levi-Montalcini, R.: Effects of mouse tumor transplantation on the nervous system. Ann. N.Y. Acad. Sci. 55: 330-343 (1952).
64a.Levi-Montalcini, R., Aloe, L., Mugnaini, E., Oesch, F. and Thoenen, H.: Nerve growth factor induced volume increase and enhanced tyrosine hydroxylase synthesis in the chemically axotomized sympathetic ganglia of newborn rats. Proc. Nat. Acad. Sci. (U.S.A.) 72: 595-599 (1975).
65. Levi-Montalcini, R. and Angeletti, P.U.: Nerve growth factor. Physiol. Res. 48: 534-569 (1968).
66. Levi-Montalcini, R. and Booker, B.: Destruction of the sympathetic ganglia in mammals by an antiserum to a nerve-growth protein. Proc. Nat. Acad. Sci. (U.S.A.) 46: 384-391 (1960).
67. Levi-Montalcini, R., Meyer, H. and Hamburger, V.: In vitro experiments on the effects of mouse sarcomas 180 and 37 on the spinal and sympathetic ganglia of the chick embryo. Cancer Res. 14: 49-57 (1954).
68. Levi-Montalcini, R., Revoltella, R. and Calissano, P.: Microtubule proteins in the nerve growth factor mediated response. Recent Progr. Hormone Res. 30: 635-669 (1974).
69. Liuzzi, A., Angeletti, P.U. and Levi-Montalcini, R.: Metabolic effects of a specific nerve growth factor (NGF) on sensory and sympathetic ganglia: enhancement of lipid biosynthesis. J. Neurochem. 12: 705-708 (1965).
70. Liuzzi, A. and Foppen, F.H.: Sterol-like compound from sensory ganglia. Effect of a nerve growth factor and insulin on its biosynthesis. Biochem J. 107: 191-196 (1968).
71. Longo, A.M. and Penhoet, E.E.: Nerve growth factor in rat glioma cells. Proc. Nat. Acad. Sci. (U.S.A.) 71: 2347-2349, (1974).
72. Lyon, G.M., Jr.: Growth stimulation of tissue culture cells derived from patients with neuroblastoma. Cancer Res. 30. 2521-2531 (1970).
73. Mathieu, C., Duprat, A.-M., Zalta, J.-P. and Beetschen, J.-C.: Action du facteur de croissance nerveuse (nerve growth factor) sur la differenciation de cellules embryonnaires d'amphibien. Exper. Cell Res. 68: 25-32 (1971).
74. Merrell, R., Pulliam, M.W., Randono, L., Boyd, L.F., Bradshaw, R.A. and Glaser, L.: Temporal changes in tectal cell surface specificity induced by nerve growth factor. Proc. Nat. Acad. sci. (U.S.A.) 72: 4270-4274 (1975).
75. Mizel, S.B. and Bamburg, J.R.: Studies on the action of nerve growth factor II. Neurotubule protein levels during neurite outgrowth. Neurobiol. 5: 283-290 (1975).
76. Mizel, S.B. and Bamburg, J.R.: Studies on the action of nerve growth factor III. Role of RNA and protein synthesis in the process of neurite outgrowth. Develop. Biol. 49: 20-28 (1976).
77. Monard, D., Solomon, F., Rentsch, M. and Gysin, R.: Glia-induced morphological differentiation in neuroblastoma cells. Proc. Nat. Acad. Sci. (U.S.A.) 70: 1894-1897 (1973).
78. Monard, D., Stockel, K., Goodman, R. and Thoenen, H.: Distinction between nerve growth factor and glial factor. Nature 258: 444-445 (1975).
79. Murphy, R.A., Oger, J., Saide, J.D., Blanchard, M.H., Arnason, B.G.W., Hogan, C., Pantazis, N.J. and Young, M.: Secretion of nerve growth factor by central nervous system

glioma cells in culture. J. Cell. Biol. (in press).

79a. Murphy, R.A., Pantazis, N.J., Arnason, B.G.W. and Young, M.: Secretion of nerve growth factor by mouse neuroblastoma cells in culture. Proc. Nat'l. acad. Sci. (U.S.A.) 72: 1895-1898 (1975).

80. Murphy, R.A., Saide, J.D., Blanchard, M.H. and Young, M.: Molecular properties of the nerve growth factor secreted in mouse saliva. Proc. Nat. Acad. Sci (U.S.A.) (in press).

81. Murphy, R.A., Saide, J.D., Blanchard, M.H. and Young, M.: Nerve growth factor in mouse serum and saliva. Role of the submandibular gland. Proc. Nat. Acad. Sci. (U.S.A.) (in press).

82. Murphy, R.A., Singer, R.H., Saide, J.D., Pantazis, N.J., Blanchard, M.H., Byron, K.S., Arnason, B.G.W. and Young, M.: Synthesis and secretion of a high molecular weight form of nerve growth factor by skeletal muscle cells in culture. Proc. Nat. Acad. Sci. (U.S.A.) (in press).

83. Nikodijevic, O., Nikodijevic, B., Zinder, O., Yu, M.-Y. W., Guroff, G. and Pollard, H.B.: Control of adenylate cyclase from secretory vesicle membranes by β-adrenergic agents and nerve growth factor. Proc. Nat. Acad. Sci. (U.S.A.) 73: 771-774 (1976).

84. Norr, S.C.: In vitro analysis of sympathetic neuron differentiation from chick neural crest cells. Develop. Biol. 34: 16-38 (1973).

85. Oger, J., Arnason, B.G.W., Pantazis, N., Lehrich, J. and Young, M. Synthesis of nerve growth factor by L. and 3T3 cells in culture. Proc. Nat. Acad. Sci. (U.S.A.) 71: 1554-1558 (1974).

86. O'Lague, P.H., Mac Leish, P.R., Nurse, C.A., Claude, P., Furshpan, E.J. and Potter, D.D.: Physiological and morphological studies on developing sympathetic neurons in dissociated cell culture; in The Synapse. Cold Spring Harbor Symposia on Quantitative Biology 15: 399-408 (1976).

87. Olson, L. and Malmfors, T.: Growth characteristics of adrenergic nerves in the adult rat. Acta Physiol. Scandinav., Suppl. 348, 1-112 (1970).

88. Otten, U. and Thoenen, H.: Modulatory role of glucocorticoids NGF-mediated NGF.mediated enzyme induction in organ cultures of sympathetic ganglia. Brain Res. 111: 438-441 (1976).

89. Pantazis, N.J., Blanchard, M.H., Arnason, B.G.W. and Young, M.: Molecular properties of the nerve growth factor secreted by L cells. Proc. Nat. Acad. Sci. (U.S.A.) (in press).

90. Pantazis, N.J., Murphy, R.A., Saide, J.D., Blanchard, M. H. and Young, M.: Dissociation of the 7S-nerve growth factor complex in solution. Biochemistry (in press).

91. Paravicini, U., Stoeckel, K. and Thoenen, H.: Biological importance of retrograde axonal transport of nerve growth factor in adrenergic neurons. Brain Res. 84: 279-291 (1975).

92. Partlow, J.M. and Larrabree, M.G.: Effects of a nerve growth factor, embryo age and metabolic inhibitors on growth of fibres and on synthesis of ribonucleic acid and protein in embryonic sympathetic ganglia. J. Neurochem. 18: 2101-2118 (1971).

93. Pulliam, M.W., Boyd, L.F., Baglan, N.C. and Bradshaw, R.A.: Specific binding of covalently cross-linked mouse nerve growth factor to responsive peripheral neurons. Biochem. Biophys. Res. Comm. 67: 1281-1289 (1975).

94. Purves, D. and NJA, A.: Effect of nerve growth factor on synaptic depression after axotomy. Nature 260: 535-536 (1976).

95. Radeva, V. and Taxi, J.: Influence du facteur de croissance nerveuse (NGF) sur la genése de synapses dans le tube neural d'embryons de tritons. Arch. Anat. Microscopique 64: 135-148 (1975).

96. Raisman, G. and Field, P.M.: Sexual dimorphism in the neuropil of the preoptic area of the rat and its dependence on neonatal androgen. Brain Res. 54: 1-29 (1974).

97. Revoltella, R., Bosman, C., and Bertolini, L.: Dectection of nerve growth factor binding sites on neuroblastoma cells by rosette formation. Cancer Res. 35: 890-895 (1975).

98. Revoltella, R., Bertolini, L., Pediconi, M. and Vigneti, E.: Specific binding of nerve growth

factor (NGF) by murine C1300 neuroblastoma cells. J. Exper. Med. 140: 437-451 (1974).

99. Roisen, F.J., Murphy, R.A. and Braden, W.G.: Dibutyryl cyclic adenosine monophosphate stimulation of colcemid-inhibited axonal elongation. Science 177: 809-811 (1972).

100. Roisen, F.J., Murphy R.A. and Braden, W.G.: Neurite development in vitro. I. The effects of adenosine 3'5'-cyclic monophosphate (cyclic AMP). J. Neurobiol. 4: 347-368 (1972).

101. Roisen, F.J., Murphy, R.A., Pichichero, M.A. and Braden, W.G.: Cyclic adenosine moophosphate stimulation of axonal elongation. Science 175: 73-74 (1972).

102. Saide, J.D., Murphy, R.A., Canfield, R.E., Skinner, J., Robinson, D.R., Arnason, B.G.W. and Young, M.: Nerve growth factor in human serum and its secretion by human cells in culture. J. Cell. Biol. 67: 376a (1975).

103. Schenkein, I., Bueker, E.D., Helson, L., Axelrod, F. and Dancis, J.: Increased nerve-growth stimulating activity in disseminated neurofibromatosis. New Engl. J. Med. 290: 613-614 (1974).

104. Schwab, M.E. and Thoenen, H.: Early effects of nerve growth factor on adrenergic neurons: an electron microscopic morphometric study of the rat superior cervical ganglion. Cell Tissue Res. 158: 543-553 (1975).

105. Scott, D., Jr. and Lui, C.N.: Effect of nerve growth factor on regeneration of spinal neurons in the cat. Exp. Neurol. 8: 279-289 (1963).

106. Server, A.C., Herrup, K., Shooter, E.M., Hogue-Angeletti, R.A., Frazier, W.A. and Bradshaw, R.A.: Comparison of the nerve growth factor proteins from cobra venom (Naja naja) and mouse submaxillary gland. Biochemistry 15: 35-39 (1976).

107. Siggers, D.C.: Nerve growth factor and some inherited neurological conditions. Proc. Roy. Soc. Med. 69: 183-184 (1976).

108. Siggers, D.C., Rogers, J.G., Boyer, S.H., Margolet, L., Dorkin, H., Banerjee, S.P. and Shooter, E.M.: Increased nerve-growth-factor β-chain cross-reacting material in familial dysautonomia. New Engl. J. Med. 295: 629-634 (1976).

109. Silberstein, S.D., Johnson, D.G., Jacobowitz, D.M. and Kopin, I.J.: Sympathetic reinnervation of the rat iris in organ culture. Proc. Nat. Acad. Sci. (U.S.A.) 68: 1121-1124 (1971).

110. Stach, R.W., Server, A.C., Pignatti, P.-F., Piltch, A. and Shooter, E.M.: Characterization of the γ subunits of the 7S nerve growth factor complex. Biochemistry 15: 1455-1461 (1976).

111. Stach, R.W. and Shooter, E.M.: The biological activity of cross-linked β nerve growth factor protein. J. Biol. Chem. 249: 6668-6674 (1974).

112. Stahn, R., Rose, S., Sanborn, S., West, G. and Herschman, H.: Effects of nerve growth factor administration on N-ethyl-N-nitroso-urea carcinogenesis. Brain Res. 96: 287-298

113. Steiner, G. and Schönbaum, Eds.: Immunosympathectomy (Elsevier, New York, N.Y. 1972).

114. Stoeckel, K., Guroff, G., Schwab, M. and Thoenen, H.: The significance of retrograde axonal transport for the accumulation of systemically administered nerve growth factor (NGF) in the rat superior cervical ganglion. Brain Res. 109: 271-284 (1976).

115. Stoeckel, K., Schwab, M. and Thoenen, H.: Specificity of retrograde transport of nerve growth factor (NGF) in sensory neurons: A biochemical and morphological study. Brain Res. 89: 1-14 (1975).

115a.Stöckel, K., Solomon, F., Paravicini, U. and Thoenen, H.: Dissociation between effects of nerve growth factor on tyrosine hydrolase and tubulin synthesis in sympathetic ganglia. Nature 250: 150-151 (1974).

116. Szutowicz, A., Frazier, W.A. and Bradshaw, R.A.: Subcellular localization of nerve growth factor receptors. J. Biol. Chem. 251: 1516-1523 (1976).

117. Szutowicz, A., Frazier, W.A. and Bradshaw, R.A.: Subcellular localization of nerve growth factor receptors. J. Biol. Chem. 251: 1524-1528 (1976).

118. Thoenen, H.: Comparison between the effect of neuronal activity and nerve growth factor on the enzymes involved in the synthesis of norepinephrine. Pharmacol. Reviews 24: 255-

267 (1972).

119. Thoenen, H., Angeletti, P.U., Levi-Montalcini, R. and Kettler, R.: Selective induction of tyrosine hydroxylase and dopamine hydroxylase in the rat superior cervical ganglia by nerve growth factor. Proc. Nat. Acad. Sci. (U.S.A.) 68: 1598-1602 (1971).

120. Thoenen, H., Saner, A., Angeletti, P.U. and Levi-Montalcini, R.: Increased activity of choline acetyltransferase in sympathetic ganglia after prolonged administration of nerve growth factor. Nature New Biol. 236: 26-27 (1972).

121. Thoenen, H., Saner, A., Kettler, R. and Angeletti, P.V.: Nerve growth factor and preganglionic cholinergic nerves; their relative importance to the development of the terminal adrenergic neuron. Brain Research 44: 593-602 (1972).

122. Tischler, A.S.: Quantitation of nerve-growth stimulating activity. N. Engl. J. Med. 290: 1203 (1974).

123. Tischler, A.S. and Greene, L.A.: Nerve growth factor-induced process formation by cultured rat pheochromocytoma cells. Nature 258: 341-342 (1975).

124. Varon, S., Nomura, J. and Shooter, E.M.: Reversible dissociation of the mouse nerve growth factor protein into different subunits. Biochemistry 7: 1296-1303 (1968).

125. Varon, S., Raiborn, C. and Tyszka, E.: In vitro studies of dissociated cells from newborn mouse dorsal root ganglia. Brain Res. 54: 51-63 (1973).

126. Virolainen, M. and Defendi, V.: Growth regulating substances for animal cells in culture; in Defendi, V. and Stoker, M., eds., Wistar Inst. Monograph No. 7, p. 37 (The Wistar Institute Press, Philadelphia 1967).

127. Waddell, W.R., Bradshaw, R.A., Goldstein, M.N. and Kirsch, W.M.: Production of human nerve-growth factor in a patient with a liposarcoma. Lancet 1: 1365-1367 (1972).

128. Waghe, M., Kumar, S. and Steward, J.K.: Nerve growth factor in human sera. J. Pediat. Surg. 5: 14-18 (1970).

129. Wiklund, L. and Bjerre, B.: Histofluorescence observations on the sympathetic nervous system after removal of the salivary glands in adult mice. Brain Res. 96: 161-168 (1975).

130. Winick, M. and Greenberg, R.E.: Appearance and localization of a nerve growth-promoting protein during development. Pediatrics 35; 221-228 (1965).

131. Young, M., Oger, J., Blanchard, M.H. Asdourian, H., Amos, H. and Arnason, B.G.W.: Secretion of a nerve growth factor by primary chick fibroblast cultures. Science 187: 361-362 (1975).

132. Young, M., Saide, J.D., Murphy, R.A. and Arnason, B.G.W.: Molecular size of nerve growth factor in dilute solution. J. Biol. Chem. 251: 459-464 (1976).

133. Young, M., Murphy, R.A., Saide, J.D., Pantazis, N.J., Blanchard, M.H. and Arnason, B.G.W.: Studies on the molecular properties of nerve growth factor and its cellular biosynthesis and secretion; in Bradshaw, Frazier, Merrell, Gottlieb and Hogue-Angeletti, **Surface Membrane Receptors, pp. 247-267 (Plenum Press, New York 1976).**

134. Zaimis, E., ed.: Nerve growth factor and its antiserum, pp. 273-000 (Athlone Press, London 1972).

CHAPTER FOURTEEN

VIRAL INFECTIONS AND BRAIN DEVELOPMENT*

RICHARD T. JOHNSON, M.D.

The extraordinary diversity of cerebral lesions that can be induced by viruses is explicable primarily by two basic principles: 1) varied cell populations of the central nervous system have different susceptibility to different viruses and, 2) viral infections can have varied effects on cells. During development the complexity of these virus-host interrelationships are compounded by changes in the cell susceptibility and response attendant on cell generation, migration, and differentiation; and by the simultaneous maturation of humoral and cell-mediated immune systems in the fetus and newborn which may modify the infectious process.

In man, a variety of viruses has been associated with congenital or neonatal neurologic diseases (table 1). However, only cytomegalovirus and rubella virus infections of the fetus and herpes simplex virus infection of the newborn are common problems. The other associations are rare, and in some cases may represent chance associations of malformations with common infections. In cytomegalovirus, rubella, and herpes simplex virus infections acute or chronic inflammatory cerebral lesions are associated with gliosis and mineralization. These neuropathologic findings are now regarded as the primary features of antecedent or ongoing viral infection of the immature nervous system. In animals, where experimental infections of the developing nervous system can be systematically studied, viruses have been found capable of producing symmetrical non-inflammatory malformations of the brain and spinal cord which lack the neuropathologic stigmata suggesting infection (30). These cerebral lesions may resemble malformations of man which are assumed to be of toxic, genetic, or vascular origin. These lesions include abnormalities of closure of the neural tube, cerebellar hypoplasia, defective myelin formation, hydranencephaly, porencephaly and hydrocephalus (table 2). Three examples will be discussed to exemplify several principles of viral

* Studies supported by grants from the United Cerebral Palsy Research and Educational Foundation (R-230-74) and U.S. Public Health Service (NS 10920).

Table 1. Viral infection of the nervous system of human fetus or neonate.

Virus	Time of infection	Neurologic disease	Reference
Cytomegalovirus	During gestation	Cytomegalic inclusion disease with microcephaly, retardation, motor visual and/or auditory deficits	19
Rubella	1st trimester	Minor mental and auditory deficits	20, 58
		Chronic encephalitis, microcephaly, retardation, diplegia, visual and auditory deficits	10
Herpes simplex	Later gestation	Hearing loss and minor retardation	23, 47
	During gestation	Encephalitis with congenital microcephaly	13, 67
	During parturition	Severe encephalitis	21, 50
Herpes zoster	1st half of gestation	Cicatricial scar, limb hypoplasia and encephalomyelitis	59, 68
Coxsackie, group B	Neonatal (? late gestation)	Encephalitis associated with myocarditis	15
Polioviruses	Late gestation	Congenital or neonatal paralytic poliomyelitis	66
Arboviruses	Late gestation	Congenital or neonatal encephalitis	7, 65, 73
Influenza	1st trimester	Doubtful relation to malformations	6, 17, 22

Modified from ref. 32.

teratogenesis: 1) defects of neural tube closure induced by myxoviruses show-ing that the malformation may be determined by the stage of development rather than by which cells are infected, 2) effect of blue tongue virus infections in sheep showing that the same agent may cause different malformations at different stages of development, and 3) induction of aqueductal stenosis by a variety of viruses demonstrating that lesions, even after cellular differen-tiation, may mimic developmental abnormalities.

EFFECTS ON ORGANOGENESIS

Both influenza A and Newcastle disease virus have been found to cause ab-normalities of closure of the neural tube when inoculated into chick embryos. Embryos inoculated at 48 hours of incubation show subsequent collapse of the primitive brain (micrencephaly), abnormalities of neural tube flexion, and failure of closure of segments of the neural tube. Histologic studies of this defect in organogenesis have failed to show histologic abnormalities of the neural ectoderm, notochord or surrounding mesenchymal cells. The mitotic activity along the luminal surface of the developing neural tube appears nor-mal (18, 60, 61).

Studies of the pathogenesis of this defect utilizing fluorescent antibody staining have indicated that in Newcastle disease virus infections, a natural virus of chickens, antigen develops within cells of the neural tube (74). In contrast, with influenza A virus infections, infection appears to be limited to the chorionic and amnionic membranes, the non-neural ectoderm and focal areas of the primitive myocardium and gut; there is no evidence of infection of the neural ectoderm, the notochord, or the contiguous mesenchymal tissue (29). Thus, in the case of these myxovirus infections of the developing chick the same teratogenic effect results even though different cells are infected by the different viruses. As long recognized with the teratogenic effects of irra-diation and chemicals, the nature of the malformation may be more de-pendent on the ontogenic stage at the time of the insult than on the specific nature of the inciting factor.

INFECTIONS OF IMMATURE CELL POPULATIONS

A variety of factors may make mitotic, undifferentiated, or migrating cells susceptible to infection and destruction. For example, parvoviruses, which

Table 2. Experimental malformations induced by viruses.

Malformation	Virus	Host species	Reference
Defects in neural tube	Myxoviruses		
	Influenza	Chick embryo	18, 29, 61
	Newcastle disease	Chick embryo	60, 74
Encephalocoele	Togavirus		
	St. Louis encephalitis	Mouse	1
Cerebellar hypoplasia	Parvoviruses		
	Rat virus	Hamster, rat, cat, ferret	36, 37, 39
	Feline panleukopenia	Cat, ferret	38
	Minute virus of mice	Mouse	41
	Arenaviruses		
	Lymphocytic choriomeningitis	Rat	49
	Tamiami	Mouse	16
	Unclassified		
	Bovine viral diarrhea	Cow, sheep	5, 62, 72
	Hog cholera	Pig	11, 28
Hypomyelinization	Unclassified		
	Hog cholera	Pig	11, 28
	Border disease	Sheep	2
Hydranencephaly	Orbivirus		
	Blue tongue	Sheep	54, 75
	Togavirus		
	Akabane	Cow	43
Porencephaly	Orbivirus		
	Blue tongue	Sheep, mouse, hamster	3, 51, 54
	Togavirus		
	Venezuelan equine encephalitis	Monkey	45
Hydrocephalus	Myxoviruses		
	Mumps	Hamster	12, 33, 35, 42
	Parainfluenza, Type 1 & 2	Hamster	14, 34
	Influenza	Hamster, mouse, monkey	34, 44
	Measles (mutant)	Hamster	24
	Reovirus		
	Type 1	Mouse, hamster, rat, ferret	40, 46
	Type 3 (mutant)	Mouse	57
	Togaviruses		
	Japanese encephalitis	Pig	64
	Ross River	Mouse	48
	St. Louis encephalitis	Mouse	1

Modified from ref. 32.

are single-stranded DNA viruses, appear incapable of replication except in cells generating host cell DNA (39). Therefore, these infections are limited to cells in mitosis as exemplified by the granuloprival cerebellar hypoplasia induced in a variety of fetal or neonatal animals with these agents in contrast to the non-neurologic disease with gastroenteritis and leukopenia induced in mature cats infected with one of these agents – feline panleukopenia virus (38). Undifferentiated migrating cells may also be susceptible because of their undifferentiated cytoplasmic membranes. Undifferentiated cells are multi-potential, and this may include the potential to replicate virus. As the developing neuron reaches the cortical plate, intense specialization of cytoplasmic membranes occurs and metabolic functions of the cell become specialized. The cell may then become non-susceptible to infection. Blue tongue virus infections of sheep or mice provide an example of this latter type of selective vulnerability and exemplify how the same agent can cause different malformations at different stages of development.

Blue tongue virus is a double-stranded RNA virus which causes an acute respiratory and gastrointestinal disease of mature sheep (27). With the introduction of a live attenuated virus-vaccine against bluetongue, ewes immunized during pregnancy were found to deliver lambs with malformed brains (62). Subsequent experimental studies showed that fetuses infected at 50 days of their 150 gestation period develop hydranencephaly, those inoculated at 75 days of gestation develop porencephalic cysts, and those inoculated at 100 days of gestation have no gross malformation of the brain but only microscopic microglial nodules (54).

Sequential studies of this infection both in the developing sheep fetus and in neonatal mice, using a mouse-adapted strain of virus, showed selective infection of the germinal cells of the subventricular zone of the forebrain (51, 52, 55). This cell population along the ventricular wall proliferates and then migrates outward forming neurons during early gestation and glia during later gestation. Infection of ovine fetuses at 50 days of gestation produces massive necrosis with destruction of the subventricular cells, but at birth the inflammatory response and necrotic tissue have cleared leaving a non-inflammatory cavity resembling hydranencephaly. By mid-gestation most of the neurons have migrated into the cortical plate, but infection of the remaining germinal cells leads to focal acute encephalitis which, at birth, appear as glial walled cystic lesions, with no inflammation, primarily in the white matter. Thus, the age-dependency of the lesions apparently is determined by the availability of vulnerable, immature cells.

Rat virus, which infects and destroys the granular cells of the cerebellum causing cerebellar hypoplasia, also infects the germinal cells of the subventricular zone of the forebrain yet in these animals no gross forebrain lesions are seen. Both rat virus and bluetongue virus initially infect the same cell population in the forebrain of neonatal hamsters and comparable amounts of virus are replicated (2). The difference in sequelae can be explained on both cellular and viral factors. Rat virus infections are limited to cells in mitosis: Therefore, a brief period of infection occurs, limited to the subventricular zone. Cells are destroyed but apparently not in sufficient numbers to lead to gross malformation. On the other hand, blue tongue virus does not require a cellular DNA synthesis replication, and this virus is not biologically restricted to the zone of mitotic activity. Although the virus appears incapable of infecting mature neurons or glia, after germinal cells are infected the infection spreads from the subventricular zone wall along the migratory paths leading to sufficient cell destruction to grossly affect forebrain development.

HYDROCEPHALUS

Some viruses selectively infect the ependymal cells and aqueductal stenosis develops; others cause chronic infection of the parenchyma giving rise to hydrocephalus ex vacuo; others may involve primarily the meninges and result in communicating hydrocephalus (31). It is the aqueductal stenosis following neonatal infections of rodents that resembles a developmental anomaly.

A number of viruses cause acute selective infection of ependymal cells associated with inflammation and loss of ependyma. Subsequent to this, for reasons as yet unclear, in the immature animal midbrain tissue closes the aqueduct (34, 35). Inflammation is resolved, gliosis is minimal, normal brain stem occupies the area previously taken up by the aqueduct, and remaining ependymal cells forming small aqueductules. Thus, in this case, all the histologic criteria for a developmental agenesis of the aqueduct of Sylvius can be fulfilled (33), yet the anomaly can be induced by infection of a fully formed aqueduct in the newborn rodent.

There is evidence, both from case reports of aqueductal stenosis following mumps virus infection (4, 25, 53, 56, 69, 71) and the observation of ependymal cells with presumed mumps virus nucleocapsids within the spinal fluid during mumps meningitis (25), that mumps may cause aqueductal stenosis in man. It is yet undetermined whether mumps virus or any of the other viruses producing this lesion in neonatal or fetal animals is capable of crossing the human placenta, causing ependymal cell destruction, and leading to hydrocephalus in the newborn.

A FURTHER EXAMPLE OF SELECTIVE VULNERABILITY

Selective vulnerability of cells during development is not limited to the central nervous system but is recognized in other organs such as the inner ear. In this organ viral infections are thought to be major causes of congenital deafness, are blamed for some cases of acquired deafness, and Meniere's disease. Furthermore, on clinical grounds viruses are generally regarded as being the major cause of acute labyrinthitis. Clinically, rubella and cytomegaloviruses are clearly associated with congenital deafness, and epidemiologic studies have associated measles and mumps virus with acquired deafness. However, inner ear infections have been the subject of few virologic studies (70). No virus has yet been recovered from the inner ear of animal or man.

Using the inner ear of the neonatal hamster, which is developmentally analogous to the 15 week-old fetus, we found that virus could be injected directly into the endolymph and perilymph by percutaneous inoculation or placed solely in the perilymph by intracerebral inoculation, since virus passes from the subarachnoid space into the perilymph via the cochlear aqueduct. Selective vulnerability of inner ear structures was shown with a variety of viruses. Influenza virus infected only the cells of the perilymphatic channels. In contrast, mumps virus infected only endolymphatic structures sparing, in large part, the sensory end organs. Herpes simplex virus and measles virus showed striking involvement of the sensory end organs – the organ of Corti, the crista and the macula (8, 9). Thus, by analogy one can envision how viral infections might produce varied cochlear and labyrinthine abnormalities or malformations in man.

SUMMARY

The recognition of viral teratogens in man is limited. We recognize those which are common teratogens, such as cytomegalovirus, or those which occur in epidemic form, such as rubella. We suspect viruses only when we see inflammation, calcification or gliosis, yet the potential for viruses to cause a wide variety of non-inflammatory defects is evident from animal studies.

Fetal infections if persistent, may give rise to elevated serum IgM antibody levels in newborns. Probably transient infections early in gestation do not. Nevertheless, considering only that group of children with elevated IgM in the newborn period 3% of newborns have greater than 22 mg% IgM and half of these exhibit major clinical abnormalities in the first year of life. Conversely, only 20% of these children with increased IgM can be accounted for on the basis of cytomegalovirus, rubella, herpes, toxoplasmosis, or syphilis, the five infectious agents suspected of being the major infectious teratogens (63). The remaining 80% are unexplained.

REFERENCES

1. Anderson, A.A. and Hanson, R.P.: Intrauterine infection of mice with St. Louis encephalitis virus: immunological, physiological, neurological and behavioral effects on progeny. Infect. Immun. 12: 1173-1183 (1975).
2. Barlow, R.M., Gardiner, A.C., Storey, I.J. and Slater, J.S.: Experiments in Border disease. II. Some aspects of the disease in the foetus. J. Comp. Path. 80: 635-643 (1970).
3. Becker, L.E., Narayan, O. and Johnson, R.T.: Comparative studies of viral infections of the developing forebrain. I. Pathogenesis of rat virus and bluetongue vaccine virus infections in neonatal hamsters. J. Neuropath. Exp. Neurol. 33: 519-529 (1974).
4. Bray, P.F.: Mumps – a cause of hydrocephalus? Pediat. 49: 446.449 (1972).
5. Brown, T.T., deLahunta, A., Bistner, S.I., Scott, K.W. and McEntee, K.: Pathogenetic studies of infection of the bovine fetus with bovine viral diarrhea virus. I. Cerebellar atrophy. Vet. Path. 11: 486-505 (1974).
6. Coffey, V.P. and Jessop, W.J.E.: Maternal influenza and congenital deformities. Lancet 1: 748-751 (1963).
7. Copps, S.C. and Giddings, L.E.: Transplacental transmission of western equine encephalitis. Pediatrics 24: 31-33 (1959).
8. Davis, L.E., Shurin, S. and Johnson, R.T.: Experimental viral labyrinthitis. Nature 254: 329-331 (1975).
9. Davis, L.E. and Johnson, R.T.: Experimental viral infections of the inner ear. I. Acute infections of the newborn hamster labyrinth. Lab. Invest. 34: 349-356 (1976).
10. Desmond, M.M., Wilson, G.S., Melnick, J.L., Singer, D.B., Zion, T.E., Rudolph, A.J., Pineda, R.G., Ziai, M.H. and Blattner, R.J.: Congenital rubella encephalitis. J. Pediat. 71: 311-331 (1967).
11. Emerson, J.L. and Delez, A.L.: Cerebellar hypoplasia, hypomyelinogenesis and congenital tremors in pigs, associated with prenatal hog cholera vaccination of sows. J. Amer. Vet. Med. Assoc. 147: 47-54 (1967).
12. Ennis, F.A., Hopps, H.E., Douglas, R.D. and Meyer, H.M., Jr.: Hydrocephalus in hamsters: induction by natural and attenuated mumps viruses. J. Inf. Dis. 119: 75-79 (1969).
13. Florman, A.L., Gershon, A.A., Blackett, P.R. and Nahmias, A.J.: Intrauterine infection with herpes simplex virus. Resultant congenital malformations. J.A.M.A. 225: 129-132 (1973).
14. Friedman, H.M., Gilden, D.H., Lief, F.S., Rorke, L.B., Santoli, D. and Koprowski, H.: Hydrocephalus produced by the 6/94 virus. Arch. Neurol. 32: 408-413 (1975).
15. Gear, J.H.S. and Measroch, V.: Coxsackie virus infections of the newborn. Prog. Med. Virol. 15: 42-62, 1973.
16. Gilden, D.H., Friedman, H.M. and Nathanson, N.: Tamiami virus induced cerebellar heterotopia. J. Neuropath. Exp. Neurol. 33: 29-41 (1974).

17. Hakosalo, J. and Saxen, L.: Influenza epidemic and congenital defects. Lancet 2: 1346-1347 (1971).
18. Hamburger, V. and Habel, K.: Teratogenetic and lethal effects of influenza-A and mumps viruses on early chick embryos. Proc. Soc. Exp. Biol. 66: 608-617 (1947).
19. Hanshaw, J.B.: Congenital cytomegalovirus infection: a fifteen year perspective. J. Inf. Dis. 123: 555-561 (1971).
20. Hanshaw, J.B., Scheiner, A.P., Moxley, A.W., Gaev, L., Abel, V. and Scheiner, B.: School failure and deafness after "Silent" congenital cytomegalovirus infection. New Eng. J. Med. 295: 468-470 (1976).
21. Hanshaw, J.B.: Herpesvirus hominis infections in the fetus and the newborn. Am. J. Dis. Child. 126: 546-555 (1973).
22. Hardy, J.B.: Fetal consequences of maternal viral infections in pregnancy. Arch. Otolaryngol. 98: 218-227 (1973a).
23. Hardy, J.B.: Clinical and developmental aspects of congenital rubella. Arch. Otolaryngol. 98: 230-236 (1973b).
24. Haspel, M.V. and Rapp. F.: Measles virus: an unwanted variant causing hydrocephalus. Science 187: 450-451 (1975).
25. Herndon, R.M., Johnson, R.T., Davis, L.E. and Descalzi, L.R.: Ependymitis in mumps virus meningitis. Electron microscopical studies of cerebrospinal fluid. Arch. Neurol. 30: 475-479 (1974).
26. Hower, J., Clar, H.E. and Duchting, M.: Aquaduckt verschluss nach Mumps-Meningitis. Dtsch. med. Wschr. 97: 43-44 (1972).
27. Howell, P.G. and Verwoerd, D.W.: Bluetongue virus. Virol. Mono. 9: 35-74 (1971).
28. Johnson, K.P., Ferguson, L.C., Byington, D.P. and Redman, D.: Multiple fetal malformations due to persistent viral infection. I. Abortion, intra-uterine death and gross abnormalities in fetal swine infected with hog cholera vaccine virus. Lab. Invest. 30: 608-617 (1974).
29. Johnson, K.P., Klasnja, R. and Johnson, R.T.: Neural tube defects of chick embryos: an indirect result of influenza A virus infection. J. Neuropath. Exp. Neurol. 30: 68-74 (1971).
30. Johnson, R.T.: Effects of viral infections on the developing nervous system. New Eng. J. Med. 287: 599-604 (1972).
31. Johnson, R.T.: Hydrocephalus and viral infections. Develop. Med. Child. Neurol. 17: 807-816 (1975).
32. Johnson, R.T.: The teratogenic effect of viruses; in Vinken and Bruyn, Handbook of Clinical Neurology, Vol. 33 (North Holland Publishing Co., Amsterdam, in press).
33. Johnson, R.T. and Johnson, K.P.: Hydrocephalus following viral infection: the pathology of aqueductal stenosis developing after experimental mumps virus infection. J. Neuropath. Exp. Neurol. 27: 591-606 (1968).
34. Johnson, R.T. and Johnson, K.P.: Hydrocephalus as a sequela of experimental myxovirus infections. Exp. Molec. Path. 10: 68-80 (1969).
35. Johnson, R.T., Johnson, K.P. and Edmonds, C.J.: Virus-induced hydrocephalus: development of aqueductal stenosis in hamsters after mumps infection. Science 157: 1066-1067 (1967).
36. Kilham, L. and Margolis, G.: Cerebellar ataxia in hamsters inoculated with rat virus. Science 143: 1047-1048 (1964).
37. Kilham, L. and Margolis, G.: Cerebellar disease in cats induced by inoculation of rat virus. Science 148: 244-246 (1965).
38. Kilham, L. and Margolis, G.: Viral etiology of spontaneous ataxia of cats. Amer. J. Path. 48: 991-1011 (1966).
39. Kilham, L. and Margolis, G.: Spontaneous hepatitis and cerebellar "hypoplasia" in suckling rats due to congenital infections with rat virus. Amer. J. Path. 49: 457-475 (1966).
40. Kilham, L. and Margolis, G.: Hydrocephalus in hamsters, ferrets, rats and mice following inoculations of reovirus type 1. I. Virologic studies. Lab. Invest. 21: 183-188 (1969).

41. Kilham, L. and Margolis, G.: Pathogenicity of minute virus of mice (MVM) for rats, mice and hamsters. Proc. Soc. Exp. Biol. 133: 1447-1452 (1970).
42. Kilham, L. and Margolis, G.: Induction of congenital hydrocephalus in hamsters with attenuated and natural strains of mumps virus. J. Inf. Dis. 132: 462-466 (1975).
43. Kurogi, H., Inaba, Y., Goto, Y., Miura, Y., Takahashi, H., Sato, K., Omori, T. and Matumoto, M.: Serologic evidence for etiologic role of Akabane virus in epizootic abortion-arthrogryposis-hydranencephaly in cattle in Japan, 1972-1974, Arch. Virol. 47: 71-83 (1975).
44. London, W.T., Fuccillo, D.A., Sever, J.L. and Kent, S.G.: Influenza virus as a teratogen in rhesus monkeys. Nature 255: 483-484 (1975).
45. London, W.T. and Sever, J.L.: Personal communication.
46. Margolis, G. and Kilham, L.: Hydrocephalus in hamsters, ferrets, rats and mice following inoculations with reovirus type 1. II. Pathologic studies. Lab. Invest. 21: 189-198 (1969).
47. Menser, M.A. and Forrest, J.M.: Rubella-high incidence of defects in children considered normal at birth. Med. J. Australia 1: 123-126 (1974).
48. Mims, C.A., Murphy, F.A., Taylor, W.P. and Marshall, I.D.: The pathogenesis of Ross River virus infection in mice. I. Ependymal infection, cortical thinning and hydrocephalus. J. Inf. Dis. 127: 121-128 (1973).
49. Monjan, A.A., Gilden, D.H., Cole, G.A. and Nathanson, N.: Cerebellar hypoplasia in neonatal rats caused by lymphocytic choriomeningitis virus. Science 171: 194-196 (1970).
50. Nahmias, A.J., Alford, C.A. and Korones, S.B.: Infection of the newborn with herpesvirus hominis. Adv. Pediat. 17: 185-226, 1970.
51. Narayan, O. and Johnson, R.T.: Effects of viral infection on nervous system development. I. Pathogenesis of bluetongue virus infection in mice. Amer. J. Path. 68: 1-14 (1972).
52. Narayan, O., McFarland, H.F. and Johnson, R.T.: Effects of viral infection on nervous system development. II. Attempts to modify bluetongue virus-induced malformations with cyclophosphamide and antithymocyte serum. Amer. J. Path. 68: 15-22 (1972).
53. Naske, R. and Poustka, K.: Eine seltine Komplikation nach Mumpsmeningoenzephalitis. Wein. Med. Wochenschr. 123: 221-224 (1973).
54. Osburn, B.I., Silverstein, A.M., Prendergast, R.A., Johnson, R.T. and Parshall, C.J.: Experimental viral-induced congenital encephalopathies. I. Pathology of hydranencephaly and porencephaly caused by bluetongue vaccine virus. Lab. Invest. 25: 197-205 (1971).
55. Osburn, B.I., Johnson, R.T., Silverstein, A.M., Prendergast, R.A., Jochim, M.M. and Levy, S.E.: Experimental viral-induced congenital encephalopathies. II. The pathogenesis of bluetongue vaccine virus infection in fetal lambs. Lab. Invest. 25: 206-210 (1971).
56. Paraicz, E.: Angaben zum membranosen Verschluss des Aqueductus Sylvii. Acta Paediat. Acad. Sci. Hung. 11: 121-133 (1970).
57. Raine, C.S. and Fields, B.N.: Neurotropic virus-host relationship alterations due to variation in viral genome as studied by electron microscopy. Amer. J. Path. 75: 119-132 (1974).
58. Reynolds, D.W., Stagno, S., Stubbs, G., Dahle, A.J., Livingston, M. M., Saxon, S.S. and Alford, C.A.: Inapparent congenital cytomegalovirus infection with elevated cord IgM levels. Causal relation with auditory and mental deficiency. New Eng. J. Med. 290: 291-296 (1974).
59. Rinvik, R.: Congenital varicella encephalomyelitis in surviving newborn. Amer. J. Dis. Child. 117: 231-235 (1969).
60. Robertson, G.G., Williamson, A.P. and Blattner, R.J.: A study of abnormalities in early chick embryos inoculated with Newcastle disease virus. J. Exp. Zool. 129: 5-43 (1955).
61. Robertson, G.G., Williamson, A.P. and Blattner, R.J.: Origin of myeloschisis in chick embryos infected with influenza-A virus. Yale J. Biol. Med. 32: 449-463 (1960).
62. Schultz, G. and Delay, P.D.: Losses in newborn lambs associated with bluetongue vaccination of pregnant ewes. J. Amer. Vet. Med. Assoc. 127: 224-226 (1955).

63. Scott, F.W., Kahrs, R.F., DeLahunta, A., Brown, T.T., McEntee, K. and Gillespie, J.H.:
Virus induced congenital anomalies of the bovine fetus. I. Cerebellar degeneration (hypoplasia), ocular lesions and fetal mummification following experimental infection with
bovine viral diarrhea-mucosal disease virus. Cornell Vet. 63: 536-560 (1973).
64. Sever, J.L.: Viral teratogens; in Perrin and Finegold, Pathobiology of Development, pp. 97-
104 (Williams & Wilkins, Baltimore 1973).
65. Shimizu, T., Kawakami, Y., Fukuhara, S. and Matumoto, M.: Experimental stillbirth in
pregnant swine infected with Japanese encephalitis virus. Jap. J. Exp. Med. 24: 363-375
(1954).
66. Shinefield, H.R. and Townsend, T.E.: Transplacental transmission of Western equine encephalomyelitis. J. Pediat. 43: 21-25 (1953).
67. Siegel, M. and Greenberg, M.: Poliomyelitis in pregnancy: effect on fetus and newborn
infant. J. Pediat. 49: 280-288 (1956).
68. South, M.A., Tompkins, W.A.F., Morris, C.R. and Rawls, W.E.: Congenital malformation
of the central nervous system associated with genital (type 2) herpesvirus. J. Pediat. 75: 13-18
(1969).
69. Srabstein, J.C., Morris, N., Larke, R.P.B., de Sa, D.J., Castelino, B.B. and Sum, E.: Is there a
congenital varicella syndrome? J. Pediat. 84: 239-243 (1974).
70. Steigman, A.J. quoted by Bray, P.F.: Letter to Editor. Pediatrics 50: 349 (1972).
71. Strauss, M. and Davis, G.L.: Viral disease of the labyrinth. I Review of the literature and
discussion of the role of cytomegalovirus in congenital deafness. Ann. Otol. Rhinol.
Laryngol. 82: 1-7 (1973).
72. Timmons, G.D. and Johnson, K.P.: Aqueductal stenosis and hydrocephalus after mumps
encephalitis. New Eng. J. Med. 283: 1505-1507 (1970).
73. Ward, G.M.: Experimental infection of pregnant sheep with bovine viral diarrhea – mucosal
disease virus. Cornell Vet. 61: 179-191 (1971).
74. Wenger, F.: Necrosis cerebral masiva del feto en casos de encefalitis equina Venezolana.
Invest. Clin. 21: 13-31 (1967).
75. Williamson, A.P., Blattner, R.J. and Robertson, G.G.: The relationship of viral antigen to
virus-induced defects in chick embryos. Newcastle disease virus. Develop. Biol. 12: 498-519
(1965).
76. Young, S. and Cordy, D.R.: An ovine encephalopathy caused by bluetongue vaccine virus. J.
Neuropath. Exp. Neurol. 23: 635-659 (1964).

CHAPTER FIFTEEN

INFECTIOUS AND IMMUNOLOGIC DISEASES AFFECTING THE DEVELOPING NERVOUS SYSTEM*

ELLSWORTH, C. ALVORD, JR., M.D. AND CHENG-MEI SHAW, M.D.

In the preceding chapter Dr. Johnson has illustrated the variety of disorders of the developing nervous system which can be produced experimentally with several viral infections. In order to keep his presentation in some perspective related to human diseases, we shall review briefly what is known about infections of the human in utero or perinatally.

First, we must point out that grossly similar lesions may have many different causes, and we must not jump to the conclusion that any, much less every, human case of the disorders he has shown – of hydranencephaly or porencephaly, of cerebellar hypoplasia, of aqueductal stenosis or of nonclosure of the neural tube – is due to a viral infection. His is a challenge which must be met with more clear-cut evidence than is now available. Even electron microscopic claims of virus-like particles must be viewed with considerable scepticism, as Shaw and Sumi (1975) have recently reported.

For example, the spectrum of porencephaly-hydranencephaly may well be due to vascular insufficiency in the distribution of one or more branches of the internal carotid arteries, perhaps produced by intermittent strangulation due to wrapping of the umbilical cord around the neck. Fig. 1 illustrates a specimen seen 20 years ago, but which we failed to recognize for several years for what it almost undoubtedly was. We had photographed it as an example of the normal straight branches of the middle cerebral artery, freely communicating over the as-yet-unfissured cerebral surface in simple arcades with similar branches of the anterior and posterior cerebral arteries, as Vander Eecken (1959) described. Only after several years of demonstrating the photograph to students did we realize that the whiteness of the arteries was not an artifact of partial replacement of blood with air but was probably an extensive platelet thrombosis, evidence of a watershed distribution of inadequate ar-

* Supported in part by research grants number GM 13543-12 and HD 02274-11 from the National Institutes of Health, U.S. Public Health Service.

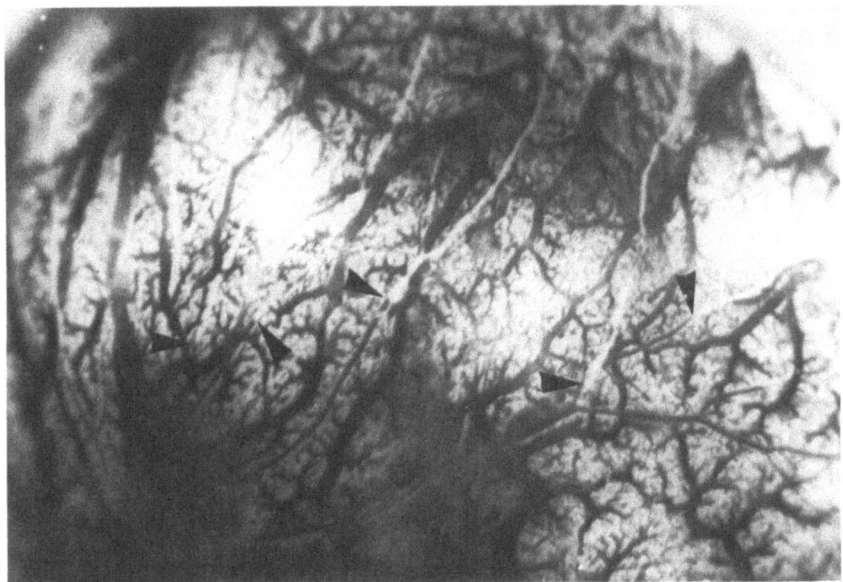

Fig. 1. Lateral view of the left cerebral hemisphere of a fetus of about 5 months gestation showing abrupt changes in the contents of each of the branches of the middle cerebral artery from obvious blood to white material (? platelet thrombi).

terial perfusion. The specimen was disposed of without microscopic examination long ago and we have not found another similar example, until recently, but microscopic studies are still not complete. This is probably the best evidence for a vascular cause of massive porencephaly-hydranencephaly that we could hope to find.

In hydranencephaly practically all of the telencephalon is absent. Usually this results from the destruction of previously formed brain. Within the skull there is a fluid-filled cavity surrounded by leptomeninges, usually with a gliotic molecular layer and sometimes with small islands of partly preserved cerebral cortex. If any cortex persists, it is usually inferior temporal and occipital. The deep cerebral nuclei are usually at least partially destroyed, but the striatum may even be hypertrophied. The diencephalon is generally somewhat better preserved and the midbrain, cerebellum, hindbrain and cranial nerves are usually normal except for the atrophy of various descending tracts (Crome and Sylvester 1958; Hamby et al. 1950).

The child may appear clinically normal at birth and for several months thereafter, especially if the hypothalamus is intact. If not, the infant usually dies in a few days. In some cases abnormal head enlargement (hydrocephalus) may become apparent after several weeks or months. The diagnosis of hydranencephaly is easily made by transillumination of the skull, and EEG ordinarily shows an absence of electric activity.

Hydranencephaly has occurred in consecutive pregnancies (Hamby et al. 1950) and in twins (Haque et al. 1969), but it usually presents as an isolated case. Environmental insults have been described in pregnancies that have resulted in a child with hydranencephaly. These include syphilis, toxoplasmosis, influenza, cytomegalic inclusion disease, attempted abortion, anoxia and trauma, including needle tracks (Crome 1972; Navin and Angevine 1968; Williams 1969). Postnatal onset has been reported (Lindenberg and Swanson 1967; Weiss et al. 1970), but the absorption of the cerebral lesion is clearly less complete than in those developing prenatally. While there may be no single etiologic agent, be it genetic or environmental, the usual pattern of destruction suggests bilateral carotid artery insufficiency with preservation of the vertebrobasilar circulation. This hypothesis has some experimental evidence to support it (Crome 1972). Halsey et al. (1971) relate the pathogenesis of hydranencephaly to a variety of processes that can reduce cerebral mass (e.g., vascular lesions, multiple cystic encephalomalacia, schizencephaly). A spectrum of anomalies from hydranencephaly to porencephaly has been produced in offspring to pregnant sheep vaccinated with live bluetongue virus at different stages in gestation (Osburn et al. 1971a, b), and the brain destruction could be secondary to the virus acting on the vascular system. Muir (1959) believes that hydranencephaly and porencephaly fall within the same spectrum of encephaloclastic disorders, and we agree that most such cases are qualitatively similar, the usual porencephaly being restricted to the distribution of only one or a few branches of the middle cerebral artery.

Reviews of earlier studies and hypotheses of the etiology of porencephaly are given by Globus (1921), Jaffe (1929) and Patten et al. (1937). In 1859, Heschl coined the term porencephalus to include any cavities within the cerebral hemispheres which communicate with either the ventricular or subarachnoid space or with both (Globus 1921). Confusion is increased if within such a broad definition are included cases with any hole within the wall of the cerebral hemisphere, whether communicating or not with the ventricular or subarachnoid space.

Yakovlev and Wadsworth (1941, 1946a, b) provide considerable clarification by subdividing cases of porencephaly into encephaloclastic and dysplastic types, the former occurring secondary to hemorrhage, infarction, inflammation or degenerative processes. These processes may begin during the fetal period, at birth or after birth; the lesions arise primarily as a result of necrotic

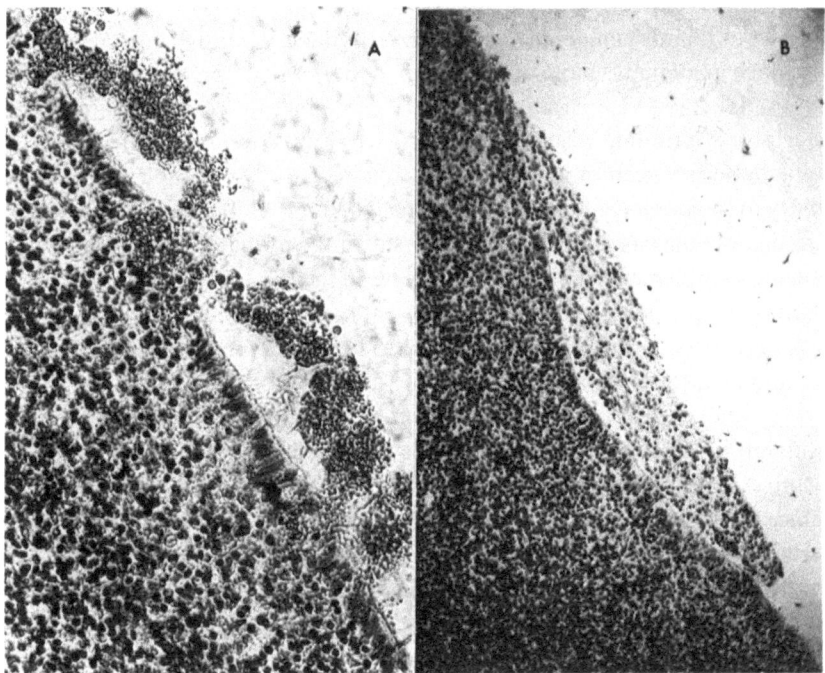

Fig. 2.A. Old petechial hemorrhages with hemosiderin- and lipid-filled macrophages rupturing through the ependyma of the lateral ventricle (Np 5271). This was an infant born at 28 weeks gestation and dying 28 days later. *B.* Granular ependymitis with astrocytic proliferation through breaks in the ependyma of the lateral ventricle of an infant born at 28 weeks gestation and dying 8 days later (Np 5122). More recent hemorrhages, similar to those shown in Fig. 2.A, were also abundant. Oil-red-O and hematoxylin.

softening and destruction of brain wall. Microscopic examination reveals proliferation of astrocytes and blood vessels around the edge of the defect. Such lesions can be unilateral but are more commonly bilateral and slightly asymmetric in both position and degree of severity. When encephaloclastic porencephaly is associated with moderate hydrocephalus, a cone-shaped defect passes through the cortex and white matter, whereas in cases of severe hydrocephalus the lesion has a pouched-out appearance. Regardless of the age at onset, a loss of nerve cells, myelinated nerve fibers and ependyma is typical, as is ulegyria (sclerotic microgyria) at the edges of the defect, especially if the onset is prenatal. Another characteristic of encephaloclastic porencephaly is that it is not commonly associated with other CNS malformations or with malformations of other organs.

The astrocytic gliosis and persistence of macrophages at the surfaces of the deep cerebral nuclei and at the edges of the primary hydranencephalic defect are much greater and more prolonged than those occurring secondarily in the pyramidal tracts (Leech and Alvord 1975). While these observations may suggest a continuing reaction, such as a viral infection could produce, the same pattern is seen in immature rats following a single simple mechanical injury to the brain (Sumi and Hager 1968a, b; Sumi 1970, pers. comm. 1976). Presumably the vascular damage at the site of the primary lesion contributes potentiating elements which do not appear in the less complicated Wallerian degeneration of the pyramidal tracts.

As another example, possibly related to hydrocephalus due to stenosis of the aqueduct of Sylvius induced by viruses (Johnson 1975; Herndon et al. 1974), we should point out that ependymal and subependymal lesions are common in the autopsied brains of premature infants (Leech and Kohnen 1974). As in fig. 2A, some of these lesions are hemorrhagic, probably petechiae etiologically similar to the more grossly apparent subependymal hemorrhages so commonly seen in the germinal matrix and rupturing into the ventricles. Others appear to be the healed stages (granular ependymitis) of old hemorrhages (fig. 2B). Again, the etiology can be quite diverse, and we know of no evidence that these may not be of viral etiology. Most commonly we attribute them to the metabolic hypoxia-acidosis which accompanies the infantile respiratory distress syndrome (IRDS) that causes practically all of the deaths of premature infants. Whether the aqueduct of Sylvius can be directly affected by comparable lesions is doubtful. Certainly blood clots or debris from these lesions in the lateral ventricles could easily block the aqueduct, destroy more or less of its ependyma and evoke secondary ependymal granulations which may later show little, if anything, of the original cause. Furthermore, the aqueduct can be secondarily compressed, even occluded, secondary to the hydrocephalus, which could be due to a variety of causes.

So much for speculation. What about the known?

Infections, especially viral, are relatively common causes of human CNS anomalies. Numerous studies over the past decade implicate several viruses, previously unsuspected of being teratogenic, as being associated with congenital malformations.

The most common type of viral involvment in fetal tissue is a direct effect on the cells. There are two possible routes by which the virus can reach the fetus: 1) by way of the amniotic cavity following a vaginal infection and entrance through the cervix, and 2) transplacentally through the bloodstream following a maternal viremia. The latter route is probably the most common since maternal viremia, increased antibody titers and isolation of viruses from the placenta (especially in rubella) are the rule. Johnson (1972) has recently reviewed the mechanisms by which some viral infections cause CNS malformations in experimental animals. He found evidence for 1) indirect effects on organogenesis, 2) selective vulnerability of specific immature cell populations and 3) primary agenesis resulting from host reactions to destruction.

One of the most distressing problems in the consideration of viruses as teratogenic agents is the fact that many of the mothers infected do not exhibit any clinical symptoms; i.e., they have subclinical or silent infections. This is well known in the case of cytomegalovirus and rubella virus and lends more credence to the possibility that several other viruses may be responsible for malformations. In addition, there is concern that some CNS diseases found in adults may be the result of slow virus infections acquired in utero (Katz, 1973).

Rubella virus was first identified as a significant cause of congenital malformations by Gregg (1941). During the past 30 years this virus has been cultured in vitro, and a vaccine developed against it. Some may regard such an interval as long, but when one considers the numerous other factors as yet unknown, this is a remarkably short chapter in medical history.

The probable mechanism by which rubella virus causes damage is by the direct invasion of sensitive organs, where it multiples and produces a cytopathologic, necrotic and inflammatory response. The effect on the developing embryo can result in numerous anomalies, some of them in the CNS. These include mental retardation, microcephaly, spasticity, deafness and blindness. Blindness may be secondary to optic nerve atrophy, cataracts or both.

Neuropathologic findings (Rorke 1973; Rorke and Spiro 1967) in cases of the rubella syndrome include degeneration of blood vessels associated with ischemic necrosis of tissue (present in 65% of cases). This vascular degeneration involves one or more layers of the vessel wall and replacement of these layers by an amorphous granular material. In some cases the vessel walls are less severely affected and masses of the amorphous material are found in the pericapillary areas. These deposits, which are PAS- and Feulgen-positive, are sometimes not associated with the vessels but may appear as discrete masses in the brain. Deposits of fibrin and calcium are occasionally found in subintimal and pericapillary spaces. Micrencephaly is present in 70% of cases and is due to a decreased number of cells. Retardation of myelinization has been noted in over half of the cases that have survived beyond one month postnatally. Other cases have shown hydrocephalus, gliosis (perivascular and marginal), leptomeningitis, cysts in the white matter lined by glia, necrosis of subcortical and periventricular white matter, calcification and subependymal cysts overlying the medio-ventral aspect of the head of the caudate nucleus (Shaw 1973; Shaw and Alvord 1974).

Although virtually every part of the CNS can be affected by the rubella virus, the eye is of special interest, since anomalies in the eyes of children were first related to the fact that the mother had had rubella early in pregnancy (Gregg 1941). Boniuk and Zimmerman (1967) studied the eyes in 19 such affected infants at autopsy. Clinical information on 17 cases revealed bilateral cataracts in 8, unilateral cataract with microphthalmia in 1, bilateral glaucoma in 1, and bilateral corneal opacifications in 1. In 6 cases the eyes were regarded as normal from the clinical viewpoint, but cataractous lens, iridocyclytis, vacuolization of iris pigment epithelium, hypoplasia of the iris and iris atrophy were found consistently microscopically. Necrosis of the ciliary body was found to be pathognomonic for the rubella virus. Other developmental defects of the eye, such as colobomata, retinal dysplasia and severe microphthalmia, were not found.

Rubella virus has its most devastating effects when the maternal infection occurs early in pregnancy. The cytopathologic effects in embryos have been studied by Tondury and Smith (1966), who established the following temporal relationships following early exposure to rubella:

Gestational Age (weeks)	Abnormal Lens (%)	Abnormal Inner Ear (%)
0–4	35	12
4–8	48	13
8–11	66	0

In cases in which the maternal infection occurs after the 12th week, structural abnormalities are more difficult to detect since the major portion of organogenesis has been completed, but Hardy et al. (1969) found a high incidence of communication deficits (decreased auditory sensitivity and delayed language development) and motor or mental retardation. These functional defects have not been analyzed morphologically.

Cytomegaloviruses (CMV) are increasingly of concern to clinicians dealing with newborns (Hanshaw 1970; Weller 1971). Recent studies have shown that between 1% and 3% of normal newborns excrete CMV at birth (Hanshaw 1970). Thus, CMV is probably the most common of the known fetal infections. The incidence of women infected during pregnancy is even higher, ranging between 2% and 5%.

The epidemiology and pathogenesis of CMV infections are now reasonably well defined. Several different strains exist, belonging as a group to the herpesvirus family. Although the prevalence of CMV varies in different areas of the world, many adults have been infected, usually through an inapparent infection. The subclinical form is especially true in women who acquire the infection during pregnancy. Except for the tendency to be of younger age, no other factors can be defined clinically in these women.

There are two mechanisms by which the infection can be transmitted during the prenatal period: natural and iatrogenic. Iatrogenic infections, by transfusions of infected blood, are infrequent and of less concern. The fetus seems most prone to adverse results during the second and third trimesters. Once a woman gives birth to an infant with congenital abnormalities caused by CMV, a second such episode is rare (Embil et al. 1970). After CMV gains access to the maternal bloodstream, transplacental transmission to the fetus occurs and may or may not cause damage. When CMV does gain access to the fetal cells, it forms intranuclear and cytoplasmic inclusion bodies, cytomegaly and foci of cell death. Time-lapse cinematography in vitro has shown that such cells can literally explode. Cell counts in infected organs reveal an absolute decrease in number of cells.

Cytomegalic inclusion disease (CID) is the term commonly used for the diagnosis of anomalies in newborns resulting from a CMV infection in utero. Characteristically these infants are "small for dates" and jaundiced, with evidence of hepatosplenomegaly, petechiae, pneumonitis and radiographic changes in the long bones. These same features can also be found in infants with the congenital rubella syndrome.

Not all of the CNS manifestations of CID are present at birth, some developing over a period of several months. Hanshaw (1970) emphasizes the following cerebral abnormalities: microcephaly, microgyria, dilated ventricles, periventricular calcifications, status spongiosus, hydrocephalus, encephalomalacia, calcified cerebral arteries, cerebral cysts, cerebral cortical immaturity, cerebellar aplasia and dolichocephaly. Subependymal cysts are common (Shaw 1973; Shaw and Alvord 1974).

In addition, deafness and ocular abnormalities have occurred in the CID syndrome, with chorioretinitis, microphthalmia, abnormal or atrophic optic discs, malformed anterior chambers, retinal calcifications and vestigial pupillary membranes. In contrast to infants with the congenital rubella syndrome, cataracts are unusual but have been reported in CID. Seizures, blindness, motor changes and mental retardation are common manifestations of CNS involvement.

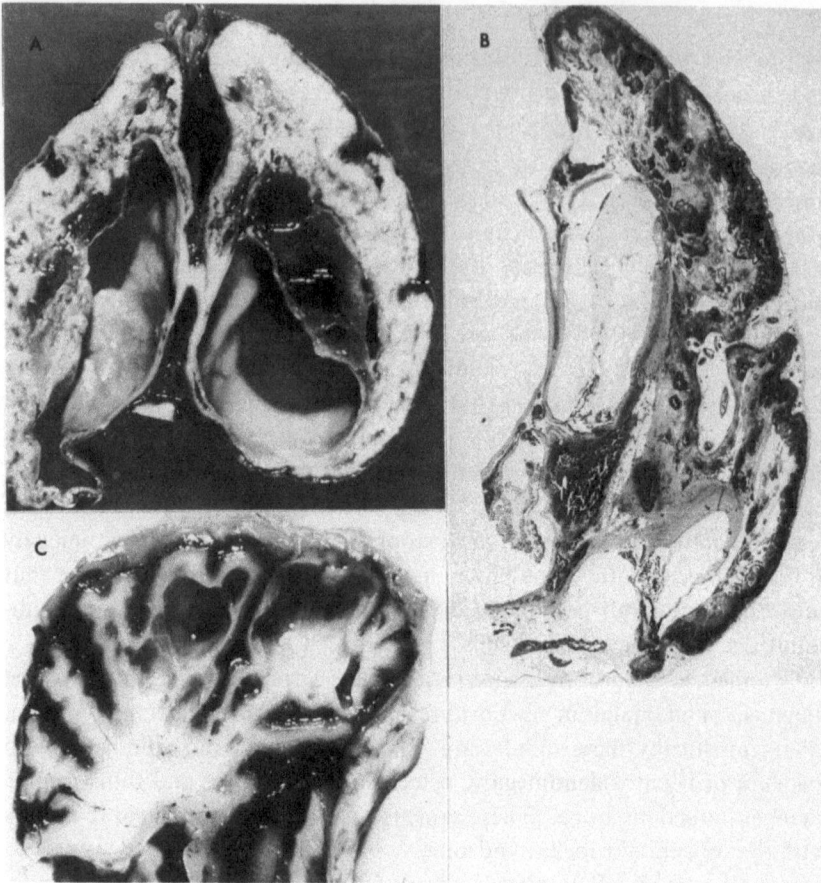

Fig. 3. Congenital herpes simplex with diffuse necrotizing encephalitis and death at 20 days (Np 4262). *A*, coronal section of the cerebral hemispheres just posterior to the corpus callosum; *B*, section of "*A*" stained with gallocyanin and Darrow red; *C*, parasagittal section of cerebellum.

The periventricular calcifications of congenital CID are almost pathognomonic. CMV has a particular affinity for the rapidly growing subependymal germinal matrix cells, especially those around the lateral ventricles. Calcium is subsequently deposited in these areas of necrosis in such a manner that an external cast of the ventricle is produced, easily seen on skull radiographs (Dennis and Alvord 1961).

At the present time serologic testing and viral cultures are available for confirmation of the diagnosis of CID, but there are no known preventive

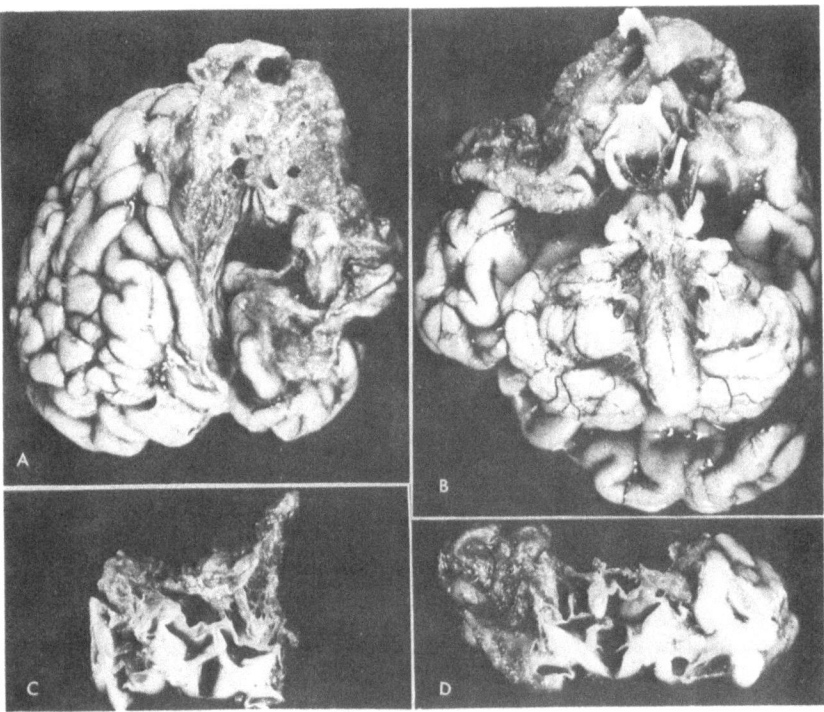

Fig. 4. Neonatal herpes simplex with multifocal necrotizing encephalitis and death at 3.5 years (Np 2895). *A*, frontal view; *B*,basal view; *C* and *D*, coronal sections through the third ventricle.

measures or therapy. Not all infants with CID have clinically detectable neurologic deficits. Berenberg and Nankervis (1970) provide follow-up data on 12 patients ranging from 4 to 12 years: 5 had severe mental retardation, 3 moderate and 1 mild; 3 had normal intelligence. Of the 3 who were normal, 1 was described as being "awkward" and the other 2 had entirely normal neuromuscular findings.

Other viruses may damage the developing brain. South et al. (1969) report an infant whose mother was infected with Type 2 herpes virus during the first few weeks of pregnancy. The infant had marked microcephaly with over-lapping cranial bones, microphthalmia, cloudy lenses, seizures and hypo-tonia. Although radiographs of the skull taken at two weeks of age were negative for intracranial calcifications, those taken at five weeks revealed periventricular casts similar to those seen in congenital CID. Microcephaly, seizures and calcification (periventricular and hippocampal) were reported by

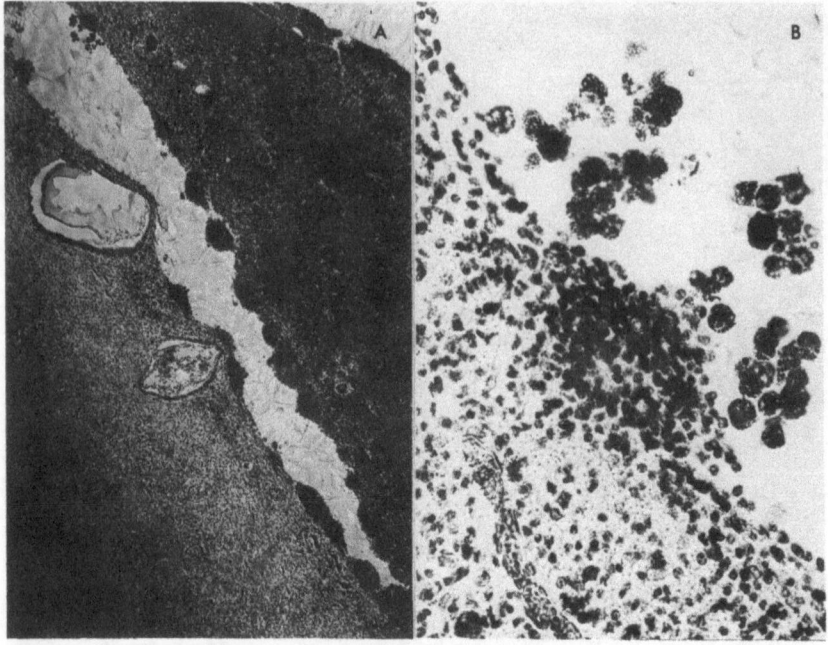

Fig. 5. Post hemorrhagic cyst in the germinal matrix of the same premature infant illustrated in fig. 2*A* (Np 5271). Oil-red-O and hematoxylin.

Florman et al. (1973) in a patient with Type 1 herpes virus.

In our laboratory we have seen 2 cases (1 definite and 1 probable) of congenital herpes encephalitis, 4 cases (3 definite and 1 probable) of neonatal herpes encephalitis and 1 definite case of neonatal Coxsackie encephalitis. These cases show more destructive and more diffuse lesions (figs. 3 and 4) than most of the diseases discussed so far.

Other viruses (e.g., Western equine, Saint Louis, influenza) have been shown to cross the placenta and infect the fetus, as reviewed by Brown (1966). The question of varicella being teratogenic has been raised by Savage et al. (1973).

Subependymal cysts have been emphasized by Shaw (1973) and Shaw and Alvord (1974), but one should not overemphasize the cysts themselves except as obvious markers of a disease acquired in utero and as an indicator of more subtle and diffuse changes that are probably responsible for the mental retardation characteristic of practically every patient who survives long enough to be tested clinically. Although 7 of the original 30 cases of subependymal cysts had obvious congenital viral infections (3 rubella and 4 CMV), none of our more recent 22 cases could be associated with a proven viral infection. Six of the original 23 cases and 3 of the 22 new cases of unknown etiology seemed likely to be of viral etiology (4 rubella, 1 CMV and 4 flu). One of the new cases could be due to repeated episodes of hypoxia, and two are definitely due to old subependymal hemorrhage (fig. 5), so common in premature infants with IRDS. There remain, however, at least 17 of the original and 16 of the new cases with no clues as to the etiology of the cysts.

The evidence for a role for other non-viral infectious agents, except for toxoplasmosis and syphilis, in the productions of CNS malformations is negligible when compared to viruses.

Toxoplasmosis is most commonly an asymptomatic infection of adults, but as an intrauterine infection it is very destructive, more so in the fetal than in the embryonic period. The affected infant can have chorioretinitis, seizures and intracranial calcifications. Warkany (1971) states that approximately one-half of the cases of toxoplasmosis have hydrocephalus and microcephaly. Hydranencephaly has also resulted from in utero infection by toxoplasmosis (Altshuler 1973). Although there may be cerebral atrophy, the frontal lobes seem to be less affected than the rest of the cortex. Toxoplasmosis has also been reported as a coexistent infection with CMV in a microcephalic stillborn (Demian et al. 1973).

Congenital syphilis has almost disappeared with penicillin treatment of primary and secondary syphilis, but occasional cases unfortunately still occur. Various types are recognized with leptomeningitis, pachymeningitis, endarteritis, optic atrophy, general paresis, tabes dorsalis and hydrocephalus due to chronic basilar leptomeningitis and/or cerebral atrophy with the meningo-vascular or paretic forms of neurosyphilis. The rarer a disease becomes, the less likely the younger physicians will recognize it, and one must turn to some of the older texts (e.g., Ford 1945) for authoritative descriptions of many of the classical varieties.

We know of no immunologic disorders which affect the developing human nervous system except for kernicterus resulting from hyperbilirubinemia usually due to Rh-incompatibility between fetus and mother (Weiser et al. 1969). Experimentally, nephrotoxic antisera have been shown to be teratogenic, probably indirectly through effects on the yolk sac (Bragonier et al. 1970). Leung et al. (1973) separated the teratogenic antibodies evoked by phosphate-buffered-saline homogenates of kidneys from the nephrotoxic ones evoked by trypsin-solubilized glomerular basement membranes. Recent reviews of the complex immunobiology of reproduction have appeared (Breyere and Shulman 1971; Beer and Billingham 1971).

SUMMARY

In order to supplement Dr. Johnson's review we wish to make two points: 1) not all grossly similar lesions should be considered to be due to the same etiology, even if experimentally documented, and 2) relatively few infections (syphilis, toxoplasmosis, cytomegalovirus, herpesvirus and rubella) and no immunologic disorders (except for kernicterus due to hyperbilirubinemia) are known to affect the developing human nervous system. Admittedly, since most human malformations are of unproven cause, these two points are not mutually exclusive. Rather, the challenge remains to show that any of the experimental models relates to any of the human diseases. In particular, the porencephaly-hydranencephaly spectrum could be due to ischemia in the watershed distribution of one or more branches of the internal carotid arteries; and aqueductal stenosis could be due to secondary blockage by blood clots or debris related to hemorrhage or necrosis in the germinal matrix or periventricular white matter, lesions most likely due to hypoxia-acidosis in premature infant, who die most often of the concomitant respiratory distress syndrome.

REFERENCES

Altshuler, G.: Toxoplasmosis as a cause of hydranencephaly. Am. J. Dis. Child. 125: 251-252 (1973).
Beer, A.E. and Billingham, R.E.: Immunobiology of mammalian reproduction. Adv. Immunol. 14: 1-84 (1971).
Berenberg, W. and Nakervis, G.: Long-term follow-up of cytomegalic inclusion disease of infancy. Pediatrics 46: 403-410 (1970).
Boniuk, M. and Zimmerman, L.E.: Ocular pathology in the rubella syndrome. Arch. Ophthalmol. 77: 455-473 (1967).
Bragonier, J.R., Frank, M.M. and Brent, R.L.: Production of congenital malformations using tissue antisera. VIII. Effectiveness of structurally modified anti-kidney antibodies. Fed. Proc. 29: 828 (1970).
Breyere, E.J. and Shulman, S.: Immunology of reproduction. In Amos, Progress in Immunology, First International Congress of Immunology, pp. 1227-1232 (Academic Press, New York 1971).
Brown, G.C.: Recent advances in the viral aetiology of congenital anomalies. Adv. Teratol. 1: 55-80 (1966).
Crome, L.: Hydrencephaly. Dev. Med. Child. Neurol. 14: 224-234 (1972).
Crome, L. and Sylvester, P.E.: Hydranencephaly (hydrencephaly). Arch. Dis. Child. 33: 235-245 (1958).
Demian, S.D.A., Donnelly, W.H., Jr. and Monif, G.R.G.: Coexistent congenital cytomegalovirus and toxoplasmosis in a stillborn. Am. J. Dis. Child. 125: 420-421 (1973).
Dennis, J.P. and Alvord, E.C., Jr.: Microcephaly with intracranial calcification and subependymal ossification: Radiologic and clinicopathologic correlation. J. Neuropathol. Exp. Neurol. 20: 412-426 (1961).

Embil, J.A., Ozere, R.L. and Haldane, E.V.: Congenital cytomegalovirus infection in two siblings from consecutive pregnancies. J. Pediatr. 77: 417-421 (1970).

Ford, F.R.: Diseases of the Nervous System in Infancy, Childhood and Adolescence. (Charles C. Thomas, Publ., Springfield 1945).

Florman, A.L., Gershon, A.A., Blackett, P.R. and Nahmias, A. J.: Intrauterine infection with herpes simplex virus: Resultant congenital malformations. JAMA 225: 129-132 (1973).

Globus, J.H.: A contribution to the histopathology of porencephalus. Arch. Neurol. Psychiatr. 6: 652-668 (1921).

Gregg, N.M.: Congenital cataract following German measles in the mother. Trans. Ophthalmol. Soc. Austral. 3: 35-46 (1941).

Halsey, J.H., Jr., Allen, N. and Chamberlin, H.R.: The morphogenesis of hydranencephaly. J. Neurol. Sci. 12: 187-217 (1971).

Hamby, W.B., Krauss, R.F. and Beswick, W.F.: Hydranencephaly: clinical diagnosis. Presentation of seven cases. Pediatrics 6: 371-383 (1950).

Hanshaw, J.B.: Developmental abnormalities associated with congenital cytomegalovirus infection. Adv. Teratol. 4: 64-93 (1970).

Haque, I.U. and Glassauer, F.E.: Hydranencephaly in twins. N.Y. State J. Med. 69: 1210-1214 (1969).

Hardy, J.B., McCracken, G.H. Jr., Gilkeson, M.R. and Sever, J. L.: Adverse fetal outcome following maternal rubella after the first trimester. JAMA 207: 2414-2420 (1969).

Herndon, R.M., Johnson, R.T., Davis, L.E., and Descalzi, L.R.: Ependymitis in mumps virus meningitis. Electron microscopical studies of cerebrospinal fluid. Arch. Neurol. 30: 475-479 (1974).

Jaffe, R.H.: Traumatic porencephaly. Arch. Pathol. 8: 787-799 (1929).

Johnson, R.T.: Effects of viral infection on the developing nervous system. N. Engl. J. Med. 287: 599-604 (1972).

Johnson, R.T.: Hydrocephalus and viral infections. Dev. Med. Child Neurol. 17: 807-816 (1975).

Katz, S.L.: Slow Virus Infections (64th Ross Conference on Pediatric Research, Columbus 1973).

Leech, R.W., and Alvord, E.C., Jr.: Wallerian degeneration in the premature human. J. Neuropathol. Exp. Neurol. 34: 92 (1975).

Leech, R.W., and Kohnen, P.: Subependymal and intraventricular hemorrhages in the newborn. Am. J. Pathol. 77: 465-476 (1975).

Leung, C.C.K., Urdaneta, A., Jensen, M., and Brent, R.L.: The production of congenital malformations and nephrotoxic serum nephritis. Fed. Proc. 32: 982 (1973).

Lindenberg, R., and Swanson, P.D.: Infantile hydranencephaly – A report of five cases of infarction of both cerebral hemispheres in infancy. Brain 90: 839-850 (1967).

Muir, C.S.: Hydranencephaly and allied disorders: a study of cerebral defects in Chinese children. Arch. Dis. Child. 34: 231-246 (1959).

Navin, J.J., and Angevine, J.M.: Congenital cytomegalic inclusion disease with porencephaly. Neurology (Minneap.) 18: 470-472 (1968).

Osburn, B.I., Johnson, R.T., Silverstein, A.M., Prendergast, R.A., Jochim, M.M., and Levy, S.E.: Experimental viral-induced congenital encephalopathies. II. The pathogenesis of blue tongue vaccine virus infection in fetal lambs. Lab. Invest. 25: 206-210 (1971a).

Osburn, B.I., Silverstein, A.M., Prendergast, R.A., Johnson, R. T., and Parshall, C.J., Jr.: Experimental viral-induced congenital encephalopathies. I. Pathology and hydranencephaly and porencephaly caused by blue tongue vaccine virus. Lab. Invest. 25: 197-205 (1971b).

Patten, C.A., Grant, F.C., and Yaskin, J.C.: Porencephaly: diagnosis and treatment. Arch. Neurol. Psychiatr. 37: 108-136 (1937).

Rorke, L.B.: Nervous system lesions in the congenital rubella syndrome. Arch. Otolaryngol. 98: 249-251 (1973).

Rorke, L.B., and Spiro, A.J.: Cerebral lesions in congenital rubella syndrome. J. Pediatr. 70: 243-

255 (1967).

Savage, M.O., Mossa, A., and Gordon, R.R.: Maternal varicella infection as a cause of fetal malformations. Lancet 1: 352-354 (1973).

Shaw, C.M.: Subependymal germinolysis. J. Neuropathol. Exp. Neurol. 32: 153 (1973).

Shaw, C.M., and Alvord, E.C., Jr.: Subependymal germinolysis. Arch. Neurol. 31: 374-381 (1974).

Shaw, C.M., and Sumi, S.M.: Nonviral intranuclear filamentous inclusions. Arch. Neurol. 32: 428-432 (1975).

South, M.A., Tomkins, W.A.F., Morris, C.R., and Rawls, W.E.: Congenital malformation of the central nervous system associated with genital (type 2) herpesvirus. J. Pediatr. 75: 13-18 (1969).

Sumi, S.M.: Reaction of the immature brain to injury. pp. 177-189 in Angle and Bering, Physical Trauma as an Etiological Agent in Mental Retardation (U.S. Government Printing Office, Washington, D.C. 1970).

Sumi, S.M., and Hager, H.: Electron microscopic study of the reaction of the newborn rat to injury. Acta Neuropathol. 10: 324-355 (1968a).

Sumi, S.M., and Hager, H.: Electron microscopic features of an experimentally produced porencephalic cyst in the rat brain. Acta Neuropathol. 10: 336-346 (1968b).

Tondury, G., and Smith, D.W.: Fetal rubella pathology. J. Pediatr. 68: 867-879 (1966).

Vander Eecken, H.M.: The Anastomoses Between the Leptomeningeal Arteries of the Brain. (Charles C. Thomas, Publ., Springfield 1959).

Warkany, J.: Congenital Malformations. (Year Book Med. Publ., Chicago 1971).

Weiser, R.S., Myrvik, Q.N., and Pearsall, N.N.: Fundamentals of Immunology for Students of Medicine and Related Sciences, pp. 263-272. (Lea & Febiger, Philadelphia 1969).

Weiss, M.H., Young, H.F., and McFarland, D.E.: Hydranencephaly of postnatal origin. J. Neurosurg. 32: 715-720 (1970).

Weller, T.H.: The cytomegaloviruses: ubiquitous agents with protean clinical manifestations. N. Engl. J. Med. 285: 203-214, 267-274 (1971).

Williams, H.J.: Skull erosion complicating traumatic porencephaly in infancy. Am. J. Roentgenol. Radium Ther. Nucl. Med. 106: 129-132 (1969).

Yakovlev, P.I., and Wadsworth, R.C.: Double symmetrical porencephalies (schizencephalies). Trans. Am. Neurol. Assoc. 67: 24-29 (1941).

Yakovlev, P.I., and Wadsworth, R.C.: Schizencephalies: a study of congenital clefts in the cerebral mantle. I. Clefts and fused lips. J. Neuropathol. Exp. Neurol. 5: 116-130 (1946a).

Yakovlev, P.I., and Wadsworth, R.C.: Schizencephalies: a study of congenital clefts in the cerebral mantle. II. Clefts with hydrocephalus and lips separated. J. Neuropathol. Exp. Neurol. 5: 169-206 (1946b).

CHAPTER SIXTEEN

BIOCHEMICAL ASPECTS OF THE MATURATION OF SEROTONINERGIC NEURONS IN THE RAT BRAIN*

M. HAMON, PH.D. AND S. BOURGOIN, PH.D.

INTRODUCTION

At birth, the new-born rat is very immature and the organization of its central nervous system is far from complete. Thus, the cerebellum is hardly visible behind the colliculi, the electrical activity of the cerebral cortex is discrete, myelination has not begun. The poor development of the central nervous system of the neonatal rat can be quantified by comparing its biochemical composition with that of the adult brain. Thus, the water content represents about 90% of the wet weight of the brain at birth whereas it is less than 80% in the adult. By contrast, both lipids and proteins increase significantly in % of wet weight from birth to adulthood (Winick and Noble, 1965; Davis and Himwich, 1973; Agrawal and Davison, 1973). The brain levels of other components, notably amino acids (Lajtha and Toth, 1973; Agrawal and Davison, 1973), monoamines are also strikingly different in the new-born and the adult rat. Since several of these small molecules are considered as putative neurotransmitters in the central nervous system of the adult rat, their presence suggests that they may also have a physiologic role in the immature brain. Indeed, monoamines, namely norepinephrine, dopamine and serotonin (5-HT) are already present inside particular neurons in neonatal rat brain. Histofluorescence techniques show that neuroblasts in the medial zone of the brain-stem, i.e. the raphe system, exhibit typical yellow 5-HT fluorescence as early as the 13th day following conception (Olson and Seiger, 1972). These cells continue to multiply and migrate for a few days during fetal life so that the typical organization of serotoninergic neurons in the raphe nuclei is reached before birth (Seiger and Olson, 1973). However, their fibers are still poorly developed and the density of 5-HT terminals in forebrain areas is much lower in the new-born than in the adult rat (Loizou, 1972). The pro-

* Supported by grants from INSERM (ATP 6.74.27), DGRST, DRME and la Société des Usines Chimiques Rhône-Poulenc.

gressive invasion of the various central areas by serotoninergic fibers and terminals proceeds for several weeks after birth so that the typical serotoninergic innervation of the central nervous system of the adult rat can be observed only at the end of the 5th-6th postnatal week. The present paper describes our studies on the metabolism (synthesis, release, catabolism. . .) of 5-HT in the brain when serotoninergic neurons are growing i.e.. during the first 6 postnatal weeks.

1. EVOLUTION OF 5-HT, 5-HYDROXYINDOLE ACETIC ACID AND TRYPTOPHAN LEVELS IN THE CENTRAL NERVOUS SYSTEM DURING DEVELOPMENT

At birth, the mean levels of 5-HT in various areas of the central nervous system of the rat are about one fourth those found in the adult. Such a difference between the adult and neonatal values has been observed in another species: the cat. By contrast, the levels of the indole-amine in the central nervous system of the newborn guinea pig are close to the adult values. On the 2nd postnatal day, they reach 80% of those found in 3 month old guinea pigs (unpublished observations). These preliminary data led us to choose the rat (Sprague-Dawley, Charles River strain) for further studies on development since 1: the guinea pig is almost mature at birth and 2: the rat is much more convenient than the cat for biochemical studies. The results described in this paper are from young rats kept with their mothers until weaning on the 22nd postnatal day. Controlled environmental conditions (Bourgoin et al., 1974), food and water ad libitum, were maintained until death.

Within the 12 hours following birth, the concentration of 5-HT in the brain stem of pups equals 32% of that found in adults. In the forebrain, this ratio equals 22%, suggesting that the differentiation of 5-HT neurons is more neurons in the brain stem than in anterior brain regions. This observation agrees with previous findings using histofluorescence techniques (Loizou, 1972). During the first three postnatal weeks, the concentration of the indoleamine in the brain stem increases rapidly so that the adult level is reached at the end of this period (fig. 1A). The week after, the increase still proceeds and 5-HT levels in the brainstem of 4 week-old rats are significantly higher than those found in adults. These latter levels are reached definitely at the end of the 5th postnatal week (fig. 1A); in the forebrain, 5-HT levels increase much more slowly. At the end of the 5th postnatal week, 5-HT levels in this region are only 75% of those found in adult rats (fig. 1A).

The developmental changes in the concentration of 5-hydroxyindole acetic

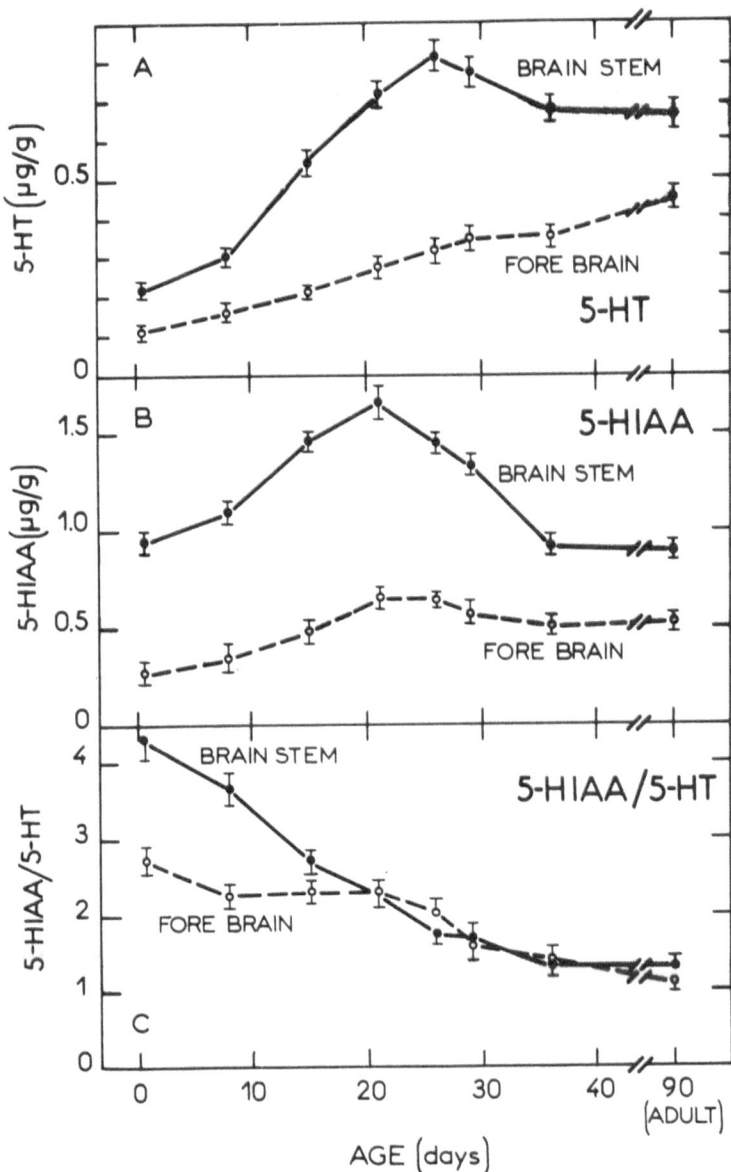

Fig. 1. Developmental changes in the levels of 5-HT (A) and 5-HIAA (B) and in the 5-HIAA/HT ratio (C) in the brain-stem and the forebrain of the rat- Time 0 corresponds to birth. Each point is the mean ± SEM of 6-8 determinations.

acid (5-HIAA), the almost exclusive product of 5-HT catabolism in brain, do not follow the same pattern as those of the indoleamine. In the brain stem of 12 h-old are already as high as in adult animals (fig. 1 *B*). Then, they become significantly higher than the adult value between the first and the 4th post-natal week with a peak around the end of the third postnatal week. At this time, 5-HIAA levels exceedthose found in the brain-stem of adult rats by 80%. A quite different picture is observed in the forebrain since during the two first postnatal weeks, 5-HIAA levels remain significantly lower than the adult values in this region (fig. 1 *B*). As in the brain stem however, levels higher than the adult ones occur in the forebrain during the 4th postnatal week (fig. 1 *B*). One week later, in the brain stem and in the forebrain, 5-HIAA levels reach their definitive values (fig. 1 *B*).

Comparison between the evolution of 5-HT and that of 5-HIAA levels in the brain stem and the forebrain can be visualized by considering the changes in 5-HIAA/5-HT ratio as a function of age. As shown in fig. 1 *C*, a clearcut difference between young and adult rats should concern the catabolism of 5-HT in brain since, for the first four postnatal weeks, both in the brain stem and the forebrain, this 5-HIAA/5-HT ratio is significantly higher than in adults. This problem will be discussed later.

Immediately after birth, the concentration of tryptophan in brain is 5-10 times higher than in adult rats. Then, it decreases progressively to reach the adult level at the end of the 4th postnatal week (Tyce et al., 1964; Bourgoin et al., 1974). The same pattern of evolution is observed in the forebrain and the brain stem (Bourgoin, 1976).

2. EVOLUTION ON 5-HT SYNTHESIS IN THE RAT BRAIN DURING ONTOGENESIS

Often when both the concentrations of the 5-HT precursor, tryptophan, and that of the 5-HT metabolite, 5-HIAA, are elevated in brain, it is considered that the turnover rate of 5-HT is increased (Tagliamonte et al., 1971). Since this particular situation occurs during early life, one might conclude that the turnover (i.e. the synthesis and the utilisation) of the indoleamine is more rapid for this period than during the adult life. To check this hypothesis, we studied the evolution as a function of age of various factors known to mo-dulate the rate of 5-HT synthesis in adult rats.

2.1. *The particular state of peripheral tryptophan.* Several observations established that only that fraction of peripheral tryptophan not bound to serum albumin can enter the brain and be used as a precursor for 5-HT synthesis. Gessa and Tagliamonte (1974) showed that the levels of tryptophan and the rate of 5-HT synthesis in brain increase when the free fraction of tryptophan in serum is enhanced. Since brain tryptophan levels are particularly high in the neonatal period it was of interest to consider the state of tryptophan in serum at this age. As shown in fig. 2, free tryptophan represents more than 95% of the total amino acid in the serum of the new born rat. This is in sharp contrast with adult rats in which free tryptophan accounts for only 5-10% of

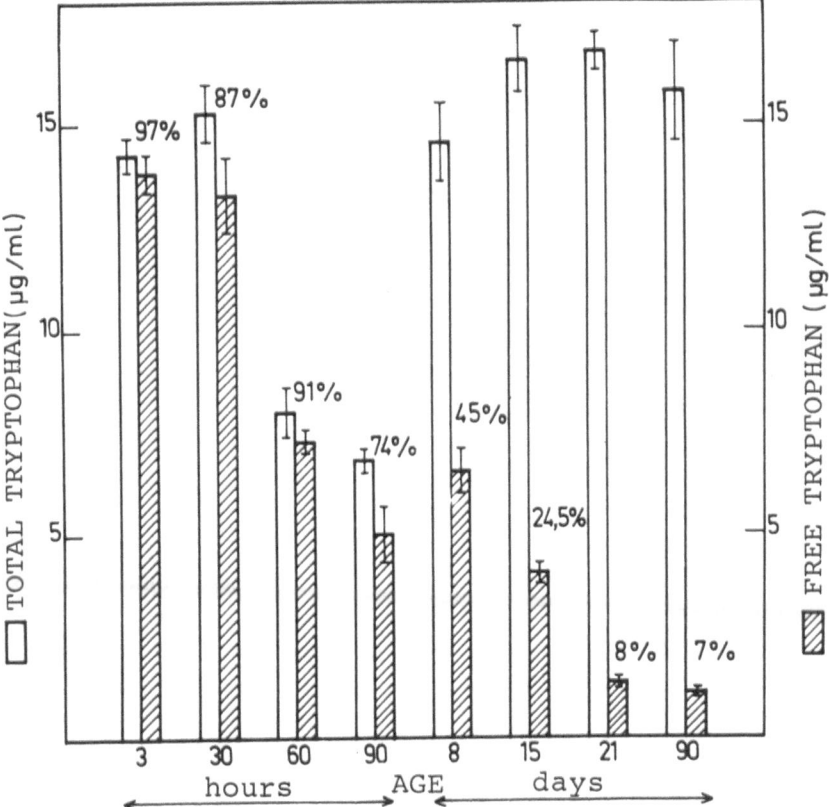

Fig. 2. Tryptophan concentrations in rat serum at various ages. Ages are given in hours or days after birth. The empty bars represent total tryptophan and the hatched ones, the free form of the amino acid. Percentage on bars express the ratio of the concentration of free tryptophan over that of total aminoacid. Each value is the mean ± SEM of 8 determinations.

the total amino acid in serum (fig. 2). The bound fraction of the indolic amino acid increases progressively as a function of age so that the adult situation is reached at the end of the third postnatal week (fig. 2).

The parallel changes of tryptophan levels in brain and the concentration of free tryptophan in serum during ontogenesis suggest that only the unbound amino acid in serum can enter brain tissues. This observation confirms previous findings (Gessa and Tagliamonte, 1974; Curzon and Knott, 1974). Its importance must be emphasized since, in contrast to these previous findings, the particular role of free tryptophan has been observed in normal physiologic conditions.

The particular state of serum tryptophan in the new born rat led us to investigate the characteristics of the aminoacid binding on plasma proteins. By isolating serum albumin from other proteins by Sephadex G 100 filtration, it can be shown that this protein binds much less tryptophan in the new born than in the adult serum (fig. 3). Bourgoin et al. (1977c) three convergent factors which explain the lack of binding of tryptophan in the serum of the new-born rat:

a) The concentration of the binding protein itself, serum albumin, in neonatal serum is about half that found in adult serum.

b) The concentration of unesterified fatty acids, natural competitive inhibitors of tryptophan binding onto serum albumin, is about three times higher in the serum of new born rats as in adults. Their level remains significantly higher than the adult value until weaning, i.e. as long as the young rat is fed with a high fat milk diet.

c) The intrinsic capacity to bind tryptophan of purified and defatted serum albumin extracted from the new born rat serum is lower than that of the adult protein. In Scatchard plots, it appears that the binding affinity (Kd = 2-2.5 × 10⁻⁴M) of both proteins is similar but the number of binding sites is lower in the neonatal serum albumin. Since there is only one binding site for tryptophan on the adult protein, this might suggest that the serum albumin extract from the neonatal serum is not homogeneous: that some of the protein(s) in this extract do not bind tryptophan at all. Attempts to differentiate several kinds of serum albumin(s) in the neonatal serum by immuno-precipitation with a specific antibody have failed. Therefore, it can be proposed that a neonatal protein unable to bind tryptophan but precipitable by a specific antibody against rat serum albumin probably exists in the serum of the new born rat. Since the possible contamination of our preparation by α-fetoprotein has been ruled out, the presence of a neonatal serum albumin was assumed. In humans, Miyoshi et al. (1966) proposed the exis-

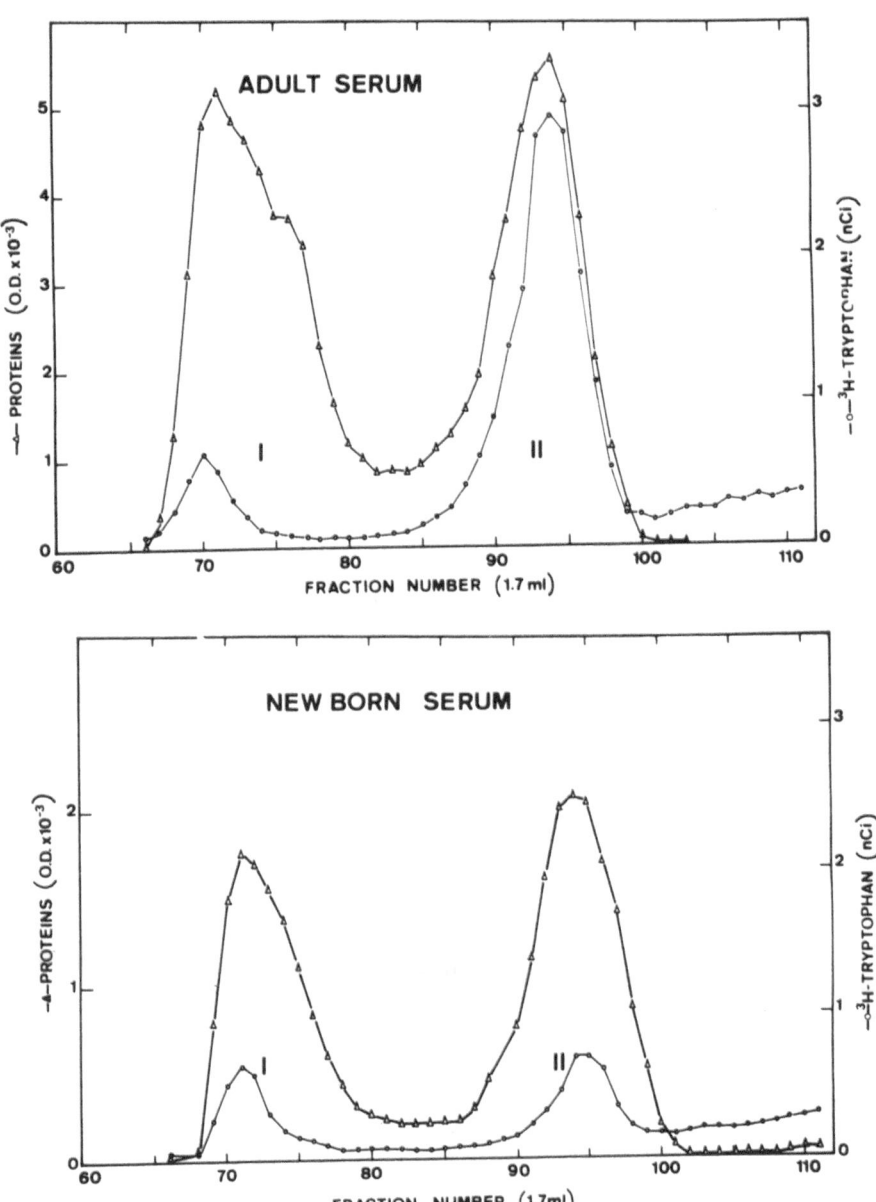

Fig. 3. Sephadex G 100 gel filtration of adult and new-born rat serum previously incubated with
³H-tryptophan. With adult serum, most of the bound ³H-tryptophan comes out of the gel in peak
II corresponding to the monomer of serum albumin. In similar experimental conditions (see
Bourgoin et al., 1977c), the serum albumin in the new born serum binds much less ³H-
tryptophan.

tence of a fetal (F) serum albumin with different characteristics than those of the adult protein. According to Tuilié and Lardinois (1972), the capacity of alb. F to bind bilirubin is lower than that of the adult serum albumin.

The lack of binding of tryptophan in serum involves important consequences on 5-HT metabolism in the brain of the new-born rat:

a) Drugs such as salicylate and benzodiazepines which alter central 5-HT metabolism in adults by increasing the concentration of free tryptophan in serum are inactive in new born rats (Bourgoin et al., 1974; 1975). This is not surprising since, potent competitive inhibitors of tryptophan binding onto serum albumin (McArthur and Dawkins, 1969; Bourgoin et al., 1975), they have nothing to inhibit in the serum of the new-born rat.

b) The availability of almost all peripheral tryptophan for brain (and other organs) during the neonatal period leads to a high concentration of this amino acid in tissues. Thus, in contrast to the well-known situation in adults (Hamon and Glowinski, 1974), tryptophan hydroxylase is almost saturated by its substrate during the early life period. As a consequence, physiologic fluctuations in the concentration of tryptophan in brain should exert less dramatic changes in the rate of 5-HT synthesis in new born than in adult rats. The lack of binding of tryptophan in serum leads to a "buffered" synthesis of 5-HT in brain during the immediate postnatal period.

2.2. *The uptake of tryptophan by brain tissues.* Since the synthesis of 5-HT is dependent on the concentration of tryptophan in brain, the uptake of tryptophan in tissues, as the concentration of free tryptophan in serum (see above), may play a major role in its control. Indeed, several authors (Hery et al., 1972, 1974; Hamon et al., 1974a, b) have shown that changes in the activity of the tryptophan uptake process in brain tissues lead to parallel modifications in the rate of 5-HT synthesis.

In order to analyze the biosynthetic capacity of the neonatal brain for 5-HT, it was therefore of interest to study the evolution of tryptophan uptake in brain during development. This study was conducted in vitro with brain slices and synaptosomes. In brain slices from adult rat, the uptake of tryptophan exhibits important regional changes. It is most active in hypothalamus slices and decreases in the following order: hypothalamus > striatum > cortex > brain-stem > cerebellum. These regional differences are not yet observed in the neonatal brain. Except in the hypothalamus, tryptophan uptake in slices from various structures of the neonatal brain is more rapid than in corresponding tissues from the adult brain. A kinetic analysis with slices from the brain stem revealed that this difference results from a better affinity of the

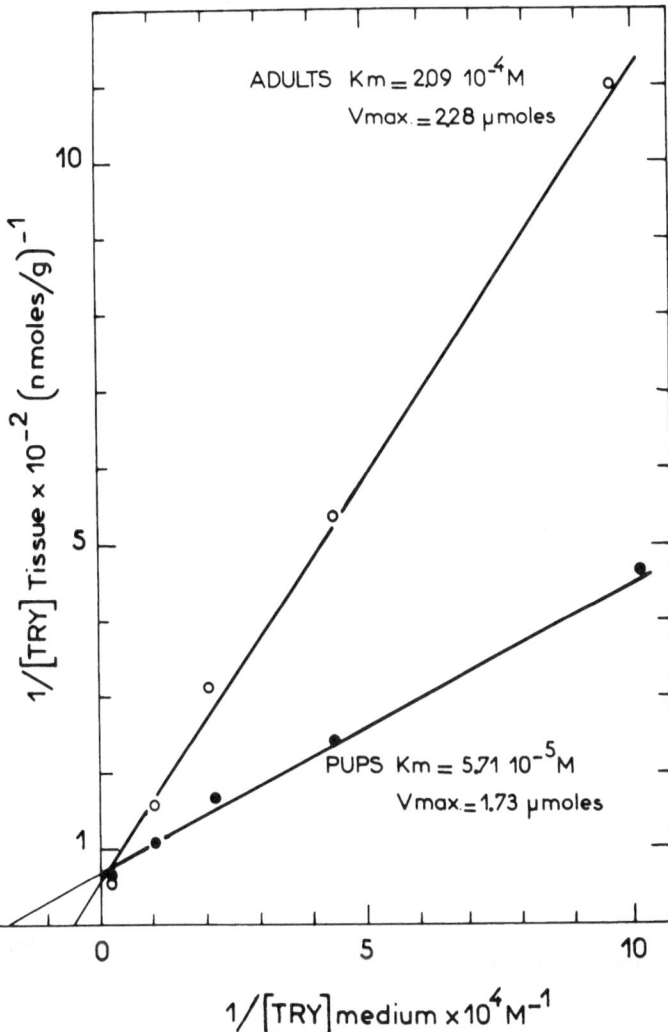

Fig. 4. Double reciprocal plots of tryptophan accumulation in brain-stem slices of new born or adult rats vs the extracellular amino acid concentration. Each point is the mean of six determinations. Vmax are expressed in µmoles tryptophan taken up in 1 g of fresh tissue for 1 hour.

carrier for tryptophan in new-born tissues (fig. 4). Similar findings were obtained by Vahvelainen and Oja (1972) using slices of cerebral cortex. As shown in table 1, differences observed with brain stem slices are also seen in synaptosomes from this region. Indeed, the rapid tryptophan uptake in syn-

Table 1. Changes as a function of age of kinetic characteristics of tryptophan uptake process in synaptosomes from the rat brain-stem. The crude synaptosomal preparation (P_2) was used. Each value is the mean of two experiments. Double reciprocal plots ($1/V$ vs $1/S$) were calculated with 6 different concentrations of tryptophan (triplicate determinations for each concentration).

Age (days)	Km (μM)	Vmax (nmol./mg prot/min)	(%)
1.5	8.42	1.14	(181)
9	8.90	1.92	(305)
16	13.60	2.08	(330)
24	12.45	1.33	(211)
90	13.68	0.63	(100)

aptosomes from the new-born rat is related to both a better affinity and a higher Vmax of the specific transport process in neonatal than in adult synaptosomes (table 1). During development, the Km reaches the adult value at the end of the 2nd postnatal week. At this age, the Vmax is maximum, being 2.3 times higher than that of adult rats (table 1).

These differences between the uptake processes for tryptophan in brains of new born and adult rats are also shown by analyzing the effects of altering the Na^+ concentration in the incubating medium of tissues. When the extracellular Na^+ concentration is reduced from 153.4 mM to 17.4 mM, only the uptake of tryptophan in brain-stem slices from new born rats is significantly decreased (-12%). In addition, probenecid (1 mM) stimulates tryptophan uptake by a Na^+-dependent mechanism only in tissues from young animals (Bourgoin, 1976). Whether these differences in tryptophan uptake by tissues of new born and adult rats are related to different carriers or to the same carrier with an activity closely dependent on developmental changes in membrane composition remains to be elucidated. However that may be, the particularities of the tryptophan uptake process during the neonatal period, as those concerning the peripheral amino acid (see above), add to the accumulation of this essential amino acid in the cells of the growing organism. In brain, this induces both a high rate of protein synthesis and a biosynthetic pathway for 5-HT near working at its maximal elocity.

2.3. *Tryptophan hydroxylase activity.* In adult rats, tryptophan hydroxylase (EC 1.14.16.4) catalyses the rate limiting step in the biosynthesis of 5-HT. The comparison of the developmental changes in the rate of 5-HT synthesis measured in vivo (fig. 5A) with those of tryptophan hydroxylase activity (fig. 5B) in the brain stem and forebrain areas strongly suggests that this enzyme begins this function even in early life. Thus, at birth, both the in vivo rate of 5-HT synthesis and the activity of tryptophan hydroxylase in the brain stem are about half the corresponding values. During the first three postnatal weeks, parallel increases in each parameter are observed despite the large changes in tryptophan levels occurring during this period. Although the total amount of tryptophan hydroxylase (per brain region) increases regularly in the brain-stem until the end of the 4th postnatal week (Bourgoin, 1976), the developmental changes in its specific activity (per mg prot.) is rather complex. As shown in fig. 5B, it first increases up to a maximum equal to 180% of the adult value at the end of the 3rd postnatal week and then decreases progressively. Kinetic analyses show that Vmax changes only account for the age-dependent variations of tryptophan hydroxylase specific activity. The same pattern, with a peak around the end of the 3rd postnatal week, is also observed for the rate of conversion of tryptophan into 5-HT and the velocity of the high affinity 5-HT uptake in synaptosomes from the brain stem (Bourgoin, 1976). These findings suggest that the density of 5-HT terminals in the brain stem is maximum between the 2nd and the 4th postnatal week. This probably results from the faster growth of serotoninergic neurons, as compared with other neuronal systems in this region. This conclusion is supported by histofluorescence data showing that serotoninergic neurons are among the first to mature in the central nervous system of the rat (Olson and Seiger, 1972; Seiger and Olson, 1973).

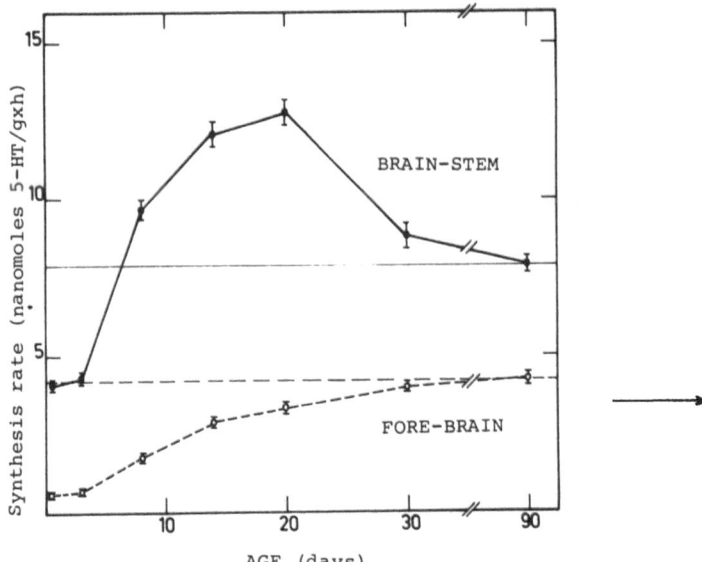

3. EVOLUTION OF 5-HT RELEASE IN THE RAT BRAIN DURING ONTOGENESIS

In vivo studies on the release of neurotransmitters in brain encounter many technical problems. No attempt has yet been made to measure the release of 5-HT in new born rat brain. However, using brain slices, the characteristics of the in vitro release of 5-HT can be studied even in neonatal tissues. Bourgoin et al. (1976b) observed that ^{3}H-5-HT newly synthesized from ^{3}H-tryptophan can be released by a depolarizing concentration of K$^+$ (30 mM) from brain stem slices of new born as well as from those of adult rats. The regulation of this K$^+$-evoked release of the indole amine by presynaptic receptors (Farnebo and Hamberger, 1974; Hamon et al., 1974a) can be detected even at birth. In

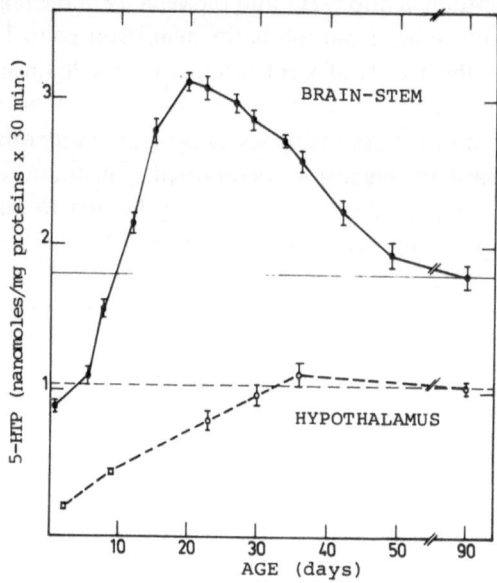

Fig. 5. Developmental changes in tryptophan hydroxylase activity and in the in vivo rate of HT synthesis in the brain stem and forebrain areas in the rat. *A*. The rate of 5-HT synthesis was calculated by measuring the initial accumulation of 5-HT in tissues after the blockade of MAO with pargyline (75 mg/kg, ip). It is expressed in nanomoles 5-HT formed per g of fresh tissue and per h. *B*. The activity of tryptophan hydroxylase was determined in the 30,000 x g supernatant of tissue homogenates made in 0.05 M Tris-acetate, pH 7.6, 2 mM β-mercaptoethanol. The assay was performed in the presence of 0.2 mM tryptophan and 0.16 mM 6-methyl-tetrahydropterin. Tryptophan hydroxylase activity is expressed in nanomoles 5-HTP formed per mg prot. and per 30 min. Each point is the mean of 6-8 determinations.

brain-stem slices of 12h-old rats, the K^+-evoked release of newly synthesized ^3H-5-HT is partially antagonized by LSD (a 5-HT agonist) whereas it is greatly enhanced in the presence of methiothepin (a 5-HT antagonist). Therefore, these data show that the 5-HT terminals in new born rats have the same potential for releasing indole amine as in adults.

According to Andén (1974), the rate of 5-HT decrease after inhibition of tryptophan hydroxylase by α-propyldopacetamide or p-chlorophenylalanine is highly dependent on the nerve impulse flow in 5-HT neurons. When parachlorophenylalanine was administered to adult and 8h-old rats three hours before death, 5-HT levels in whole brain were significantly reduced (-16%) only in adult rats (Bourgoin, 1976). Since parachlorophenylalanine is equipotent in vitro in inhibiting tryptophan hydroxylase from adult and new born rats, the lack of effect of this drug on the endogenous 5-HT levels in 8h-old rats may be the consequence of a low nerve impulse flow in 5-HT neurons at this age. Another biochemical parameter which depends on the nerve impulse flow in 5-HT neurons is the 5-HIAA/5-HT ratio. Thus the administration of LSD in adult rats induces a simultaneous reduction in the nerve impulse flow in 5-HT neurons, a decreased release of 5-HT in vivo and a diminished 5-HIAA/5-HT ratio in brain tissues (Hamon et al., 1974; Gallager and Aghajanian, 1975). The latter effect of LSD treatment also occurs in new born rats (Bourgoin et al., 1976). Therefore it may be hypothesized that 5-HT neurons discharge in physiologic conditions even at birth, but the frequency is probably lower than during the adult life. Although indirect, this evidence strongly suggests that 5-HT is released from 5-HT terminals in vivo in the neonatal rat brain.

4. EVOLUTION OF 5-HT RECEPTORS DURING ONTOGENESIS IN THE RAT

Von Hungen et al. (1975) were the first to demonstrate that 5-HT is able to stimulate adenylate cyclase in the neonatal rat brain. The Km of this enzymic activation by 5-HT is equal to $1.4 \times 10^{-6}M$ 5-HT (fig. 6). The sensitivity of this enzyme to various pharmacologic agents means that this adenylate cyclase may be associated with 5-HT receptors in neuronal membranes. Thus, putative agonists like LSD and 5-methoxy-N, N-dimethyltryptamine stimulate this enzyme (Hamon et al., 1976) whereas 5-HT antagonists such as metergoline, methiothepin, mianserin, cinanserin, cyproheptadine... counteract the stimulating effect of 5-HT on this enzyme (Enjalbert et al., unpublished observations). In addition, the regional distribution of 5-HT sensitive adenylate cyclase closely resembles that of 5-HT terminals: in particular, its activity is maximal in the hypothalamus, the colliculi and the spinal cord, all regions containing large populations of serotoninergic terminals. The 5-HT sensitive adenylate cyclase is probably associated with postsynaptic 5-HT receptors since the degeneration of serotoninergic neurons induced by a midbrain raphe lesion in the newborn rat does not alter its activity in forebrain areas (Bourgoin et al., 1977b).

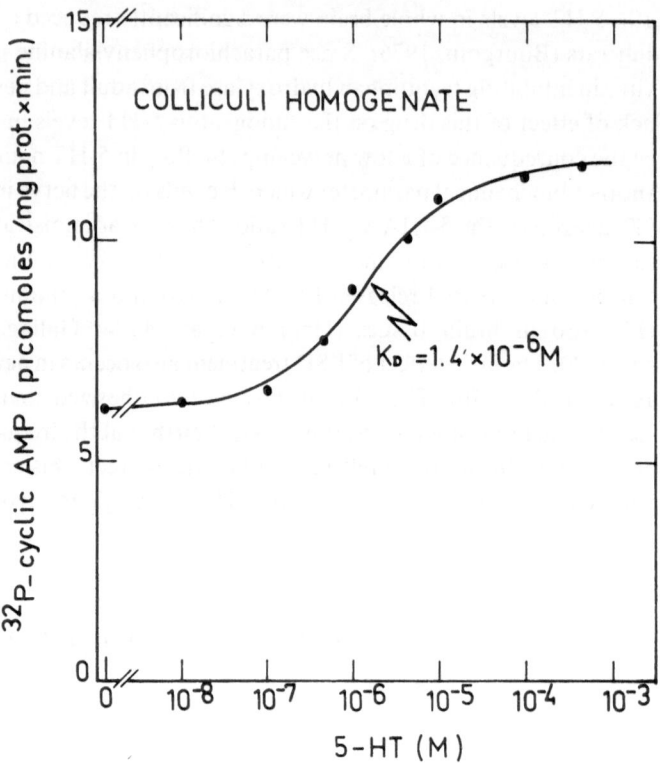

Fig. 6. Stimulating effect of 5-HT on adenylate cyclase activity in colliculi homogenate from new-born rats. Adenylate cyclase activity was measured by the conversion of α-^{32}P-ATP into ^{32}P-cyclic AMP (see Hamon et al., 1976). Each point is the mean of triplicate determinations.

During ontogenesis, the absolute stimulating effect of 5-HT on adenylate cyclase activity (in △ picomoles cAMP formed/mg prot.) remains constant. However, since the basal adenylate cyclase activity increases, the relative stimulation induced by 5-HT decreases progressively as a function of age (Von Hungen et al., 1975). This should not be interpreted in terms of 5-HT receptor desensitization since only the percent stimulation, not the absolute stimulating effect ($\triangle$) of the indole amine decreases during development.

Another way to study the receptors of neurotransmitters consists of looking for high affinity bindings of these ligands onto synaptic membranes. Bennett and Snyder state (1976) that the binding site for 5-HT which is characterized by a dissociated constant of $\sim$ 10 nM may be that of specific 5-HT receptors. Indeed, the sensitivity of this binding to various 5-HT agonists and antagonists fits with this concept. Finally, convincing evidence for the postsynaptic localization of these high-affinity binding sites particularly in regions rich in 5-HT terminals was presented in the same paper.

The comparison of data obtained with these two experimental approaches suggests that the 5-HT receptor which is characterized by a Kd of $\sim$ 10 nM is not related to the 5-HT sensitive adenylate cyclase described above. First, their regional localizations do not superimpose exactly: for instance, the number of high affinity binding sites for 5-HT is much higher in the striatum than in the spinal cord (Bennett and Snyder, 1976; Hamon and Bourgoin, unpublished observations) whereas the reverse is true for the activity of 5-HT sensitive adenylate cyclase (Enjalbert et al., unpublished observations). Second, the concentration of 5-HT receptors (per g of fresh tissue) characterized by a high affinity binding progressively increases as a function of age whereas the activity of 5-HT sensitive adenylate cyclase remains constant throughout the life span. Third, the Kd of the 5-HT binding ($\sim$ 10 nM) is about two orders of magnitude lower than the Kd of the 5-HT sensitive adenylate cyclase (1.4 μM, fig. 6).

No evidence is available as to the physiologic role of these 5-HT receptors in the central nervous system. In particular, although several behavioural observations show that some 5-HT receptors exhibit hypersensitivity following the degeneration of 5-HT neurons (Stewart et al., 1976; Trulson et al., 1976), neither the receptor with a high affinity binding for 5-HT (Bennett and Snyder, 1976) nor that which is associated with the adenylate cyclase (Bourgoin et al., 1977b) becomes hypersensitive in these conditions. In addition, the specificity of these two receptors is questionable. Fluphenazine, a dopamine antagonist in brain, also inhibits the 5-HT sensitive adenylate cyclase (Enjalbert et al., unpublished observations) and the high affinity binding of 5-HT onto synaptic membranes (Bennett and Snyder, 1976). Although this may well occur in vivo also, no evidence for a 5-HT receptor blocking action of this drug has been obtained on the basis of 5-HT turnover studies.

In conclusion, the biochemical data available favour at least two kinds of 5-HT postsynaptic receptors present at birth in rat brains. Gerschenfeld and Paupardin-Tritsch (1974) have distinguished 6 kinds of 5-HT receptors in the ganglionic nervous system of Aplysia. Therefore, in addition to the presynaptic 5-HT receptor and the two postsynaptic ones described above, one may conjecture whether the mammalian central nervous system contains other types of 5-HT receptors, also involved in synaptic transmission and modulation.

5. FORMATION AND ELIMINATION OF 5-HIA IN THE BRAIN DURING RAT DEVELOPMENT

The results discussed in part 3, concerning the nerve impulse flow in 5-HT neurons and the in vivo release of the indoleamine in brain led to the conclusion that the neuronal activity of serotoninergic systems is much lower in new born than in adult rats. In contrast with this conclusion, we observed that the 5-HIAA/5-HT ratio in brain for the neonatal period is significantly higher than in adult tissues (fig. 1), suggesting that the turnover of the indoleamine is accelerated in new born rats. In vitro studies with brain stem slices, in which neuronal activity has disappeared, demonstrate that the catabolism of newly synthesized ^{3}H-5-HT is faster in neonatal than in adult tissues (Bourgoin et al., 1974, 1977a). Therefore, in contrast to observations made on adult rats, the high 5-HIAA/5-HT·ratio occurring in tissues of new born animals is not related to the nerve impulse flow within 5-HT neurons. Several hypotheses were then proposed to explain this paradoxical high value of 5-HIAA/5-HT ratio occurring in the brain at birth:

a) During the neonatal life, the blood-brain barrier for 5-HT is not yet operating (Loizou, 1972) and 5-HT synthesized at the periphery may well enter the brain and be catabolised into 5-HIAA. This would lead to a high 5-HIAA/5-HT ratio in the brain since this exogenous 5-HT would be completely converted into 5-HIAA in the non-serotoninergic cells. However, raphe lesions in the new born rat result in parallel decreases in 5-HT and 5-HIAA levels in the brain, suggesting that 5-HIAA is formed only from 5-HT synthesized inside serotoninergic neurons (Bourgoin et al., 1977b).

b) Karki et al. (1962) hypothesized that the binding of 5-HT in vesicular organelles inside serotoninergic neurons may be less efficient in immature neurons than in the adult brain. According to this hypothesis, newly synthesized 5-HT would be partially stored in vesicles and the most part catabolised into 5-HIAA in the brain of the new born rat. In this case, since reserpine inhibits completely the storage capacity in young as well as in adult brain, the difference concerning the catabolism of the indoleamine in both tissues should be no longer found after its administration. Indeed, reserpine treatment (5mg/kg i.p.) results in an acceleration of the conversion of 5-HT into 5-HIAA in tissues from adult and new born rats but the rate of 5-HT catabolism is still much higher in the neonatal brain (Bourgoin et al., 1977a). Therefore, a possible deficiency in 5-HT storage capacity of serotoninergic neurons cannot completely account for the high rate of 5-HIAA formation occurring in neonatal tissues.

c) The enzyme responsible for the catabolism of 5-HT is monoamine oxidase A (MAO A). It was therefore important to analyze the developmental changes of this enzymatic activity in the brain to seek a possible correlation with those of the 5-HIAA/5-HT ratio. The specific activity of MAO A, both in the brain stem and the fore brain, is significantly higher for the first three postnatal weeks than later on (Bourgoin et al., 1977a). Similar findings were reported recently by Blatchford et al. (1976) and Suzuki et al. (1976) about MAO A activity in various fore brain areas and spinal cord. Kinetic analyses show that the differences observed between adult and new born enzyme activities are exclusively due to changes in Vmax (fig. 7).

d) Previous observations by Bourgoin et al. (1974) suggested that the output of 5-HIAA from brain tissues is delayed in the new born rat. This could also account for the high 5-HIAA/5-HT ratio occurring in early life. Indeed, more recent studies by Bourgoin et al. (1977a) demonstrate that delayed elimination of 5-HIAA occurs in vitro in brain slices but not in vivo. This conclusion is in agreement with data of Bass and Lundborg (1976) showing that the mechanisms involved in the output of 5-HIAA from brain, although different in new born and adult rats, are equipotent at each age.

In conclusion, the high 5-HIAA/5-HT ratio found in new born rat brains seems to result only from the developmental changes in the enzyme activity responsible for the catabolism of 5-HT: MAO A. Since phospholipids play a major role in inducing the specificity of the different forms of MAO (Houslay and Tipton, 1973), it can be assumed that this high MAO A activity results from a particular phospholipid composition of mitochondrial membranes in the new born rat. Whatever it may be, the physiologic meaning of this high rate of 5-HT catabolism in early life is still unknown.

CONCLUSION

Serotoninergic neurons are among the first to differentiate in the rat central nervous system. The yellow fluorescence typical of 5-HT appears as early as the 13th day of gestation in midbrain raphe nuclei (Olson and Seiger, 1972).

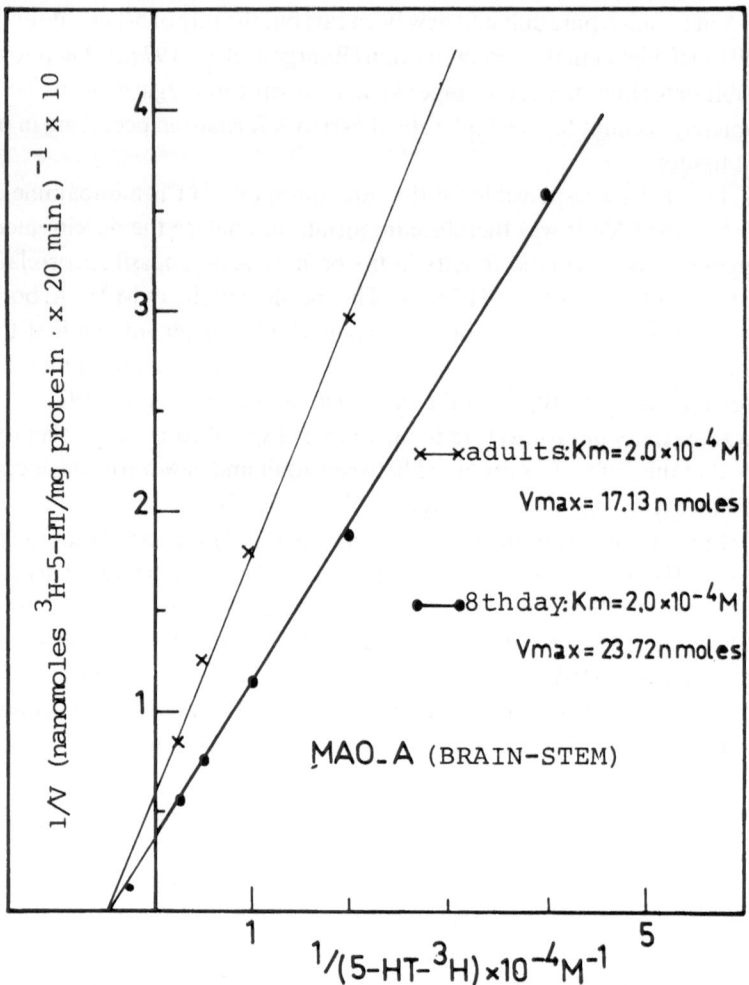

Fig. 7. Double reciprocal plots of MAO A activity in the brainstem of adult or 8-day-old rats versus the 5-HT concentration. MAO A activity was measured in whole homogenates with ^{3}H-5-HT as the substrate (Hamon et al., 1976). Each point is the mean of triplicate determinations. Vmax are expressed in nanomoles ^{3}H-5-HT catabolized per mg prot. and per 20 min.

According to Lauder and Bloom (1974), the differentiation of serotoninergic neurons occurs well before the end of cellular proliferation in the raphe nuclei, suggesting that differentiated 5-HT neurons are still able to multiply for a few days during fetal life.

Differentiated serotoninergic neurons are characterized by several metabolic peculiarities: 1) they synthesize 5-HT from tryptophan. The enzyme catalysing the first step, i.e. tryptophan hydroxylase, is present only in these neurons in brain: 2) they store 5-HT in specific organelles (synaptic vesicles) which protect the indoleamine from the intraneuronal catabolism by MAO; 3) owing to a specific permease with a high affinity for 5-HT they take up the indoleamine which has been released in the synaptic cleft.

In normal rats, the developmental changes in tryptophan hydroxylase activity, 5-HT storage capacity and 5-HT synaptosomal uptake in a given brain area are parallel (Bourgoin, 1976). Therefore, this suggests that all the specific properties of serotoninergic neurons are set up in a well coordinated way in the course of cell differentiation. However, several observations have shown that this differentiation pattern can be permanently altered by particular treatments during the perinatal period. Tonge (1974) reported that the perinatal administration of phencyclidine, methylamphetamine or chlorpromazine induced permanent alterations in the concentration and metabolism of 5-HT in various regions of the rat brain. Even more dramatic changes in serotoninergic innervation of the central nervous system are produced by the administration of 5,6 or 5,7-dihydroxytryptamine at birth. Indeed, the peripheral injection of these neurotoxic agents results in the specific degeneration of numerous 5-HT terminals in the spinal cord and forebrain areas (Sachs and Jonsson, 1975; Bourgoin and Hamon, unpublished observations). By contrast, these treatments lead to a supra-normal proliferation of 5-HT terminals in the brain stem (Sachs and Jonsson, 1975; Bourgoin and Hamon, unpublished observations). Little is known about the physiologic and behavioural consequences of these permanent alterations of serotoninergic innervation induced by perinatal pharmacologic treatments. The only obvious difference concerns the body weight of 5,7-dihydroxytryptamine treated rats which remains significantly lower than that of pair-aged animals for several weeks after the treatment. The behavioural alterations resulting from these pharmacologic treatments are probably very discrete since Adrien et al. (1976) have noted that the almost complete degeneration of 5-HT terminals in the forebrain areas after midbrain raphe lesion in early life does not alter the ontogeny of slow-wave and paradoxical sleep in the rat. Since 5-HT neurons play a major role in sleep mechanisms during the adult life (Jouvet, 1969) one may propose that compensatory mechanisms (supersensitivity of some 5-HT receptors?) develop in the brain after such manipulations in early life.

The present data show that the metabolism of 5-HT in brain during the immediate postnatal period is characterized by striking differences when compared to that occurring in the adult life. They consist mainly in the greater availability of tryptophan for brain tissue and an increased rate of catabolism of the indoleamine in the new borns. Since both peculiarities do not depend on 5-HT neurons but on the periphery (tryptophan availability) and on other cells in the brain (the lesion of serotoninergic neurons does not alter the development of MAO A in brain, Bourgoin et al., 1977a) this demonstrates that the metabolism of 5-HT largely depends on extraneuronal events even at birth. Therefore, alterations in the peripheral metabolism of tryptophan for instance during development (by food deprivation) may perturb the metabolism of 5-HT in the neonatal rat as pharmacologic treatments do. Since permanent changes in the metabolism of 5-HT in brain can be induced by pharmacologic manipulations during the perinatal period, why not suppose that these alterations in peripheral tryptophan, if too dramatic, would be also able to cause long lasting modifications in central 5-HT metabolism? However it may be, data obtained with pharmacologic treatment (Tonge, 1974) underline the critical role of the early life period in the maturation of serotoninergic neurons on the central nervous system.

SUMMARY

In the rat, most of axonal growth and synaptogenesis of serotoninergic neuronal systems in the brain occur after birth. This offers an opportunity to study the metabolic changes of a neurotransmitter, serotonin (5-HT), for the maturation of a given group of neurons in the central nervous system.

The developmental changes of various biochemical markers of 5-HT neurons: endogenous 5-HT levels, synaptosomal high affinity 5-HT uptake, tryptophan hydroxylase activity, indicate that the maturation of these neurons is faster in the brain stem than in forebrain areas. The values of these markers reach adult levels in less than 1 month after birth in the brain stem whereas the corresponding time for forebrain areas is about 5-6 weeks.

Although the neuronal activity of 5-HT systems is probably low in the new born rat, the turnover of 5-HT is relatively more rapid than in adults. This results from two characteristics: 1) the concentration of the precursor of 5-HT synthesis in brain, tryptophan, is higher in the new born than in the adult rat; 2) the activity of the enzyme responsible for the catabolism of 5-HT, the A form of monoamineoxidase in the brain, is significantly higher in neonates than in adults.

The first characteristic, which concerns the concentration of tryptophan in brain is the consequence of differences in peripheral tryptophan in adult and new born rats. In adult rats, tryptophan in serum is almost totally bound to serum albumin whereas it is more than 90% free in the serum of neonates. Therefore, drugs known to alter the metabolism of 5-HT in brain by displacing bound tryptophan from serum albumin in adult animals (such as salicylate) are ineffective in new borns.

The second characteristic leads one to wonder about the physiologic significance of the high rate of 5-HT catabolism in brain during ontogenesis. As suggested previously, this could be related to a postulated critical role of 5-HT as a chemotactic agent involved in the normal growth of various neuronal systems. This point is discussed on the basis of behavioural data obtained in rats whose serotoninergic neurons were lesioned in early life.

REFERENCES

Adrien, J., Bourgoin, S. and Hamon, M.: Midbrain raphe lesion in the new-born rat. I. Neurophysiological aspects of sleep. Brain Res. (in press, 1976).

Agrawal, H.C. and Davison, A.N.: Myelination and amino-acid imbalance in the developing brain; in Biochemistry of the developing brain," ed. Himwich, W.A.; Dekker, M., vol. 1, pp. 143-186 (N.Y., 1973).

Anden, N.E.: Effect of acute axotomy (spinal cord transection) on the turnover of 5-hydroxytryptamine; in "Serotonin-New-Vistas," eds. Costa, E., Gessa, G.l. & Sandler, M., Adv. Biochem. Psychopharmacol., 10, 35-43 (Raven Press, N.Y., 1974).

Bass, N.H. and Lundborg, P.: Transport mechanisms in the cerebrospinal fluid system for removal of acid metabolites from developing brain; in "Transport phenomena in the nervous system. Physiological and pathological aspects," eds. Levi, G., Battistin, L. and Lajtha, A. Adv. Exp. Med. Biol., 69: 31-40 (Plenum Press, N.Y., 1976).

Bennett, J.P. Jr. and Snyder, S.H.: Serotonin and lysergic acid diethylamide binding in rat brain membranes: relationship to postsynaptic serotonin receptors. Mol. Pharmacol., 12, 373-389 (1976).

Blatchford, D., Holzbauer, M., Grahame-Smith, D.G. and Youdim, M.B.H.: Ontogenesis of enzyme systems deaminating different monoamines. Brit. J. Pharmacol., 57: 279.293 (1976).

Bourgoin, S.: "Evolution du métabolisme cérébral de la sérotonine chez le rat au cours du développment." Thèse de Doctorat es sciences, Université Paris VII, pp. 1-303 (1976).

Bourgoin, S., Artaud, F., Adrien, J., Hery, F., Glowinski, J. and Hamon, M.: 5-hydroxytryptamine catabolism in the new-born rat brain. J. Neurochem. 28: 415-422 (1977a).

Bourgoin, S., Artaud, F., Enjalbert, A., Hery, F., Glowinski, J. and Hamon, M.: Acute effects of the stimulation or blockade of 5-HT receptors on 5-HT metabolism in the rat brain during ontogenesis. J. Pharmacol. exp. Ther. in press (1976).

Bourgoin, S., Enjalbert, A., Adrien, J., Hery, F. and Hamon, M.: Midbrain raphe lesion in the new-born rat. II. Biochemical alterations in serotoninergic innervation. Brain Res. 127 no. 1: 11-126 (1977b).

Bourgoin, S., Faivre-Bauman, A., Benda, P., Glowinski, J. and Hamon, M.: Plasma tryptophan and 5-HT metabolism in the CNS of the new-born rat. J. Neurochem., 23: 319-327 (1974).

Bourgoin, S., Faivre-Bauman, A., Hery, F., Ternaux, J.P. and Hamon, M.: Characteristics of tryptophan binding in the serum of the new-born rat. Biol. Neonate. 31: 141-154 (1977c).

Bourgoin, S., Hery, F., Ternaux, J.P. and Hamon, M.: Effects of benzodiazepines on the binding of tryptophan in serum. Consequences on 5-hydroxyindoles concentrations in the rat brain. Psychopharmacol. Comm., 1: 209-216 (1975).

Curzon, G. and Knott, P.J.: Effects on plasma and brain tryptophan in the rat of drugs and hormones that influence the concentration of unesterified fatty acid in the plasma. Brit. J. Pharmacol., 50: 197-204 (1974).

Davis, J.M. and Himwich, W.A.: Amino acids and proteins of developing mammalian brain; in "Biochemistry of the developing brain", ed. Himwich, W.A., Dekker, M., Vol. 1, pp. 55-110 (N.Y., 1973).

Farnebo, L.O. and Hamberger, B.: Regulation of (^{3}H)-5-hydroxytryptamine release from rat brain slices. J. Pharm. Pharmacol., 26: 642-643 (1974).

Gallager, D.W. and Aghajanian, G.K.: Effects of chlorimipramine and lysergic acid diethylamide on efflux of precursor formed ^{3}H-serotonin: correlation with serotoninergic impulse flow. J. Pharmacol. exp. Ther., 193: 785-795 (1975).

Gerschenfeld, H.M. and Paupardin-Tritsch, D.: Ionic mechanisms and receptor properties underlying the responses of molluscan neurons to 5-hydroxytryptamine. J. Physiol. (Lond.), 243: 427-456 (1974).

Gessa, G.L. and Tagliamonte, A.: Serum free tryptophan control of brain concentrations of tryptophan and of synthesis of 5-hydroxytryptamine; in "Aromatic amino acids in the brain," Ciba Foundation Symposium 22, Elsevier, Excerpta Medica, North Holland pp. 207-216 (1974).

Hamon, M., Bourgoin, S., Enjalbert, A., Bockaert, J., Hery, F., Ternaux, J.P. and Glowinski, J.: The effects of quipazine on 5-HT metabolism in the rat brain. Naunyn Schmiedeberg's Arch. Pharmacol., 294: 99-108 (1976).

Hamon, M., Bourgoin, S., Jagger, J. and Glowinski. J.: Effects of LSD on synthesis and release of 5-HT in rat brain slices. Brain Res., 69: 265-280 (1974a).

Hamon, M., Bourgoin, S., Morot-Gaudry, Y., Hery, F. and Glowinski, J.: Role of active transport of tryptophan in the control of 5-HT biosynthesis; in "Serotonin-New Vistas," eds. Costa, E., Gessa, G.L. & Sandler, M., Adv. Biochem. Psychopharmacol., 11: 153-162 (Raven Press, N.Y., 1974b).

Hamon, M. and Glowinski, J.: Regulation of serotonin synthesis. Life Sci., 15: 1533-1548 (1974).

Hery, F., Rouer, E. and Glowinski, J.: Daily variations of serotonin metabolism in the rat brain. Brain Res., 43: 445-465 (1972).

Hery, F., Rouer, E., Kan, J.P. and Glowinski, J.: The major role of tryptophan active transport in diurnal variations of 5-hydroxy-tryptamine synthesis in the rat brain; in "Serotonin-New Vistas" eds. Costa, E., Gessa, G.L. & Sandler, M., Adv. Biochem. Psychopharmacol., 11: 163-168 (Raven Press, N.Y., 1974).

Houslay, M.D. and Tipton, K.F.: The nature of electrophoretically separable multiple forms of rat liver monoamine oxidase. Biochem. J., 135: 173-186 (1973).

Jouvet, M.: Biogenic amines and the states of sleep. Science, 163: 32-41 (1969).

Karki, N., Kuntzman, R. and Brodie, B.B.: Storage, synthesis and metabolism of monoamines in the developing brain. J. Neurochem., 9: 53-58 (1962).

Lajtha, A. and Toth, J.: Perinatal changes in the free amino-acid pool of the brain of mice. Brain Res., 55: 238-241 (1973).

Lauder, J.M. and Bloom, F.E.: Ontogeny of monoamine neurons in the locus coeruleus, raphe nuclei and substantia nigra of the rat. 1. Cell differentiation. J. Comp. Neurol., 155: 469-482 (1974).

Loizou, L.A.: The postnatal ontogeny of monoamine containing neurons in the central nervous system of the albino rat. Brain Res., 40: 300-311 (1972).

McArthur, J.N. and Dawkins, P.D.: The effect of sodium salicylate on the binding of L-tryptophan to serum proteins. J. Pharm. Pharmacol., 21: 744-750 (1969).

Miyoshi, K., Saijo, K., Kotani, A., Kashiwagi, T. and Kawai, H.: Characteristic properties of fetal human (alb. F) in isomerization equilibrium. Tokushima J. exp. Med., 13: 121-128 (1966).

Olson, L. and Seiger, A.: Early prenatal ontogeny of central monoamine neurons in the rat: fluorescence histochemical observations. Z. Anat. Entwickl. Gesh. 137: 301-316 (1972).

Sachs, C. and Johnsson, G.: 5,7-dihydroxytryptamine induced changes in the postnatal development of central 5-hydroxytryptamine neurons. Med. Biol., 53: 156-164 (1975).

Seiger, A. and Olson, L.: Late prenatal ontogeny of central monoamine neurons in the rat: fluorescence histochemical observations. Z. Anat. Entwickl. Gesh., 140: 281-318 (1973).

Stewardt, R.M., Growdon, J.H., Cancian, D. and Baldessarini, R.J.: 5-hydroxytryptophan-induced myoclonus: increased sensitivity to serotonin after intracranial 5,7-dihydroxytryptamine in the adult rat. Neuropharmacol. 15: 449-455 (1976).

Suzuki, O., Noguchi, E. and Yagi, K.: Postnatal development of 5-hydroxytryptamine and monoamine oxidase in rat spinal cord. J. Neurochem., 27: 319-320 (1976).

Tagliamonte, A., Tagliamonte, P., Perez-Cruet, J., Stern, S. and Gessa, G.L.: Effect of psychotropic drugs on tryptophan concentration in the rat brain. J. Pharmacol. exp. Ther., 177: 475-480 (1971).

Tonge, S.R.: Permanent alterations in 5-hydroxytryptamine metabolism in discrete areas of rat brain following exposure to drugs during the period of development. Life Sci., 15: 245-249 (1974).

Trulson, M.E., Rose, C.A. and Jacobs, B.L.: Behavioral evidence for the stimulation of CNS serotonin receptors by high doses of LSD. Psychopharmacol. Comm., 2: 149-164 (1976).

Tuilie, M. and Lardinois, R.: The binding of unconjugated bilirubin by human sera and purified albumins. Biol. Neonate, 21: 447-462 (1972).

Tyce, G.M., Flock, E.V. and Owen, C.A. Jr.: Tryptophan metabolism in the brain of the developing rat: in "Progress in Brain Research. The developing brain," eds. Himwich, W.A. & Himwich, H.E., vol. 9: 198-203 (Elsevier, Amsterdam, 1964).

Vahvelainen, M.L. and Oja, S.J.: Kinetic of influx of phenylalanine, tyrosine, tryptophan, histidine and leucine into slices of brain cortex from adult and 7-day-old rats. Brain Res., 40: 477-488 (1972).

Von Hungen, K., Roberts, S. and Hill, D.F.: Serotonin-sensitive adenylate cyclase activity in immature rat brain. Brain Res., 84: 257-267 (1975).

Winick, A. and Noble, A.: Quantitative changes in DNA, RNA and protein during prenatal and postnatal growth in the rat. Develop. Biol., 12: 451-466 (1965).

CHAPTER SEVENTEEN

THE DEVELOPMENT OF CENTRAL CATECHOLAMINE NEURON SYSTEMS: SOME IMPLICATIONS FOR DISEASE

ROBERT Y. MOORE. M.D.. PH.D.

Two major groups of catecholamine neuron systems are present in the mammalian brain. One of these, the dopamine systems, have cell bodies which are located principally in the rostral mesencephalon and in the periventricular region of the diencephalon. The axonal projections of these systems are topographically organized and terminate in two major zones. The largest projection arises from dopamine cell bodies of the ventral tegmental area and the adjacent substantia nigra, pars compacta. It terminates in the neostriatal nuclei, in basal forebrain areas such as the septum, amygdala, and olfactory nuclei and in restricted portions of the cerebral neocortex. The dopamine cell bodies of the diencephalic periventricular system have axonal projections terminating predominantly in the median eminence and in the neuro-intermediate lobe of the pituitary. These systems appear phylogenetically stable and have the same organization in man (Nobin and Björklund, 1973; Olson, Boreus and Seiger, 1973) as in lower mammals (Moore, Björklund and Stenevi, 1974).

The second major catecholamine neuron systems are those which employ noradrenaline as a neurotransmitter. These systems have cell bodies located entirely in the pons and medulla and appear distinguishable into two major components. The first, and largest of these, is the locus coeruleus system. The locus coeruleus is a pontine nucleus composed entirely of noradrenaline neurons. Projections from locus coeruleus neurons distribute widely and diffusely throughout the neuraxis. The current evidence is that the locus coeruleus projects to the spinal cord gray matter, to the cerebellar cortex, to wide areas of the brainstem including specific nuclei such as the cochlear nuclei, to the hypothalamus, to the thalamus, to basal forebrain areas and to the entire neocortex (cf. Moore et al., 1974, for review). And, although definitive data are not available, it appears that a single locus coeruleus neuron may project to most, if not all, of these areas. As indicated above, the projection also appears to be diffuse and not topographic, in contrast to the dopamine systems. The second noradrenaline system, the lateral tegmental system, has cell bodies located in the lateral part of the reticular formation in a nearly continuous line extending from the rostral pons to caudal medulla. Within this system, however, there are relatively large accumulations of neurons in the rostral pons and in the caudal medulla. Both seem to have ascending and descending projections but the medullary neurons predominantly innervate the spinal cord and brainstem whereas the pontine neurons predominantly project to the upper brainstem, hypothalamus and basal forebrain. This system differs from the locus coeruleus system in not contributing to the noradrenaline innervation of the cerebral and cerebellar cortices.

Both the dopamine and noradrenaline system neurons are formed early in embryonic development (Lauder and Bloom 1974; Olson and Seiger, 1972) in the rat and, although data on dates of neuron origin are not available, the same appears to hold true for man (Olson et al., 1973; Nobin and Björklund, 1973). Shortly after the neurons are formed they produce axons and these along with the serotonin neuron projections, are among the earliest axonal projections to be formed in the mammalian brain (Olson and Seiger, 1972, 1973). Two examples of this will be discussed. In both the rat and human brain the nigrostriatal projection is evident very early in development, appearing very soon after the cells of origin of the projection have been formed (Olson and Seiger, 1972, 1973; Olson et al., 1973; Nobin and Björklund, 1973). Similarly, locus coeruleus axons are among the first to reach many of the areas they innervate. In the hippocampal formation, for example, locus coeruleus axons are present in Ammon's horn prenatally, prior to innervation by any other afferent system (Smith and Moore, 1976). The process of innervation of the structure is orderly, however, and the area dentata, which is formed almost exclusively postnatally in the rat, is innervated as the neurons of the structure differentiate (Smith and Moore, 1976). The pattern of locus coeruleus axon innervation of the hippocampal formation appears to follow precisely the pattern of cell formation and differentiation. The same pattern also occurs in the cerebral neocortex (Levitt and Moore, unpublished observations). Locus coeruleus noradrenaline axons appear very early in the neocortical plate and the pattern of development of the innervation quite precisely follows the development of the structure.

The question which arises from these observations is what significance they may have for disease and, particularly for the effects of brain damage in infancy and childhood? This can best be answered by reviewing some further data concerning the response of catecholamine neurons to injury and their capacity for plasticity in both the adult and developing animal. A series of studies (cf. Moore et al., 1974) has shown that noradrenaline neurons have a remarkable capacity for showing regenerative responses following injury to the nervous system. These responses may occur in two situations, one in which the axon of the noradrenaline neurons itself is injured and one in which the axon of the noradrenaline neurons itself is injured and one in which the noradrenaline neuron responds to injury to other neurons. In the first case, the axon responds by vigorous regenerative sprouting which is capable of forming an extensive terminal plexus producing functional contacts. This has been studied most extensively in the adult nervous system (cf. Moore et al., 1974, for review) but experiments on the developing nervous system have shown similar phenomena (Jonsson, Pycock, Fuxe and Sachs, 1974). The second case is, perhaps, the more interesting. In this the noradrenaline neuron responds to removal (denervation) of other afferents to the same nucleus or area by collateral sprouting and reinnervation of denervated post-synaptic sites. Dopamine neurons exhibit regenerative sprouting but have not been shown to form extensive terminal plexus nor to undergo collateral sprouting. Thus, the data now available indicate that noradrenaline neuron systems are a unique component of the catecholamine neuron systems in their formation of diffuse, non-topographic projection systems and in exhibiting vigorous and extensive responses to injury. This is of particular interest in that recent ultrastructural studies of noradrenaline terminals in cerebral cortex (Descarries, Watkins and Lapierre, 1977) have demonstrated that only about 5% of all noradrenaline terminals in cortex make conventional synaptic contacts with post-synaptic elements. Several interpretations of this are possible but it is intriguing to speculate that this indicates these terminals are poorly specified in respect to the elements with which they make synaptic contacts. Consequently, they could readily respond to either functional changes or injury by readjusting synaptic relationships.

Regardless of whether the speculation offered above is correct, it is clear that the morphology of the noradrenaline neuron systems and their demonstrated capacity to respond to injury have implications for recovery of function in the human nervous system. This would appear to be true particularly in the developmental period because of their early appearance, ubiquitous distribution and apparent lack of specificity. The role that catecholamine neurons play in the functional response to brain injury in infancy and childhood is unknown. It is interesting, however, that many of the drugs used to modify behavior in the brain injured child are ones which either mimic or alter the actions of catecholamine neurons.

REFERENCES

Descarries, L., Watkins, K.C. and Lapierre, Y. Noradrenergic axon terminals in the cerebral cortex of rat. III. Topometric ultrastructural analysis. Brain Res. In press (1977).

Johnsson, G., Pycock, Ch., Fuxe, K. and Sachs, C. Changes in the development of central noradrenaline neurones following neonatal administration of 6-hydroxydopamine. J. Neurochem. 22: 419-426 (1974).

Lauder, J.M. and Bloom, F.E. Ontogeny of monoamine neurons in the locus coeruleus, raphe nuclei, and substantia nigra of the rat. I. Cell differentiation. J. Comp. Neurol. 155: 469-482 (1974).

Moore, R.Y., Björklund, A. and Stenevi, U. Growth and plasticity of adrenergic neurons. In Schmitt, F.O. and Worden, F.G. (Eds.). The Neurosciences – Third Study Program, pp. 961-978. (M.I.T. Press, Cambridge, 1974a).

Nobin, A. and Björklund, A. Topography of the monoamine neurons systems in the human brain as revealed in fetuses. Acta physiol. scand. Suppl. 388: 1-40, (1973).

Olson, L. and Seiger, A. Early prenatal ontogeny of central monoamine neurons in the rat: Fluorescence histochemical observations. A. Anat. Entwickl.-Gesch. 137: 301-316 (1972).

Olson, L. and Seiger, A. Late prenatal ontogeny of central monoamine neurons in the rat: Fluorescence histochemical observations. A. Anat. Entwickl.-Gesch. 140: 281-318 (1973).

Olson, L., Boreus, L.O. and Seiger, A. Histochemical demonstration and mapping of 5-hydroxytryptamine- and catecholamine-containing neuron systems in the developing human fetal brain. Z. Anat. Entwickl.-Gesch. 139: 259-282 (1973).

Smith, R.L. and Moore, R.Y. Development of the noradrenergic innervation of the hippocampus, Neuroscience Abstracts 2: 228 (1976).

CHAPTER EIGHTEEN

DEVELOPMENTAL RETARDATION IN RAT BRAIN AFTER PERINATAL SEIZURES*

FRED PLUM AND CLAUDE WASTERLAIN

In human beings, generalized seizures occurring during the first two years of life imply a high risk that the child will later show neurologic abnormalities or learning problems (Rodin, 1968). Falconer (1974) believes that repetitive convulsions or an episode of severe status epilepticus in this age group poses an especial danger, and the experience of others leads to the same view (Aicardi and Chevrie, 1970).

Generalized convulsions produce significant changes in the blood respiratory gases as well as striking chemical changes in the brain of the adult

Table 1. Chemical effects of seizures in adult animals.

	Control	Single ES Unparalyzed	Paralyzed Ventilated
PaO_2mm Hg	100-120	50-60	100-60
$PaCO_2$mm Hg	38-42	45-55	38-42
Protein Synthesis[1]	100%	76%** immed. 103% 10 min.	
Polyribosome/ Ribosome Ratio	100%	83%**[2] immed. 117%** 30 min.	100%[2]
Phosphocreatine[4]	4.05 ± 0.11	3.52 ± 0.10*	3.71 ± 0.21
ATP[4]**	2.74 ± 0.01	2.56 ± 0.04**	2.70 ± 0.05
ADP[4]***	0.43 ± 0.02	0.49 ± 0.03	0.45 ± 0.04
AMP[4]***	0.028 ± 0.002	0.080 ± 0.014**	0.034± 0.008

* Value different from control with P < 0.05:
** P < 0.01 (1) Cotman, 1971; (2) MacInness, 1970; (3) Wasterlain, 1972; (4) Duffy, 1975;
*** values expressed as m mol/kg net weight ± S.E.M.

experimental animal. Table 1 lists these changes and it will be noted that most of the abnormalities accompanying cerebral seizures are accentuated when a generalized convulsion occurs. (Duffy, Howse and Plum, 1975). One finds a

* These studies were aided by Grant # NSO 4928 from NINCDS, NIH.

fall in the high energy compounds of phosphocreatine and ATP, a substantial rise in lactic acid in the tissue, inhibition of protein synthesis (Cotman, et al., 1971; Wasterlain, 1974) and alterations in cerebral ribosomes (MacInness, McConkey and Schlesinger, 1970; Wasterlain, 1972). Following brief episodes of repetitive seizures, the histologic anatomy of the brain remains undamaged (Wasterlain, 1976), but with experimental status epilepticus lasting an hour or more in the baboon, Meldrum and Brierley (1973) find anoxic cell changes in the hippocampal formation.

The inhibition of protein synthesis and change in ribosomes that follow generalized seizures last only a few minutes or so, not long enough to explain the memory impairment of epilepsy in the adult. During periods of rapid brain growth, however, impaired protein synthesis lasting even a short time would pose a greater risk, since a relatively small inhibition of protein synthesis could substantially inhibit the formation of DNA and, consequently, cell division. (Enger and Tobey, 1972).

These considerations led us to study the influence of epileptic seizures on brain and body growth in neonatal animals (Wasterlain and Plum, 1973a). We used the rat for convenience, and in a subsequent experiment examined the influence of such seizures on the animals behavioral development (Wasterlain and Plum, 1973b). In this chapter we will describe the effects of single seizures on the brain of the newborn. In the next chapter, Wasterlain will say what happens to the newborn's brain chemistry in status epilepticus. He will also describe a form of treatment that sharply reduces mortality from status epilepticus in the experimental neonate (Wasterlain and Duffy, 1976).

We studied a large number of newborn rats taking litter mates paired for sex and weight as experimental subjects and controls. The controls were equally handled except for the seizure. The animals were weighed on alternate days, weaned at 22 days and killed at 30 days of life. If an animal died prematurely, both he and his litter pair were omitted from the experiment.

The effect of seizures was studied on three stages of rat brain development (Winick and Noble, 1965). One group of 44 animals received a single daily electroconvulsive stimulus (ECS) consisting of 150 v, one sec. AC delivered to the ears between days 2 and 11. This time coincides with the maximal postnatal burst of mitotic activity in rat brain. A second group of 44 animals received one daily ECS between days 2 and 11 of life, which is the time when cell division is slowing down, but cell growth and myelination is at a maximum. A third group of 48 animals received one ECS daily between days 19 and 28 of life, when mitotic activity is low and myelination well advanced.

A minimum of four control and experimental animals in each group were killed by perfusion fixation for histologic examination of the brain. No differences between ECS animals and controls could be seen by this technique.

In the remaining animals, the cerebral hemispheres and diencephalon above the colliculi were removed, weighed, and analyzed for total protein, DNA, RNA, and cholesterol by standard procedures described elsewhere (Lowry, 1951; Zlatkis, Zak and Bogle, 1953; Webb and Levy, 1955; Enesco and Leblond, 1962; Winick and Noble, 1965; and Munro and Fleck, 1966).

ECS in these animals resulted in a tonic-clonic seizure lasting 15-20 secs., and increasing in intensity as the animals grew older. Following the onset of seizures the animals lagged in growth (table 2). Body weight was reduced

Table 2. Effects of ECS on brain development.

	Group I Control	Exper.	Group II Control	Exper.	Group III Control	Exper.
Body weight	78.96	61.47*	58.80	48.65*	61.92	52.93*
gm ± SE	± 2.86	± 4.34	± 5.25	± 3.83	± 3.58	± 2.27
Brain weight	1.150	0.985*	1.037	0.975*	1.085	1.059
gm ± SE	± 0.019	± 0.029	± 0.020	± 0.024	± 0.013	
Total brain	1599.	1377.*	1374.	1356.	1366.	1387.
DNA μg ± SE	±62.	±47.	±48.	±73.	±26.	±30.
Total brain	2606.	2227.*	2378.	2188.**	2309.	2252.
RNA μg ± SE	±59.	±54.	±77.	±55.	±64.	±73.
Brain protein	136.9	119.4*	118.5	104.1*	117.	116.8
mg SE	± 3.0	± 2.5	± 2.7	± 1.4	± 2.5	± 2.8
Brain cholesterol	15.36	12.92*	13.22	12.96	14.02	13.75
mg SE	± 0.34	± 0.72	± 0.43	± 2.11	± 0.41	± 0.39

* Difference P < 0.01
** Difference P < 0.05

compared to controls by 22% in group 1 and by 15% in the other two groups of animals. Brain weight of animals receiving seizures was affected almost as much as body weight in the first 2 groups being 14% less than controls in group 1 and 8% less in group 2. Brain weight did not fall in group 3, despite the fall in body weight and the fact that seizures appeared, if anything, more intense in these older animals.

Neonatal seizures had important effects on brain chemistries (fig. 1). In group 1 animals, all the variables, i.e., DNA, RNA, protein and cholesterol declined by an amount proportional to the fall in brain weight. In group 2, DNA was normal, but RNA and total protein both were less than in controls. In group 3 animals, none of the chemical variables differed significantly from the litter mate controls, who received no seizures.

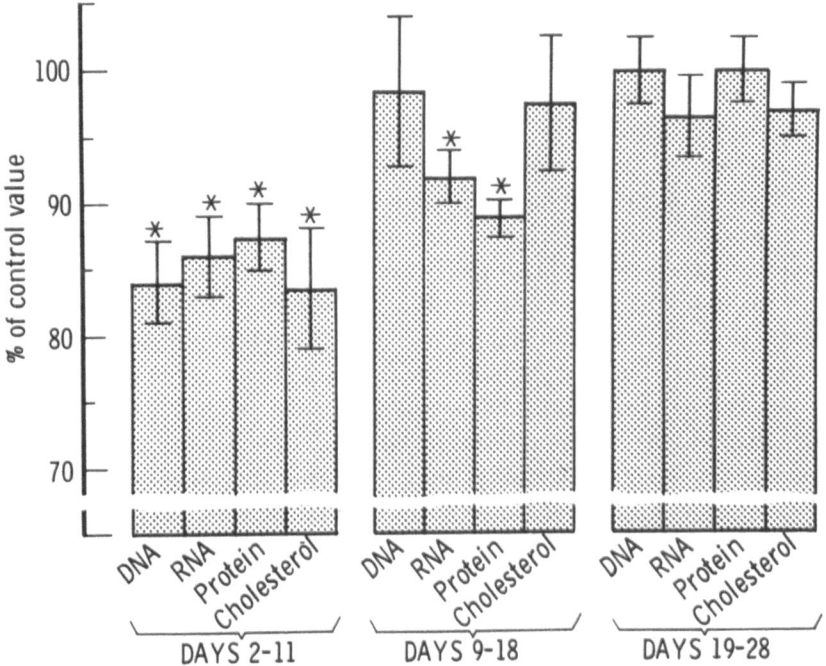

Fig. 1. Changes in brain chemistries compared to control litter mates in neonatal rats receiving one daily ECS during the time indicated. All starred values significant at least P < 0.05.

We interpret these changes as implying that seizures during the neonatal period interfered with brain cell division (total DNA) and consequently resulted in a reduced number of total cells at 30 days. The fewer number of brain cells, in turn, were reflected by reduced amounts of cellular components. During the suckling period, the second period of experimental study, daily seizures interfered with the development of a brain with a normal complement of cells, by impeding connectivity, which caused a lowered brain weight and protein content of the tissue. Once past the stage of rapid development, however, seizures no longer interfered with brain size or components, even though body size in this third group of animals was smaller than litter mates.

Before discussing the implications of these experiments, it may be useful to mention certain precautions that were taken against artifact. For one thing, histologic examination showed very little in the way of abnormality and specifically no evidence of cerebral infarction, hemorrhage or electrical damage in experimental animals. Furthermore, the electrical stimulus was not the

significant variable since, in a later experiment, seizures induced by convulsant gases or analeptics induced similar chemical changes in brain. We
cannot yet tell to what degree anoxia or circulatory changes induced by the
seizures were a factor in the outcome. Placing animals in 100% oxygen during
and after the induced seizures had no influence on the findings, however.

The most cogent question is whether these changes in brain are specific to
the epileptic insult or merely reflect the effect of malnutrition and weight loss.
In fact, these changes in brain weight are not what one would expect from
mere malnutrition. The research of Winick (Winick and Noble, 1966) and
others (Howard, 1965; Dobbing and Sands, 1971) indicates that during neonatal starvation, body weight falls drastically, yet brain weight declines com-
paratively less so that the normal ratio between brain weight and body weight
rises. By contrast, following seizures in these animals with still growing
brains, brain weight fell almost as much proportionately as body weight and
in some instances the usual ratio of brain weight to body weight was reversed.

In a subsequent experiment. Wasterlain has directly compared brain
weights at 30 days between animals with seizures and starved litter mates with
no seizures, but the same body weight. Despite similar weights, the brains of
the animals with seizures during the first 10 days of life, weighed 5% less and
had 5% less DNA than their equal-weight litter mates. Both values are statistically significant at the 5% level (Wasterlain, 1976b).

The precise mechanism of the fall in DNA, RNA and brain weight secondary to infantile seizures, is not known. One may ask what significance the
chemical findings in the brain carry for the development of the organism and
to what degree the developmental stages of the rat can be regarded as comparable to stages in human development.

First, to look at the developmental effects on the rat. In a study subsequent
to the above, we took pairs of newborn animals and subjected one of the pair
to two daily ECS-induced convulsions between days 2 and 11 of life. In
addition to non-electrically stimulated controls, we also gave subconvulsive
electrical stimulation to 15 animals to determine if the current, per se, had any
adverse effect. The animals were tested daily from days 2 to 21 of life for the
appearance of a variety of developmental responses, including the righting
reflex, free fall righting, negative geotaxis, cliff avoidance, auditory startle
response, visual placing and palmar grasp. From days 6 to 24 they also were
tested for swimming ability. All animals were weaned at 22 days of age and
killed for analysis at 30 days. Table 3 indicates the significant changes. The
brains of experimental animals were smaller than the controls, while "current
controls" were similar to full controls in all respects. The experimental ani-

mals showed significant retardation in a majority of the developmental mile-stones, as indicated (table 3).

What relevance has ECS-induced brain damage in rats to brain damage following neonatal seizures in humans? One can only speculate. The rat brain is active mitotically for the first 20 days of life, but the second half of this period is perhaps more like the early postnatal development of human brains when the rate of cell division is moderately active but myelination is very active (Dobbing, 1970). Human cerebellar DNA increases fourfold between

Table 3. Age of developmental milestones in days with neonatal seizures.

	Epilepsy	Controls
Brain weight	1.082	0.979*
Startle response	11.7	10.8*
Visual placing	17.8	16.5*
Free-fall righting	18.7	16.4*
Swimming	13.5	12.1*
Adult swimming	21.0	18.2*

 * P < 0.01
 ** P < 0.02

birth and 1 year of age and cerebral DNA rises threefold (Winick, Rosso, and Waterlow, 1970). One doesn't know whether neurons or glia are predom-inantly involved in the postnatal cell divisions in either rat or man, but the question is of secondary concern. Both cell types undoubtedly influence mature brain development, and the adverse effects of neonatal cretinism or phenylketonuria leave little doubt of the risk of temporarily exposing the early developing brain to noxious influences which selectively impair development.

Direct evidence is lacking on whether seizures during the first year of life impair the growth of the human brain, but several investigators infer that neonatal seizures carry an unfavorable prognosis. As already mentioned, we know in man that the onset of seizures before the age of 1 year correlates with learning difficulties in school, although many such patients no longer have seizures when they reach school age (Rodin, 1968). Similarly, Aicardi and Chevrie (1970) demonstrated that severe febrile seizures sometimes result in permanent and prominent intellectual retardation. But perhaps the most tell-ing study of the adverse effects of early life seizures comes from the British Epilepsy Association (Butler, 1972). That group reported that children with a history of epilepsy performed less well than the general population, both in school and on testing. The difference held true even when the child lacked any evidence of what would ordinarily be regarded as mental retardation and had no detectable evidence of organic brain illness.

We recognize that major differences between the human and the experimental situation force one to temper any conclusions from this animal study. The differences on the human side include uncertainty from the clinical reports of the nature, cause, timing and complications of the seizures that preceded learning disability. On the animal side, the differences involve the slightly greater immaturity at birth of the rat brain, as well as the fact that some non-specific effects of ECS and handling may have contributed to the adverse effects of epilepsy in our animals. Nevertheless, the experiments strongly support the view that generalized convulsions pose a considerable threat to the development of the immature brain and they imply the need to treat vigorously human infants with generalized seizures of any cause.

REFERENCES

Aicardi, J. and Chevrie, J.J.: Convulsive status epilepticus in infants and children: A study of 239 cases. Epilepsia 11: 187-197 (1970).

Butler, N.R.: Yhe pattern of convulsive disorders in Britain's children. The Second Kenneth Gibson Chevrie, Lecture, University of Bristol, Bristol, England (1972).

Cotman, C.W., Banker, G., Zornetzer, S.F. and McGaugh, J.L.: Electroshock effects on brain protein synthesis; Relation to brain seizures and retrograde amnesia. Science 173: 454-456 (1971).

Dobbing, J.: Undernutrition and the developing brain: The relevance of animal models to the human problem. Am. J. Dis. Child. 120: 411-415 (1970).

Dobbing, J. and Sands, J.: Vulnerability of developing brain: IX The effect of nutritional growth retardation on the timing of the brain growth spurt. Biol. Neonate 19: 363-378 (1971).

Duffy, T.E., Howse, D.C. and Plum, F.: Cerebral energy metabolism during experimental status epilepticus. J. Neurochem. 24: 925-934 (1975).

Enesco, M. and LeBlond, C.P.: Increase in cell number as a factor of growth in the organs of the young male rat. J. Embryol. Exp. Morphol. 10: 530-562 (1962).

Enger, M.D., and Tobey, R.A.: Effects of isoleucine deficiency on nucleic acid and protein metabolism in cultured Chinese hamster cell: Continued ribonucleic acid and protein synthesis in the absence of deoxyribonucleic acid synthesis. Biochemistry 11: 269-277 (1972).

Enger, M.D. and Tobey, R.A.: Effects of isoleucine deficiency on nucleic acid and protein metabolism in cultured Chinese hamster cell: Continued ribonucleic acid and protein synthesis in the absence of deoxyribonucleic acid and on DNA, RNA and cholesterol contents of the brain and liver in infant mice. J. Neurochem. 12: 181-191 (1965).

Lowry, O.H., et al.: Protein measurement with the Folin phenol reagent. J. Biol. Chem. 193: 265-275 (1957).

MacInness, J.W., McConkey, E.H. and Schlesinger, K.: Changes in brain polyribosomes following an electroconvulsive seizure. J. Neurochem. 17: 457-460 (1970).

Meldrum, B.S. and Brierley, J.B.: Prolonged epileptic seizures in primates: Ischemic cell change and its relation to ictal physiological events. Arch. Neurol. 28: 10-17 (1973).

Munro, H.N. and Fleck, A.: The determination of nucleic acids, in Glick, D. (ed.). Methods in Biochemical Research, pp. 113-176 (New York, Interscience, 1966, vol 14).

Rodin, E.A.: The prognosis of patients with epilepsy, pp. 277-313 Charles C Thomas, Springfield, Ill., 1968.

Wasterlain, C.G.: Breakdown of brain polysomes in status epilepticus. Brain Res. 39: 278-284 (1972).

Wasterlain, C.G.: Inhibition of brain protein synthesis by epileptic seizures without motor manifestations. Neurology 24: 175-180 (1974).

Wasterlain, C.G.: Effects of neonatal status epilepticus on rat brain development. Neurology 26: 975-986 (1976).

Wasterlain, C.G.: Developmental effects of seizures: Role of Malnutrition. Pediatrics 57: 197-200 (1976).

Wasterlain, C.G. and Duffy, T.E.: Status Epilepticus in immature rats: Protective effects of glucose on survival and brain development. Arch. Neurol. 33: 821-827 (1976).

Wasterlain, C.G. and Plum, F.: Vulnerability of developing rat brain to electroconvulsive seizures. Arch. Neurol. 29: 38-45 (1973a).

Wasterlain, C.G. and Plum, F.: Retardation of behavioral landmarks after neonatal seizures in rats. Trans. Am. Neurol. Assn. 98: 47-48 (1973b).

Webb, J.M. and Levy, H.B.: A sensitive method for the determination of deoxyribonucleic acid in tissues and microorganisms. J. Biol. Chem. 213: 107-117 (1955).

Winick, M. and Noble, A.: Quantitative changes in DNA, RNA, and protein during prenatal and methodes growth in the rat. Dev. Biol. 12: 451-466 (1965).

Winick, M. and Noble, A.: Cellular response in rats during malnutrition at various ages. J. Nutr. 89: 300-306 (1966).

Winick, M., Rosso, P. and Waterlow, J.: Cellular growth of cerebrum, cerebellum, and brain stem in normal and marasmic children. Exp. Neurol. 26: 393-400 (1970).

Zlatkis, A., Zak, B. and Bogle, A.J.: A new method for the direct determination of cholesterol. J. Lab. Clin. Med. 41: 486-492 (1953).

CHAPTER NINETEEN

PATHOPHYSIOLOGIC BASIS FOR THE SELECTIVE VULNERABILITY OF THE IMMATURE RAT BRAIN TO SEIZURES*

CLAUDE G. WASTERLAIN, M.D. AND THOMAS E. DUFFY, PH.D.

Seizures too mild to damage the adult rat brain irreversibly impair brain development when they occur during vulnerable periods (Wasterlain & Plum, 1973; Wasterlain 1976a, 1976b; Nealis et al., 1976). We are just beginning to understand the biochemical basis for this selective vulnerability of the developing brain. This understanding is of more than academic interest since we now know that development of the human brain is predominantly post-natal (Dobbing & Sands, 1973), and elucidation of pathophysiologic mechanisms offers important clues for therapy of neonatal seizures.

Our efforts to date have focused on two main areas: the role of the immature blood-brain barrier (Wasterlain & Duffy, 1976), and the effect of inhibiting brain DNA synthesis at critical periods of development (Wasterlain 1976a). The findings indicate that several mechanisms are responsible for the selective vulnerability of the developing brain to metabolic insults. First, during seizures glucose transport across the immature blood-brain barrier is unable to keep pace with the glycolytic demand, so that during repetitive seizures brain glucose falls in the presence of normal or elevated blood glucose. Second, in the developing brain, insults (e.g., seizures) too mild to cause neuronal necrosis can inhibit DNA synthesis. When this inhibition occurs at critical times of development, brain growth can be irreversibly curtailed.

A. ROLE OF THE IMMATURE BLOOD BRAIN-BARRIER

Methods. Pregnant Wistar rats (CFN strain, Carworth, NY) were housed individually in a room with a 7:00 AM to 7:00 PM light cycle. After delivery, each litter was kept with its natural mother. At the age of four days, littermates were paired by sex and body weight ($\pm$ 0.5 g). In most experiments

* Supported by NIN CDS Research Grant NS 13515 and by the Research Service of the Veterans Administration.

both animals were subjected to intermittent seizures for two hours. One animal of each pair was injected intraperitoneally (IP) with 10% of its body weight of isotonic glucose solution (0.25 M) 30 minutes prior to the initiation of seizures. The paired control animal received 10% of its body weight of isotonic saline solution (IP). In some experiments a third littermate, matched for sex and weight with the other two, received neither an injection nor seizures. In an additional series of experiments, saline-injected rats were compared with uninjected paired littermates subjected to the same seizure regimen. No differences in survival or cerebral effects of seizures were noted between these two control groups. To minimize the effects of circadian variations, all experiments were carried out between the hours of 1:00 and 4:00 PM.

Status epilepticus was induced either by a single injection of bicuculline (10 mg/kg IP, dissolved in 0.1 N HCl and brought to pH 7 immediately prior to injection), or by intermittent exposure to the volatile convulsant, flurothyl. Flurothyl seizures were induced by placing the animals in a 1-liter jar that was partially immersed in a 36° water bath; every five minutes, flurothyl was injected against the inner wall of the jar in volumes of 50 μl for the first 90 minutes of the experiment and 100 μl thereafter. For adult animals a 5-liter jar was used, and the dose of flurothyl increased proportionately. Once the convulsant was added to the jar, the lid was replaced for four minutes and then removed for one minute. The oxygen content of the jar, measured with an oxygen-sensitive electrode (Lex-O_2-Con), remained close to that of room air throughout the experiment. Chemical methods were described previously (Wasterlain & Plum, 1973; Wasterlain 1976a; Wasterlain & Duffy, 1976).

Depletion of Brain Glucose by Seizures. Exposure of four-day-old rats to flurothyl produced EEG spikes (figure 1) and tonic-clonic seizures. No difference in seizure activity or duration was noted between glucose-injected rats, controls injected with isotonic NaCl, and uninjected controls. In saline-treated controls, glucose concentrations in blood were slightly elevated before the onset of status epilepticus (perhaps reflecting the stress of injection) and remained within normal limits or slightly elevated during 50 minutes of repetitive seizures (figure 2). In contrast, brain glucose concentrations fell abruptly during the first series of convulsions, then more slowly over the next 20 minutes, but by 50 minutes of seizures had decreased to 30% of resting values. In this group of animals, the ratio of brain glucose: blood glucose declined progressively with repeated seizures (fig. 3), suggesting that the transport of glucose from blood to brain could not keep pace with glycolytic demand.

FLUROTHYL SEIZURES IN IMMATURE RATS

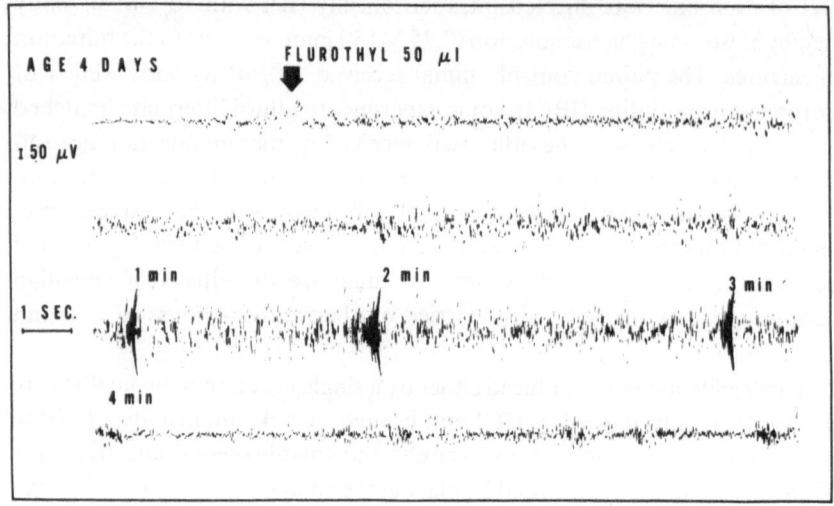

Fig. 1. Electroencephalogram during convulsive seizures induced by exposure to flurothyl (50 μl per liter) in 4-day-old rats. Exposure to flurothyl ceased after 3½ minutes.

In fact, the concentration of glucose in the intracellular fluid (ICF) of the brains of convulsing animals can be approximated from the glucose levels in blood and whole brain assuming: (1) a brain water content of 88% (Vernadakis & Woodbury, 1965); (2) a brain blood volume of 3% (Buschiazzo et al., 1970) (a minimal estimate since seizures are likely to cause vasodilation (Plum et al., 1974); (3) an extracellular fluid (ECF) space of 20% (Katzman & Pappius, 1973; Lajtha 1957; Ferguson & Woodbury, 1969); and (4) a concentration of glucose in the ECF equal to 25% of the blood glucose concentration (Meyers & Netsky, 1962). According to these assumptions, which are conservative in all instances, the concentration of glucose in the intracellular compartment of the brain is given by the expression: [Whole brain] = $0.88 (0.03$ [Blood] $+ 0.20$ [ECF] $+ 0.77$ [ICF]. Substitution into this equation of the mean glucose concentrations observed in brain (0.48 mM) and blood (5.85 mM) of saline-treated rats after 50 minutes of seizures predicts an intracellular glucose concentration of about 70 μM, a concentration at which phosphorylation of glucose by hexokinase would be expected to be rate limited (Lowry & Passonneau, 1964).

Unexpectedly, there was no indication of a progressive fall in cerebral ATP

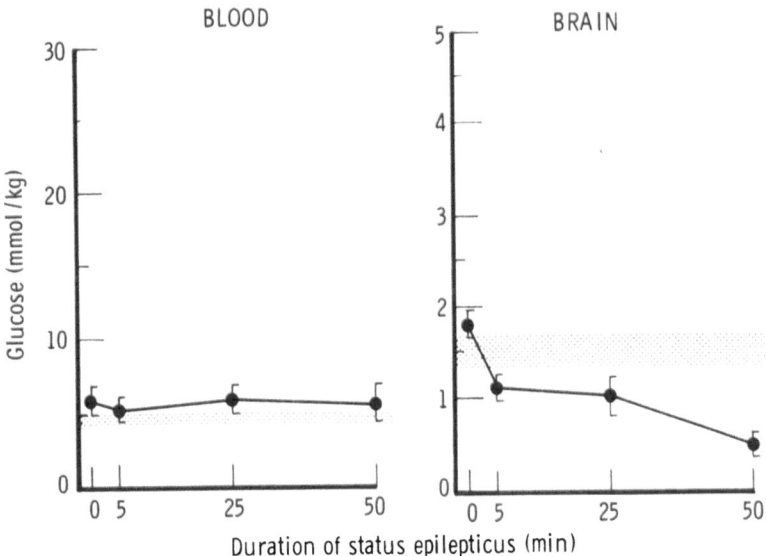

Fig. 2. Blood and brain glucose concentrations (means ± SEM) during seizures induced by flurothyl. Shaded areas represent mean control values (n = 5) ± SEM. Experimental animals received an injection of isotonic NaCl solution 30 minutes before seizures, as a control for the therapeutic injection of glucose in a parallel group. Experimental blood glucose content differed from controls at 0 and 25 minutes (P < .05), brain glucose content differed from controls after 50 minutes (P < .001).

levels with repeated seizures (fig. 4), suggesting that lack of glucose during seizures impaired brain growth by depleting the organ of its main carbon source, rather than by simply causing a cumulative energy deficit. This differed from the situation encountered during seizures in the adult (Holowach-Thurston et al., 1974; Holowach-Thurston et al., 1973) but resembled that found in neonatal hypoglycemia (Chase et al., 1973).

Other conditions that activate glycolysis in the developing brain, such as anoxia (Duffy et al., 1975; Holowach-Thurston et al., 1973) or salicylate poisoning (Holowach-Thurston et al., 1970), cause a similar fall in brain glucose at normal blood glucose levels. The finding in four-day-old rats that the brain: blood glucose ratio declined progressively during status epilepticus (fig. 3) suggests that glucose uptake by the brain could not keep pace with cerebral glycolysis, which in adult animals is greatly stimulated by seizures (Duffy et al., 1975).

In adult and immature animals alike, glucose enters the brain from the

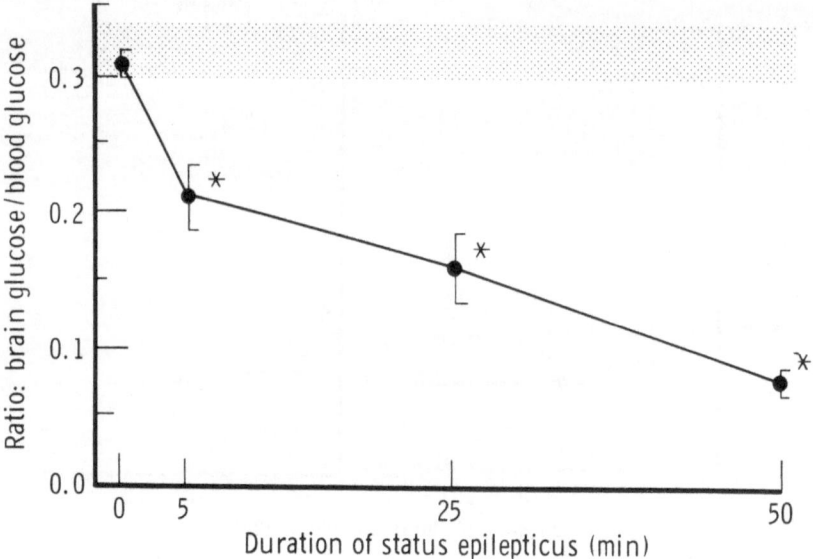

Fig. 3. Ratio of brain glucose to blood glucose concentrations in saline-treated rats. Values are calculated from data in fig. 2. Shaded area represents mean control values ± SEM. Asterisk indicates different from control at P < .01.

blood via a non-energy requiring, carrier-mediated transport system (Fishman 1964; Growdon et al., 1971), the role of diffusion being negligible. The concentration of the membrane carrier has been shown to be lower in newborn than in adult rat brain (Moore et al., 1971). We hypothesized that this low carrier capacity was responsible for the inability of glucose transport to match glycolytic demand during neonatal seizures.

Rationale for Treatment. Ordinarily, one would expect that failure of transport due to low carrier capacity would not be correctable (short of increasing carrier concentration). However, the glucose transport system present in the blood-brain barrier has a low affinity for glucose ($k_m = 7$ mM) (Moore et al., 1971). As a result, it is less than one-half saturated at physiologic blood glucose concentrations (ca 5 mM, 90 mg/100 ml). Increasing the blood glucose concentration to 20 mM would be expected to double the quantity of glucose that enters the brain.

In fact, four-day-old rats pretreated with glucose (10% of body weight IP) had blood glucose levels in the 20 mM range. The predicted initial twofold rise

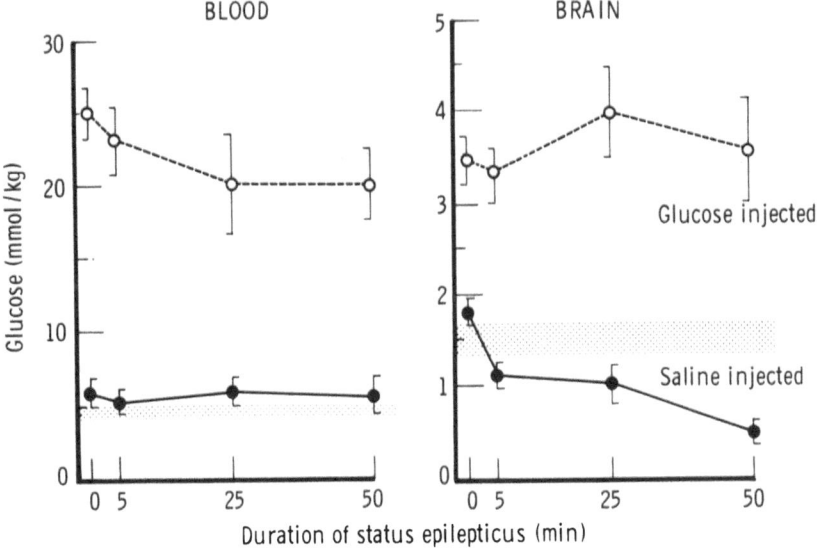

Fig. 4. Blood and brain glucose concentrations in 4-day-old rats during status epilepticus. Experimental conditions were as before except that seizures were not extended beyond 50 minutes in order to avoid data derived from moribund animals. Mortality in 4-day-old rats under these conditions was 6% in saline-treated and 0 in glucose-treated animals. Shaded area represents mean control values (n = 5) ± SEM. Each point represents mean of four to ten animals; vertical lines denote ± 1 standard error. In saline-injected rats, blood glucose content differed from control at 0 and 25 minutes (P < .05); brain glucose content differed from controls after 50 minutes (P <.001). All points pertaining to the glucose-injected group (brain and blood) differed from control values (P <.01).

in brain glucose was observed and depletion of brain glucose during seizures was completely prevented (fig. 4).

Effects of Glucose Supplements on Mortality and Cerebral Morbidity. Littermates paired for sex and weight were subjected to severe status epilepticus. One animal received a glucose load, its paired control received a similar load of isotonic NaCl. For each age group we chose a duration of status epilepticus which would kill about half of the saline-injected controls. Pretreatment of immature rats with glucose increased their survival during flurothyl-induced status epilepticus, the protection by glucose being most striking among the younger age groups (fig. 5). At one day of age, 10 of 29 saline-injected and only one of 29 glucose-injected rats died (seizures lasted three hours). Among four-day-old rats, 42 of 103 saline-treated and two of 103 glucose-treated animals died during the course of status

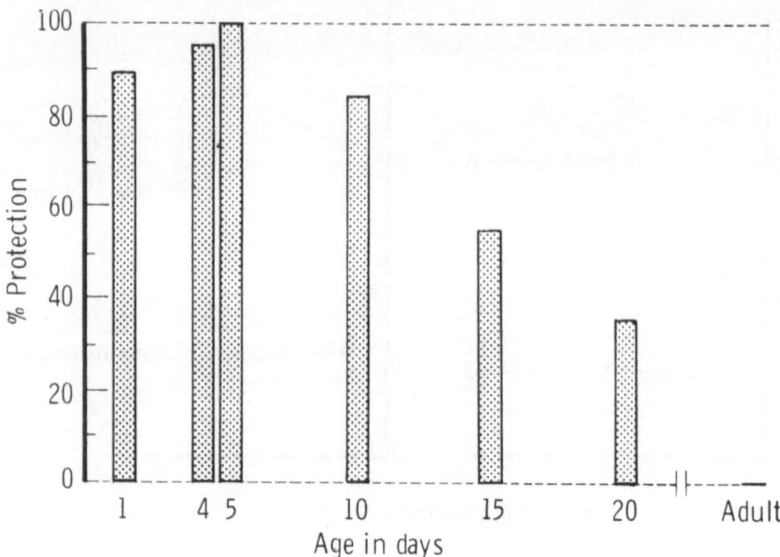

Fig. 5. Effect of glucose on mortality from repetitive seizures (see "Methods"). Percentage protection by glucose equals [(No. of saline-treated dead − No. of glucose-treated dead)/No. of saline-treated dead] × 100. Total number of pairs tested in each age group was 29 (1 day of age), 103 (4 days), 13 (5 days), 16 (10 days), 12 (15 days), 13 (20 days), and 12 (adult), respectively.

epilepticus (two hours). Among 10-day-olds, 13 of 16 saline-injected rats and two of 16 glucose-injected rats died. Among 20-day-olds, 11 of 13 saline-treated and seven of 13 glucose-treated animals died during repeated seizures. In adult rats, mean survival was 65 minutes in the saline-treated group and 72 minutes in the glucose-treated group; the difference in survival

Table 1. Effect of glucose on mortality from electroshock-induced seizures in young adult rats.*

	Saline-Treated (N / 14)	Glucose-Treated (N / 14)
Mean survival, min	34.7 ± 5.8	38.4 ± 3.0
LD_{50}, min**	31	39
LD_{100}, min**	60	50

* Animals were injected with either isotonic saline or glucose, and 30 minutes later were subjected to repeated electrically-induced (150v, ac, 1-second duration) seizures every 30 seconds. There were no significant differences (P > .05) between the two groups.

** LD_{50} indicates number of seizures sufficient to kill 50"₀ of the animals, LD_{100} number of seizures sufficient to kill all the animals.

time was not significant. Mortality was 100% in both groups, precluding comparison, so a second experiment was devised.

Adult rats were injected with glucose or saline, and 30 minutes later were subjected to supramaximal electroshocks every 30 seconds through ear clips. Again, no significant difference was found between the two groups. These data indicate that glucose administration, which reduced mortality from flurothyl-induced status epilepticus by as much as 90% during the first week of life, afforded adult rats little, if any, protection from repeated chemical (figure 5) or electrical (table 1) seizures.

It was conceivable that the difference in mortality between saline and glucose-injected immature rats during seizures represented simply an adverse effect of saline and not protection by glucose. To test this possibility, the mortality of paired uninjected and saline-injected four-day-old rats was compared during flurothyl-induced status epilepticus lasting for two hours. In each group, 11 of 20 rats (45%) died, indicating that saline pretreatment had little or no effect upon mortality from repeated seizures.

The lower mortality of glucose-treated animals during seizures compared with saline-treated rats was unrelated to serum osmolarity (table 2).

Glucose pretreatment also protected immature rats from the mortality of status epilepticus induced by other convulsive agents (fig. 6, 7).

Influence of Glucose on the Developmental Effects of Seizures. Two groups of four-day-old rats were injected with saline or glucose and subjected to two hours of flurothyl-induced seizures. Survivors were sacrificed three days later, and their brain weights and brain chemistries compared to those of untreated control littermates and to each other. Brain weight and brain DNA were significantly higher in glucose-treated rats compared with saline-treated animals (but lower than in untreated controls) (table 3). Since diploid cells of the rat contain 6.2 pg DNA per nucleus (Enesco & Leblond, 1962), the mean cell

Table 2. Serum osmolarities during flurothyl-induced status epilepticus in 4-day-old rats.*

Time After Injection, min	Saline- Injected	Glucose- Injected	Uninjected Controls
0	298	290	294
30	302	305	
60	304	306	
120	307	307	

* Animals were injected with either isotonic glucose or saline, and 30 minutes later subjected to intermittent seizures with flurothyl. Each value is the mean of three determinations of freezing point depression; sera from three animals were pooled for each measurement. All values expressed as mOsm.

BICUCULLINE SEIZURES IN IMMATURE RATS

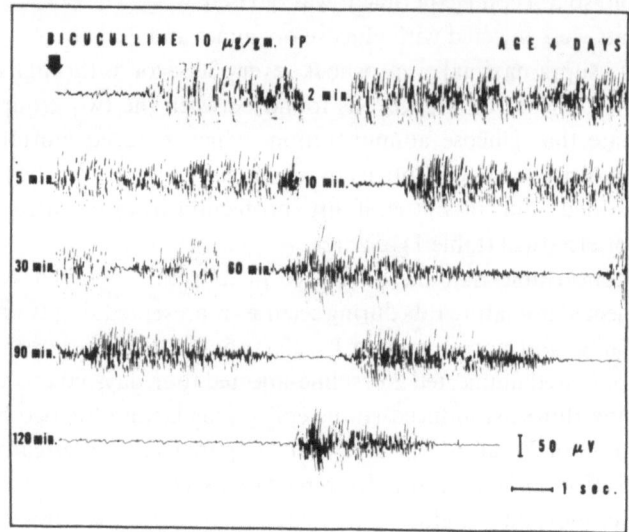

Fig. 6. Electroencephalogram of a 4-day-old rat during status epilepticus induced by intraperitoneal injection of bicuculline (10 mg per kilogram). In this animal no clinical seizure was seen
after 90 minutes.

deficit per brain was calculated to be 10.7 million cells in the saline group and
2.9 million cells in the glucose group.

The lower DNA content of brain in immature rats subjected to repeated
seizures compared with controls was found previously to reflect inhibition of
synthesis rather than enhanced breakdown (Wasterlain 1974a). Pretreatment
with glucose reduced the inhibitory effect of seizures on brain DNA synthesis
(table 4).

B. EFFECT OF INHIBITING BRAIN DNA SYNTHESIS DURING CRITICAL
     PERIODS OF DEVELOPMENT

One obvious difference between the immature and adult brain results from
the presence of active cell divisions in the former. Cell division continues until
the age of 20 days in the rat brain (Winick & Noble, 1966), two years in the
human brain (Dobbing & Sands, 1973). Many of the metabolic effects of
convulsive seizures (Duffy et al., 1975; Wasterlain 1972) would be expected to

NEONATAL S.E. : PROTECTIVE EFFECTS OF GLUCOSE

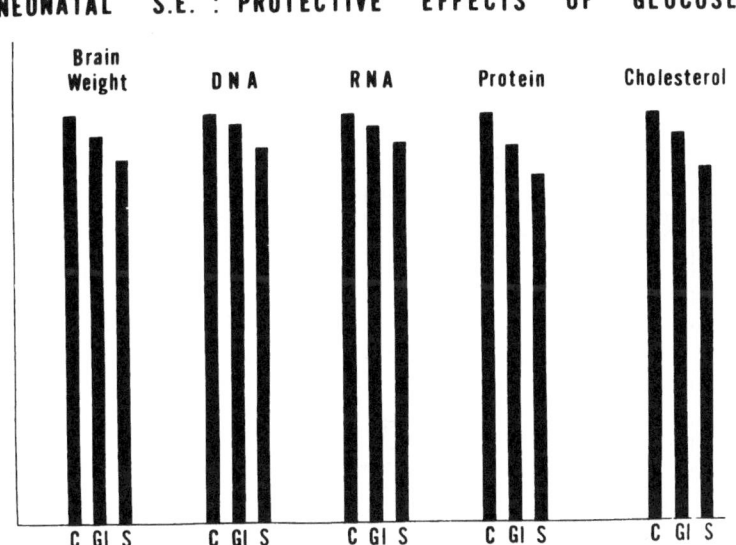

Fig. 7. Protection by glucose against the cerebral effects of seizures in 4-day-old rats. C = untreated controls, S = animals given seizures after a saline injection, G = animals given seizures after a glucose injection. 3 days after seizures glucose-injected rats fared better than saline-treated animals for brain weight, brain DNA, brain cholesterol and cell number (P < .05).

curtail cell division. To investigate how this would affect cerebral development, we designed a model of status epilepticus which produced no evidence of cell necrosis in the brain (Wasterlain 1976a). Intermittent exposure of four-day-old rats to flurothyl (two hours, 1500 μl) induced no qualitative histological changes in any brain region (fig. 8) and did not increase DNA breakdown (fig. 9), but severely curtailed brain DNA synthesis (fig. 10). Table 5 shows the short-term effects of this treatment on brain development. Forebrain weight was reduced 19.6%. Total brain DNA content was 16.8% lower than in controls. Since the number of polyploid cells in rat brain is small (Herman & Lapham, 1969), and since these cells (e.g., Purkinje cells, hippocampal pyramidal cells) did not seem affected by status epilepticus as judged from histologic sections, we can reasonably assume that the reduction of brain DNA in status-epilepticus-treated rats reflected a reduction of the number of forebrain cells. At seven days of age, the mean deficit in the number of cells in the forebrains of experimental rats was 30 million cells. Reductions of the total forebrain content of RNA, protein, and cholesterol in experimental rats were of the same order of magnitude as the changes in brain weight and brain

Table 3. Brain development of immature rats after status epilepticus: protective effect of glucose. *

	Controls	Glucose-Injected	SalineInjected
Brain weight, mg	452 ± 24°	424 ± 18**	404 ± 18***
Brain RNA, μg	1172 ± 42	1141 ± 34	1093 ± 39***
Brain DNA, μg	819 ± 30	801 ± 18	753 ± 19***
Brain protein, mg	40.8 ± 2.2°	37.6 ± 1.9**	34.8 ± 1.8***
Brain cholesterol, mg	2.37 ± 0.13	2.26 ± 0.13	2.06 ± 0.14***
Rna concentration, μg/gm	2736 ± 92	2763 ± 91	2645 ± 110
DNA concentration, μg/gm	1812 ± 64	1889 ± 42	1864 ± 44
Protein concentration, mg/gm	90.3 ± 4.5	88.7 ± 4.2	86.1 ± 4.1
Cholesterol concentration, mg/gm	5.24 ± 0.30	5.33 ± 0.29	5.10 ± 0.31
Protein-DNA ratio	49.8 ± 2.4	46.9 ± 2.1	46.2 ± 2.0
RNA-DNA ratio	1.43 & 0.40	1.42 ± 0.40	1.45 ± 0.40
Cholesterol-DNA ratio	2.89 ± 0.20	2.82 ± 0.20	2.74 ± 0.20
Cell number, X 10⁻⁶	132.1 ± 4.8	129.2 ± 2.9	121.4 ± 3.1***
Mean weight per nucleus, ng	3.31 ± 0.06	3.24 ± 0.09	3.28 ± 0.08

* Rats matched for sex, litter, and body weight (0.5 gm) were treated at 4 days of age and killed at 7 days of age. Treatment consisted of injection of isotonic glucose (10° of body weight) followed by two hours of flurothyl-induced seizures (glucose group), injection of isotonic sodium chloride solution followed by two hours of seizures (saline group), of simple handling (control group).
Different from controls: ** P <.05; *** P <.01.
Different from glucose-injected: ° P <.05; °° P <.01.

Table 4. DNA synthesis in immature rat brains during seizures.*

	DNA Synthesis	Acid-Soluble (CPM × 10³/gm brain)	Ratio**
Controls	7444 ± 248***	637 ± 55	12.33 ± 1.03***
Seizures			
No pretreatment	4016 ± 568****	720 ± 70	5.56 ± 0.43****
Saline-treated	3970 ± 539****	718 ± 67	5.62 ± 0.64****
Glucose-treated	4796 ± 531****	691 ± 85	6.93 ± 0.41****

* Littermate 4-day-old rats were subjected to flurothyl-induced seizures every five minutes. Saline or glucose pretreatment consisted of the intraperitoneal injection (10"₀ of body weight) of an isotonic solution 30 minutes prior to the initiation of seizures. At 60 minutes after the first exposure to flurothyl, all animals (control and experimental) received an intraperitoneal injection of [methyl-³H] thymidine (5 μCi/gm) and were decapitated 30 minutes later. All values represent the means of eight determinations ± 1 standard error.
** Calculated as the ratio of the mean CPM in DNA per gram of brain divided by the mean CPM X 10³ in the acid-soluble fraction per gram of brain.
*** Different from the "seizure-no pretreatment" group at P <.05.
**** Different from control at P <.01.

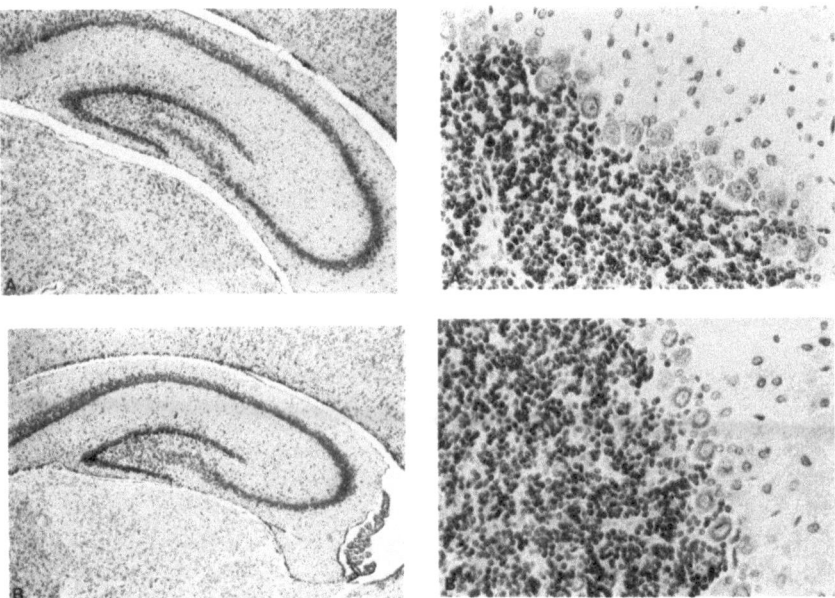

Fig. 8. (Left side). Hippocampus of 30-day-old littermates sacrificed 26 days after "severe" flurothyl status epilepticus. In experimental rats (*B*) both Sommer sector and resistant sector are indistinguishable from control (*A*).
(Right side). Cerebellar cortex of 30-day-old paired littermates. The number and appearance of both Purkinje cells and granule cells are similar in a rat subjected to "severe" flurothyl status epilepticus (*B*) and in a control animal (*A*).

Brain DNA Turnover After S.E.

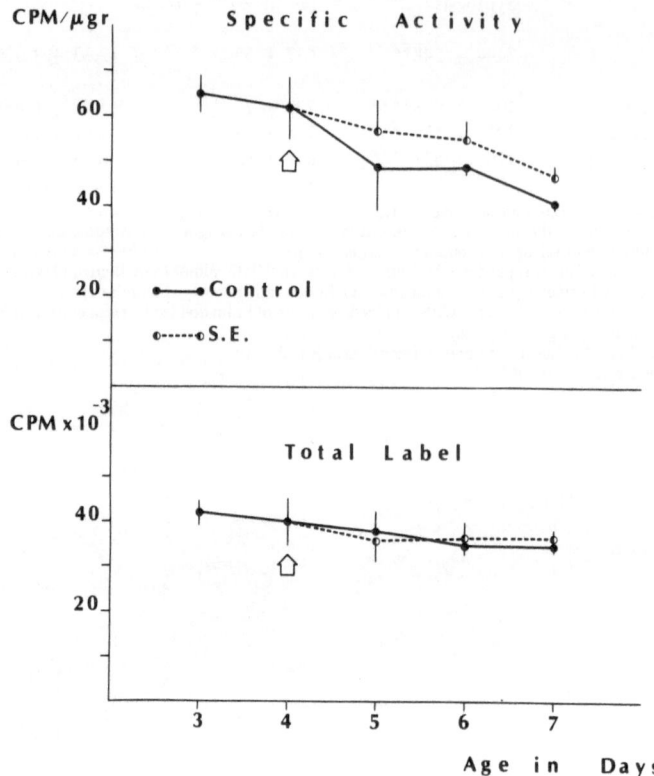

Fig. 9. Total and specific radioactivity of forebrain DNA after [³H]-thymidine injection at age 2 days. In experimental rats, convulsive seizures were induced by flurothyl (1500 μl) at age 4 days (arrow). Note the slight downward trend for total DNA label, and the lack of influence of seizures on DNA turnover, suggesting that no extensive breakdown of brain cells occurred.

DNA, as illustrated in fig. 11. Concentrations of DNA, RNA, protein, and cholesterol and measurements reflecting cell size (weight per cell, RNA per cell, protein per cell) were similar in control and experimental groups (Table 5). Neonatal status epilepticus, like most neonatal metabolic insults, reduced the number of brain cells without modifying mean cell size.

Cholesterol accumulation was reduced by status epilepticus but since both cholesterol concentrations and the amount of cholesterol per cell

DNA   Synthesis   During   S.E.

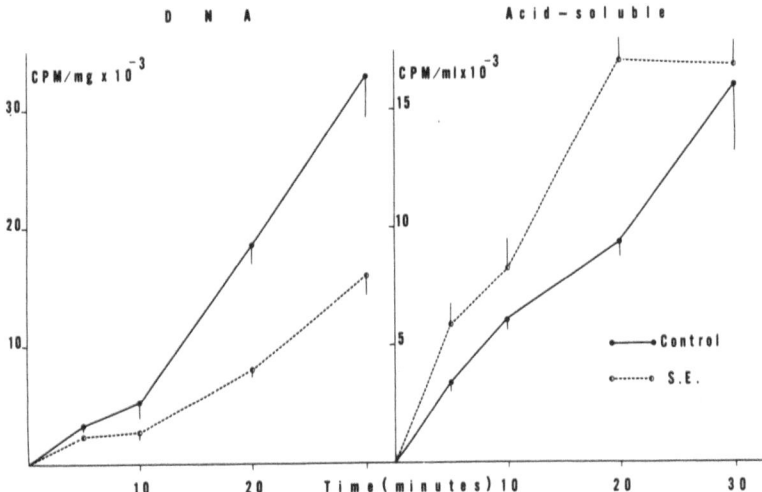

Fig. 10. Effect of neonatal status epilepticus (Flurothyl, 1,500 µl) on [³H]-thymidine incorporation into forebrain DNA. Experimental rats were subjected to status epilepticus at age 4 days. Controls were handled the same way but received no seizures. After 1 hour of status epilepticus, rats were injected with 5 µCi per gram of [³H]-methyl-thymidine intraperitoneally. The increase in radioactivity of the acid-soluble precursor pool is seen on the right. On the left, radioactivity incorporated into DNA, which is lower during status epilepticus (incorporation time 30 minutes) (P < .01).

(cholesterol/DNA ratio) were unaffected by the treatment, this fall seemed to reflect a reduction in cell number rather than a specific effect on lipid metabolism.

Twenty-six days after status epilepticus, a small but significant reduction of forebrain weight persisted (table 5). Forebrain DNA, protein, and cholesterol also were reduced (fig. 11). As expected, forebrain concentrations of DNA were lower and those of RNA much lower than in seven-day-old rats, while concentrations of protein were almost fourfold higher (a reflection of cell differentiation), and cholesterol concentrations were nearly 20-fold those in seven-day-old rats (reflecting the advanced degree of myelination in the older age group). Concentrations of DNA, RNA, protein, cholesterol, and the amounts of RNA, protein, or cholesterol per cell were not affected by status epilepticus.

In summary, a single, two-hour episode of status epilepticus at age four

Table 5. Short-term effects of severe status epilepticus induced by flurothyl (1.500 μl) for 2 hours on rat forebrain development.

No.	Controls (10)	Status (10)	Difference
Body weight, day 4 (gm)	11.7 ± 0.4	11.5 ± 0.6	NS
Body weight, day 7 (gm)	19.8 ± 0.8	12.6 ± 1.0	$p < 0.01$
Forebrain weight (mg)	669 ± 8	538 ± 20	$p < 0.01$
DNA (μg)	$1,123 \pm 67$	934 ± 38	$p < 0.02$
RNA (μg)	$2,089 \pm 28$	$1,675 \pm 55$	$p < 0.01$
Protein (mg)	23.3 ± 0.8	18.9 ± 0.4	$p < 0.01$
Cholesterol (mg)	0.45 ± 0.01	0.35 ± 0.01	$p < 0.01$
DNA concentration (μg/gm w.w.)	$1,684 \pm 114$	$1,749 \pm 86$	NS
RNA concentration (μg/gm w.w.)	$3,121 \pm 60$	$3,069 \pm 30$	NS
Protein concentration (mg/gm w.w.)	34.9 ± 1.3	35.3 ± 1.0	NS
Cholesterol concentration (mg/gm w.w.)	0.68 ± 0.02	0.65 ± 0.04	NS
Cell number X 10^{-6}	181 ± 11	151 ± 6	$p < 0.02$
Weight/cell (mμg)	3.67 ± 0.21	3.59 ± 0.18	NS
RNA/cell (μg)	11.70 ± 0.58	11.20 ± 0.59	NS
Protein/cell (μg)	131 ± 9	126 ± 5	NS
Cholesterol/cell (μμg)	2.54 ± 0.18	2.31 ± 0.10	NS

EFFECT OF NEONATAL STATUS EPILEPTICUS ON RAT BRAIN
DEVELOPMENT - INDOKLON, 1,500 µl

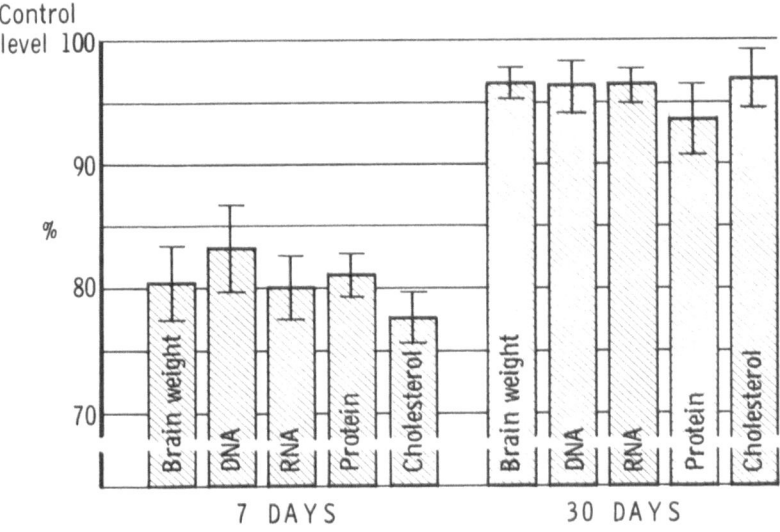

Fig. 11. Effects of severe status epilepticus induced by flurothyl at 4 days of age, on forebrain weight and forebrain chemistries 3 to 26 days later. Experimental values are represented as percent of control littermates. All experimental values shown are different from controls by a t-test for paired data (P < .05).

days was sufficient to reduce brain size and brain cell number permanently, but did not modify mean brain cell size. The cerebellum was similarly affected but recovered completely by the age of 30 days (fig. 12).

Behavioral Changes. In a previous study, maturation of swimming patterns and of free-fall righting proved very sensitive to metabolic brain damage. Status epilepticus delayed significantly the final maturation of both behavioral milestones (table 6). It also decreased the flurothyl seizure thresholds of experimental animals, as tested 24 days after seizures (table 7).

C. DISCUSSION

Mechanism of Developmental Brain Damage Induced by Status Epilepticus. A single episode of repetitive seizures, administered to rats during the vulnerable period of brain development, was sufficient to reduce brain weight and

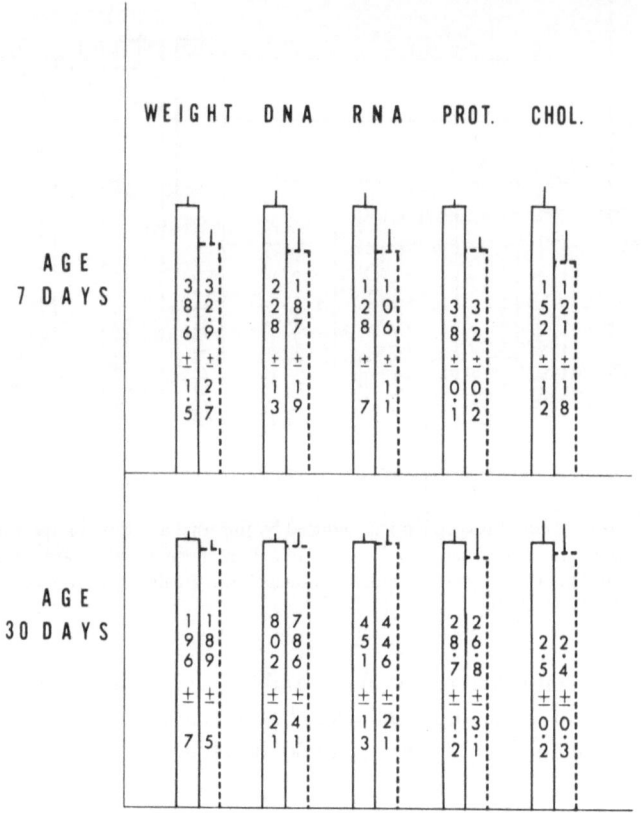

EFFECT OF NEONATAL STATUS EPILEPTICUS
ON CEREBELLAR DEVELOPMENT (INDOKLON 1500λ)

Fig. 12. Cerebellar weight, DNA, and chemistries after "severe" status epilepticus (Flurothyl 1,500 μl). Solid bars represent controls, dotted bars represent littermates subjected to status epilepticus. All differences between controls and experimentals were significant at age 7 days (P < .05); none were significant at age 30 days.

the number of brain cells permanently. It is remarkable that this permanent damage could result from an insult too mild to produce histologic or biochemical evidence of cell necrosis. Repetitive seizures are known to frequently damage the brain in all species studied, including man (Fowler 1957; Meldrum et al., 1974; Meldrum & Brierley, 1973; Spielmeyer 1927; Scholz

Table 6. Ontogeny of behavioral milestones after neonatal status epilepticus.

	Controls	Experimentals
Free-fall righting	16.5 ± 0.2	18.2 ± 0.4*
Swimming ability	12.2 ± 0.3	13.3 ± 0.2*

Values indicate the number of postnatal days required to reach the adult type of response (mean ± SEM).
* Different from controls (p < 0.01).

Table 7. Flurothyl seizure thresholds.

	Controls	Experimentals
Seizure latency	599 sec ± 14	529 sec ± 8*
Quantity of flurothyl	199 μl ± 5	176 μl ± 3*

Rats were placed into a 1 gallon jar into which flurothyl was infused at the rate of 20 μl per minute. Seizure threshold was defined as the amount of flurothyl infused into the jar at the time of onset of a full tonic seizure (mean ± SEM).
* Different from controls (p < 0.01).

1951; Norman 1964). It has always been assumed that cerebral atrophy following seizures was the result of widespread neuronal necrosis. Indeed, autopsy studies in adults dying of status epilepticus showed neuronal ischemic cell changes in cortex, cerebellum, and basal ganglia (Fowler 1957; Spielmeyer 1927; Scholz 1951; Norman 1964). Our study demonstrates that when repetitive seizures occur during the vulnerable period of cerebral mitotic activity, a more subtle and previously unrecognized form of insult can occur. Such an insult escaped histologic detection but nevertheless had profound and sometimes permanent effects on the development of brain and behavior. The profound inhibition of cerebral and cerebellar DNA synthesis observed in this study suggests a mechanism for this type of developmental brain damage and indicates that its occurrence should be restricted to the period of active cerebral mitotic activity, which covers the first three weeks of life in the rat (Winick & Noble, 1966) and, if extrapolation is possible, the first two years in man (Dobbing & Sands, 1973).

Relevance to Clinical Problems. This work should not be blindly extrapolated to clinical situations, but it is directly relevant to human problems for two reasons. First, man like the rat is a postnatal brain developer. A recent study of human brain growth, which for the first time included adequate numbers of brains in the older age groups, showed that "most of the human brain growth spurt is postnatal" (Dobbing & Sands, 1973). Forebrain DNA (reflecting the number of cells in that region) quadruples from birth (mean 0.62 mM of DNA-P per brain) to adulthood (mean 2.5 mM DNA-P per brain) (Dobbing & Sands, 1973). Active cell multiplication takes place in the human brain until at least two years of age. The bulk of mitoses are probably glial but some (e.g., in the external granular layer of the cerebellum) are almost certainly neuronal (Woodard 1960; Raaf & Kernohan, 1944; Jacobson 1970). Indeed, the marked effects of postnatal metabolic insults like hyperphenylalaninemia or hypothyroidism on children's intellectual development suggest that the human brain is particularly vulnerable during infancy and early childhood.

Second, repetitive seizures have been shown to be deleterious to the immature brain and are often followed by intellectual deterioration (Aicardi & Chevrie, 1971) and ventricular dilation (Aicardi & Baraton, 1971). Even relatively mild seizures may have subtle adverse effects on intellectual performance, as shown by a recent study of identical twins discordant for febrile seizures (Schiottz-Christensen & Bruhn, 1973).

Therapeutic Implications of the Fall in Brain Glucose During Seizures. Our experimental findings emphasize that during seizures in the immature brain the presence of a normal or even slightly elevated blood glucose concentration cannot be assumed to ensure adequate supplies of glucose to the brain. Assuming that the membrane transport system in human brain has an affinity for glucose similar to that of the rat, at blood glucose concentrations ordinarily accepted as "low normal" for human infants (1.5 mM, 27 mg/100 ml), the membrane carrier should be less than 20% saturated, possibly restricting the availability of glucose to the brain. Furthermore, increasing the blood glucose concentration would readily increase the saturation of the carrier and promote the entry of glucose into the brain. In this respect, it is important to remember that the brain of the human infant is much less mature than was previously recognized (Woodard 1960) and appears to be highly vulnerable to hypoglycemia (Raaf & Kernohan, 1944; Jacobson 1970; Schiottz-Christensen & Bruhn, 1973; Lombroso 1974). Accordingly, when treating neonatal or infantile seizures it would seem particularly desirable to prevent hypoglycemia and, perhaps, to maintain the blood sugar slightly elevated in order to maximize the availability of glucose to the brain.

REFERENCES

1. Aicardi, J., & Chevrie, Convulsive status epilepticus in infants and children: A study of 239 cases. Epilepsia 11: 187-197 (1971).
2. Aicardi, J., & Baraton, J.: A pneumoencephalographic demonstration of brain atrophy following status. Dev. Med. Child Neurol. 13: 660-667 (1971).
3. Buschiazzo, P.M., Terrell, E.B., & Regen, D.M.: Sugar transport across the blood-brain barrier. Am. J. Physiol. 219: 1505-1513 (1970).
4. Chase, H.P., Marlow, R.A., Dabiere, C.S., et al.: Hypoglycemia and brain development. Pediatrics 52: 513-520 (1973).
5. Dobbing, J., & Sands, J.: Quantitative growth and development of human brain. Arch. Dis. Child. 48: 757-767 (1973).
6. Duffy, T.E., Howse, D.C., & Plum, F.: Cerebral energy metabolism during experimental status epilepticus. J. Neurochem. 24: 925-934 (1975).
7. Duffy, T.E., Kohle, S.J., & Vannucci, R.C.: Carbohydrate and energy metabolism in perinatal rat brain: Relation to survival in anoxia. J. Neurochem. 24: 271-276 (1975).
8. Enesco, M., & Leblond, C.P.: Increase in cell number as a factor of growth in the organs of the young male rat. J. Embryol. Exp. Morphol. 10: 530-562 (1962).
9. Ferguson, R.K., & Woodbury, D.M.: Penetration of ^{14}C-insulin and ^{14}C-sucrose into brain, cerebrospinal fluid and skeletal muscle of developing rats. Exp. Brain Res. 7: 181-194 (1969).
10. Fishman, R.A.: Carrier transport of glucose between blood and cerebrospinal fluid. Am. J. Physiol. 206: 836-844 (1964).
11. Fowler, M.: Brain damage after febrile convulsions. Arch. Dis. Child. 32: 67-76 (1957).
12. Growdon, W.A., Bratton, T.S., Houston, M.C., et al.: Brain glucose metabolism in the intact mouse. Am. J. Physiol. 221: 1738-1745 (1971).
13. Herman, C.J., & Lapham, L.W.: DNA content of neurons in the cat hippocampus. Science 160: 537 (1968).
14. Holowach-Thurston, J., Hauhart, R.E. & Jones, E.M.: Anoxia in mice: Reduced glucose in brain with normal or elevated glucose in plasma and increased survival after glucose treatment. Pediatr. Res. 8: 238-243 (1974).
15. Holowach-Thurston, J., Hauhart, R.E., & Jones, E.M., et al.: Decrease in brain glucose in anoxia in spite of elevated plasma glucose levels. Pediatr. Res. 7: 691-695 (1973).
16. Holowach-Thurston, J., Pollock, P.J., Warren, S.K., et al.:Reduced brain glucose with normal plasma glucose in salicylate poisoning. J. Clin. Invest. 49: 2139-2145 (1970).
17. Jacobson, M.: Developmental Neurobiology, p. 54 (Holt, Rinehart & Winston, Inc. New York 1970).
18. Katzman, R., & Pappius, H.M.: Brain Electrolytes and Fluid Metabolism, pp. 33-43 (Williams & Wilkins, Baltimore 1973).
19. Lajtha, A.: The development of the blood-brain barrier. J. Neurochem. 1: 216-227 (1957).
20. Lombroso, C.T.: The treatment of status epilepticus. Pediatrics 53: 536-540, (1974).
21. Lowry, O.H., & Passonneau, J.V.: The relationships between substrates and enzymes of glycolysis in brain. J. Biol. Chem. 239: 31-42 (1964).
22. Meldrum, B.S., Horton, R.W., & Brierley, J.B.: Epileptic brain damage in adolescent baboons following seizures induced by Allyl-glycine. Brain 97: 407-418 (1974).
23. Meldrum, B.S., & Brierley, J.B.: Prolonged epileptic seizures in primates: Ischemic cell change and its relation to ictal physiological events. Arch. Neurol. 28: 10-17 (1973).
24. Moore, T.J., Lione, A.P., Regen, D.M., et al.: Brain glucose metabolism in the newborn rat. Am. J. Physiol. 221: 1746-1753 (1971).
25. Myers, G.G., & Netsky, M.G.: Relation of blood and cerebrospinal fluid glucose. Arch. Neurol. 6: 18-26 (1962).
26. Nealis, J.G.T., Depiero, T.J., Ouellette, et al.: Neurologic sequelae of febrile convulsions: An experimental study. Neurology 26: 363 (1976).

27. Norman, R.M.: The neuropathology of status epilepticus. Med. Sci. Law 4: 46-51 (1964).
28. Plum, F., Howse, D.G., & Duffy, T.E.: Metabolic effects of seizures; in Plum, F. (ed), Brain Dysfunction in Metabolic Disorders: Proceedings, Vol. 53, Association for Research in Nervous and Mental Disorders, pp. 141-157 (Raven Press, New York 1974).
29. Raaf, J., & Kernohan, J.W.: A study of the external granular layer of the cerebellum. Am. J. Anat. 75: 151-172 (1944).
30. Schiottz-Christensen, E., & Bruhn, P.: Intelligence, behaviour and scholastic achievement subsequent to febrile convulsions: An analysis of discordant twin-pairs. Dev. Med. Child Neurol. 15: 565-575 (1973).
31. Scholz, W.: Die Krampfschadigungen des Gehirns, p. 116 (Springer, Berlin 1951).
32. Spielmeyer, W.: Die pathogenese des epileptischen Krampfes. Z Ges. Neurol. Psychiat. 109: 501-520 (1927).
33. Vernadakis, A., & Woodbury, D.M.: Cellular and extracellular spaces in developing rat brain. Arch. Neurol. 12: 284-293 (1965).
34. Wasterlain, C.G., & Duffy, T.E.: Status epilepticus in immature rats. Arch. Neurol. 33: 821-827 (1976).
35. Wasterlain, C.G.: Effects of neonatal status epilepticus on rat brain development. Neurology 26: 975-986 (1976a).
36. Wasterlain, C.G.: Effects of neonatal status seizures on ontogeny of reflexes and behavior. An experimental study in the rat. Europ. Neurol. (in press) (1976b).
37. Wasterlain, C.G.: Inhibition of cerebral protein synthesis by epileptic seizures without motor manifestations. Neurology 24: 175-180 (1974a).
38. Wasterlain, C.G., & Plum, F.: Vulnerability of developing rat brain to electroconvulsive seizures. Arch. Neurol. 29: 38-45 (1973).
39. Wasterlain, C.G.: Breakdown of brain polysomes in status epilepticus. Brain Res. 39: 278-284 (1972).
40. Winick, M., & Noble, A.: Cellular response in rats during malnutrition at various ages. J. Nutr. 89: 300-306 (1966).
41. Woodard, J.S.: Origin of the external granular layer of the cerebellar cortex. J. Comp. Neurol. 115: 65-73 (1960).

CHAPTER TWENTY

POST-NATAL SEIZURES. CLINICAL EFFECTS

JEAN AICARDI

The possible effects of epileptic seizures on the developing brain have long been a subject of interest. Special attention has been paid to epileptic sequelae, in particular temporal lobe epilepsy (Cavanagh and Meyer 1956, Ounsted et al. 1966, Falconer 1971, Lindsay, 1971) and to hemiplegias consecutive to prolonged unilateral seizures (Gastaut et al., 1960, Norman, 1962, Isler, 1969, Aicardi et al., 1969). The pathologic effects of epileptic fits on the brain have also been extensively discussed (Scholz, 1951, Sano and Malamud, 1953, Norman, 1964, Margerison and Corsellis, 1966).

This work is limited to the study of the major clinical effects of severe and prolonged epileptic convulsions ($\geqslant$ 30 minutes) in infants and children between 1 month and 5 years of age. More subtle consequences such as behavioral and learning disturbances, and the possible effects of short repeated fits are not considered.

NATURE OF CLINICAL ABNORMALITIES FOLLOWING SEVERE POST-NATAL CONVULSIONS

Three main sequelae have been attributed to severe infantile convulsions : 1) permanent neurologic defects, especially hemiplegias; 2) mental retardation; 3) lesional epilepsies. Data describing the incidence of such sequelae are inadequate. In a series of 239 cases of convulsive status epilepticus (Aicardi and Chevrie 1970), mental and/or neurologic sequelae were present in 57% of the cases (table 1). When cases of cryptogenic status only were followed for one year the incidence of sequelae was 28%. In a series of 402 children with febrile convulsions, the incidence of neurologic or mental sequelae was 14.6% (Aicardi and Chevrie, 1976) (table 2). Epileptic sequelae in the same series were observed in 40% and 28.4%, respectively. The epilepsies observed after prolonged infantile convulsions were mostly partial or secondarily generalised ones, of a type usually associated with brain damage (table 3). These results conform with those of several series of temporal lobe epilepsy in which

Table 1. Neurologic and mental deficits following status epilepticus (≥ 1h) in 239 patients.

	No.*	acquired**
Permanent neurologic signs	88 (37%)	47 (20%)
• Hemiplegia	28	28
• Other	60	19
Mental retardation	114 (48%)	78 (33%)
Total number with any disability	136 (57%)	82 (34%)

* The number of patients in whom neurologic signs, mental retardation or any disability were found.
** Number of patients among whom neurologic signs, etc appeared "de novo" after their episode of status epilepticus.

a high incidence of antecedent severe infantile convulsions was found (Falconer et al. 1964, Ounsted et al., 1966). The significance of these figures is open to question since they were the result of retrospective studies in selected cases. They do demonstrate, however, the potential severity of infantile seizures.

The fact that mental, neurologic or epileptic sequelae can follow severe infantile convulsions is no proof of a causal relationship. Both the convulsions and the sequelae could result from an acute cerebral disorder. Or the initial convulsions could have been the acute manifestation of a pre-existing, latent, neurologic problem.

To prove that the above mentioned sequelae are a direct consequence of the convulsive seizures themselves, several conditions must be fulfilled.

1) The brain must have been normal (as far as clinical evaluation is possible) prior to the occurrence of the first severe convulsion. For this reason, we selected a population of 402 patients with febrile convulsions and discarded our initial series of 239 cases of status epilepticus, as a number of these were due to gross brain lesions. Febrile convulsions are considered as being due to a genetic propensity to seizures and it is unlikely that antecedent brain damage

Table 2. Neurologic and mental deficits following febrile convulsions in 384 patients.

Permanent neurologic signs	37	(9.6%)
• Hemiplegia	24	
• Others	13	
Mental retardation	54	(14.1%)
Total number with any disability	56	(14.6%)

Table 3. Types of epilepsy following febrile convulsions.

Type of seizures	Number	
– Partial motor	49	
– Partial complex (psychomotor)	51	81 (71%)*
– Secondarily generalized (Lennox Syndrome)	9	
– Generalized		28 (24.6%)
– Unclassified		5
– Total		114

* A number of patients had both "partial motor "and" partial complex" types.

was present in a significant proportion of our population. In addition, 18 patients in whom neurologic anomalies had been suspected prior to the first seizure were excluded, leaving a series of 384 patients with "pure" febrile convulsions. In all the patients with sequelae, these were noted only after one or several seizures.

2) The location of sequelae, when lateralised must be the same as that of the initial "causative" seizure. In our series all post-convulsive hemiplegias occurred on the side which had been exclusively or predominantly involved in the convulsion (Aicardi et al. 1969). Likewise, localized epileptic sequelae (partial motor seizures or clearly defined interictal EEG spike focus) were always attributable to the involvement of the cerebral hemisphere which had been the origin of the severe convulsion (table 4).

3) There must exist a significant relationship between the severity of the febrile convulsions and the incidence of sequelae. Table 5 shows such a relation between the incidence of sequelae and the duration and unilateral

Table 4.

	Partial motor seizures		Lateralized EEG focus		Total lateralized		
	R	L	L	R	R	L	
Left-sided fc					—		
24	0	17	0	20	0	24	
							42 (100%)
Right-sided fc							
18	15	0	17	0	18	0	
Bilateral fc							
74	10	9	3	5	11	13	24 (32%)

Table 5. Comparison between 131 patients with all types of sequelae and "controls".

Factor controls	sequelae	controls	p
Age at 1st F.C.	16.2 m	20.3 m	< 0.05*
Seizure ≥ 30 mm	66 (50.4)	15 (17.4)	< 0.0005***
Lateralized seizure	57 (43.5)	15 (17.4)	< 0.0005***
Female sex	67 (51.1)	40 (46.5)	< 0.30
Abnormal delivery	19 (14.5)	19 (22.1)	< 0.20
Birth weight	3.14 kg	3.19 kg	> 0.80
Positive family history	44 (33.6)	29 (33.7)	> 0.50
Febrile recurrences	58 (44.3)	51 (59.3)	< 0.05*

location of the febrile seizures. It has been shown previously that duration and lateralization are closely interrelated and that duration was the main responsible factor (Chevrie and Aicardi, 1975, Aicardi and Chevrie, 1976).

4) There should be no relation between the incidence of sequelae and etiologic factors other than the severity of the convulsions. This condition is only partially fulfilled (table 5) as some other factors, especially age, are important. The fact that other factors than the severity of the convulsions play a rôle, is not incompatible with the proposed hypothesis but indicates that modifying conditions may also intervene.

The conditions listed above are necessary but insufficient. Another more direct argument favoring the potential noxious role of long lasting convulsions, whatever their cause, can be drawn from the neuroradiologic study of patients with severe seizures. Among 18 infants hospitalised for an episode of status epilepticus, two pneumoencephalograms were performed consecutively, the first one within 10 days of the attack, the second during the following months (usually within two months). In 16 cases, a significant degree of ventricular dilatation appeared between the two examinations, resulting from atrophy, not from hydrocephalus (fig. 1A and B). The localisation of atrophy corresponded closely, as a rule, to that of the status epilepticus (table 6). No specific encephalopathy could be diagnosed in any of these patients. 11 of these infants had not previously convulsed. 7 had 3 or more seizures antedating their status attack. In this latter group (including 6 unilateral and 1 generalised statuses), the antecedent convulsions did not differ clinically from the severe seizure, except as to duration. While one may argue that, in the first 11 patients, an acute unrecognised encephalopathy could have been responsible for both the atrophy and status, in the latter 7 patients it is difficult to assume that the cause of the status episode was not the same as that of the antecedent, identical, brief seizures. It is difficult to attribute the post-

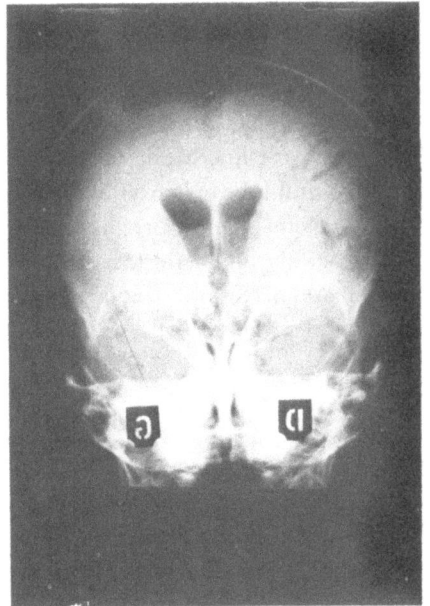   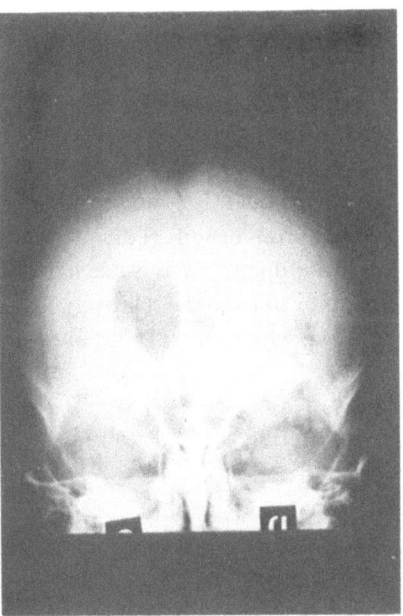

Fig. 1A. Pneumoencephalogram immediately   *Fig. 1B.* Same patient 6 weeks later.
after seizure episode.

Table 6.

			Acquired atrophy			No atrophy
R	L	B				
Right-sided status						
7			1	5		1
Left-sided status						
7			6	0	1	
Bilateral status						
4				3		1
Total 18				16		2

Acquired atrophy refers to atrophy which appeared de novo after an episode of status. Lateralisation as
indicated is either exclusive or predominant (whether of status or of atrophy).

convulsive atrophy to anything other than the long lasting seizure itself
(Aicardi and Baraton, 1971). Recently, Gastaut (1976) has confirmed these
results, using computerized tomography, repeated several times in a small
number of patients. This technique demonstrates the extent of hemispheric

edema and the rapidity with which atrophy (both cortical and periventricular) develops.

Such massive lesions result only from very severe seizures (usually several hours). It is possible, however, that more limited lesions, difficult to detect by conventional means, can be produced by less severe fits. Mesial temporal lesions, frequently found in patients with temporal lobe epilepsy following infantile convulsions (Falconer et al. 1964, Margerison and Corsellis, 1966) could be due to such a mechanism. Their generally unilateral localization might result from the relatively long lasting seizures necessary to induce brain damage, which convulsions are usually unilateral.

CONCLUSION

A clinical study cannot conclusively prove a causative relation between severe, post-natal convulsions and the neurologic sequelae which can follow them. The arguments presented suggest that such a causative link exists and that experimental data obtained in animals (Meldrum and Brierley 1973, Meldrum and Horton 1973, Meldrum et al. 1973) may be applicable to humans.

REFERENCES

Aicardi, J., Amsili, J. and Chevrie, J.J. Acute hemiplegia in infancy and childhood. Develop. Med. Child Neurol., 11, 162-173 (1969).

Aicardi, J. and Baraton, J. A pneumoencephalographic demonstration of brain atrophy following status epilepticus. Develop. Med. Child Neurol. 13, 660-667 (1971).

Aicardi, J. and Chevrie, J.J. Convulsive status epilepticus in infants and children. A study of 239 cases. Epilepsia (Amst.), 11, 187-197 (1970).

Aicardi, J. and Chevrie, J.J. Febrile convulsions: neurological sequelae and mental retardation. In Brain dysfunction in febrile convulsions, ed. by M.A. Brazier and F. Coceani, pp. 247-257 (Raven, New York, 1976).

Cavanagh, J.B. and Meyer, A. Aetiological aspects of Ammon's horn sclerosis associated with temporal lobe epilepsy. Brit. Med. J., 2, 1403-1405 (1956).

Chevrie, J.J. and Aicardi, J. Duration and lateralization of febrile convulsions. Etiological factors. Epilepsia (Amst.), 16, 781-789 (1975).

Falconer, M.A.: Genetic and related aetiological factors in temporal lobe epilepsy; a review. Epilepsia (Amst.), 13, 279-285 (1971).

Falconer, M.A., Serafetinides, E.A. and Corsellis, J.P.N.: Etiology and pathogenesis of temporal lobe epilepsy. Arch. Neurol. (Chic.), 10, 233-248 (1964).

Gastaut, H., Poirier, F., Payan, H., Salamon, G., Toga, M. and Vigouroux, M.: H.H.E.: Syndrome. Hemiconvulsions, hemiplegia, epilepsy. Epilepsia (Amst.), 1, 418-447 (1960).

Isler, W.: Acute hemiplegias and hemisyndromes in childhood (Heinemann, London, 1971).

Lindsay, J.M.M.: Genetics and epilepsy: a model from critical path analysis. Epilepsia (Amst.) 12, 47-54 (1971).

Margerison, J.H. and Corsellis, J.A.N.: Epilepsy and the temporal lobe. A clinical, electroencephalographic and neuropathological study of the brain in epilepsy with particular reference to the temporal lobes. Brain, 89, 499-530 (1966).

Meldrum, B.S. and Brierley, J.B.: Prolonged epileptic seizures in primates. Ischemic cell change and its relation to ictal physiological events. Arch. Neurol. (Chic.), 28, 10-17 (1973).

Meldrum, B.S. and Horton, R.W.: Physiology of status epilepticus in primates. Arch. Neurol. (Chic.), 28, 1-9 (1973).

Meldrum, B.S., Vigouroux, R.A. and Brierley, J.B.: Systemic factors and epileptic brain damage: prolonged seizures in paralyzed, artificially ventilated baboon. Arch. Neurol. (Chic.). 29, 82-87 (1973).

Norman, R.M.: Neuropathological findings in acute hemiplegia in childhood, with special reference to epilepsy as a pathogenic factor. In: Acute hemiplegia in childhood, edited by M. Bax and R.G. Mitchell (Heinemann, London, 1962).

Norman, R.M.: The neuropathology of status epilepticus. Med. Sci. Law, 4, 46-51 (1964).

Ounsted, C., Lindsay, J.M.M. and Norman, R.: Biological factors in temporal lobe epilepsy. (Heinemann, London 1966).

Sano, K. and Malamud, N.: Clinical significance of sclerosis of the cornu Ammonis: ictal psychic phenomena. Arch. Neurol. Psychiat. (Chic.) 70, 40-53 (1953).

CHAPTER TWENTY-ONE

SPECIFIC POSTNATAL THREATS TO BRAIN DEVELOPMENT: DENDRITIC CHANGES

MADGE E. SCHEIBEL AND ARNOLD B. SCHEIBEL

Our approach to the problem of brain damage in infancy and childhood is a bit different from that of some of the other contributors to this book. For we have had the opportunity to see what are presumed to be remote consequences, structural and behavioral, in a series of patients who have come to surgery in middle life with various patterns of intractable partial epilepsy. (8, 12) The seizure equivalent states and the electrical storms which have been shown to accompany them have been found characteristically associated with the temporal lobes of these patients (17) slightly more than half of whom have histories of difficult and prolonged deliveries or infantile febrile convulsions.

We shall describe some of the morphologic changes which we have found in the hippocampi of approximately 40 individuals, mostly young adults with clinical seizures. In the context of this book we consider them as possible models of pathology, resulting from brain damage of one sort or another, suffered in infancy or childhood. However, as we shall show, we are not convinced that these patterns of temporal lobe pathology and disordered cortical functions stand solely as monuments to discrete insults to the brain, whether genetic, traumatic, or infectious in early life. The question raised alternatively is whether these may signal, instead, an active and continuing process in which previously uninvolved neurons become involved in the degenerative sequence, culminating in cell death.

This paper is based largely on the study of Golgi impregnated specimens of temporal neocortex and archicortex. Although the limitations of the method for quantitative neurobiology are well known, its compensating features, especially the panoramic view it affords of the nerve cell in its neuropilic environment, more than compensate for this deficiency. The dendritic systems of neurons turn out to be extraordinarily sensitive indices of pathology, and it is these elements more than any other which are susceptible to visualization with the Golgi techniques. When used in conjunction with the electron microscope, the combination promotes optimal opportunity for the critical structural study of the neuron and its setting.

Our observations are based on material from two sources: a) surgical specimens removed from patients with temporal lobe epilepsy (n = 32) (8, 12), and
formalin preserved material originally obtained at autopsy and located with
the help of the Medical Center computer (n = 8). All of the former patient
group are alive and being followed periodically. They represent a fairly homogenous group, medically still in their young or middle years and, so far as we
know, not suffering from other medical diseases of great consequence. The
latter patient group are more disparate, showing a wide range of diagnosed
pathology at death, and identified primarily on the basis of a shared diagnosis
of clinical seizures. Since six of these eight patients were less than 18 years of
age at death, it is not surprising that congenital or genetic disease was found in
the majority of them.

The range of dendritic change which we have observed is great and includes, at one end of the spectrum, minimal spine loss (or non-formation) and
nodule formation on portions of a single dendrite in the dendritic domain of
one neuron, to extensive deteriorative changes in the total dendritic domains
of cell ensembles, culminating in complete disappearance of groups of neurons, often with glial replacement. In one or two cases of brain defective
epileptic children we have seen what look like outgrowths of very small super-
numerary dendrites or extraordinary hyperplastic spine systems growing
from both cell body and dendrite. While we can only guess at the significance
of these small structures, they serve to emphasize the protean nature of dendritic response to pathogenic stimuli.

The most frequent and obvious change which we have seen in our series of
resected hippocampi is an apparent diminution or loss of dendrite spines and
an increasing irregularity of dendrite silhouette. Various stages of such
changes appear in figure 1, which also indicates that patchy loss of spines may
appear anywhere along apical or basilar dendrites in either terminal or more
central position along the shaft. In most cases spine loss is accompanied by
the appearance of series of nodules scattered along the course of the dendrite
producing a "string-of-beads" type of deformity.

We have previously discussed the possible significance of these phenomena,
(12) emphasizing that absence of spines and appearance of nodules may in
themselves have no pathologic significance, depending on the circumstances.

For instance, spines are characteristically lost from the cell body and proximal segment of apical dendrite of most cortical pyramidal cells during the
immediate prenatal or perinatal period in most mammals. (11) The majority

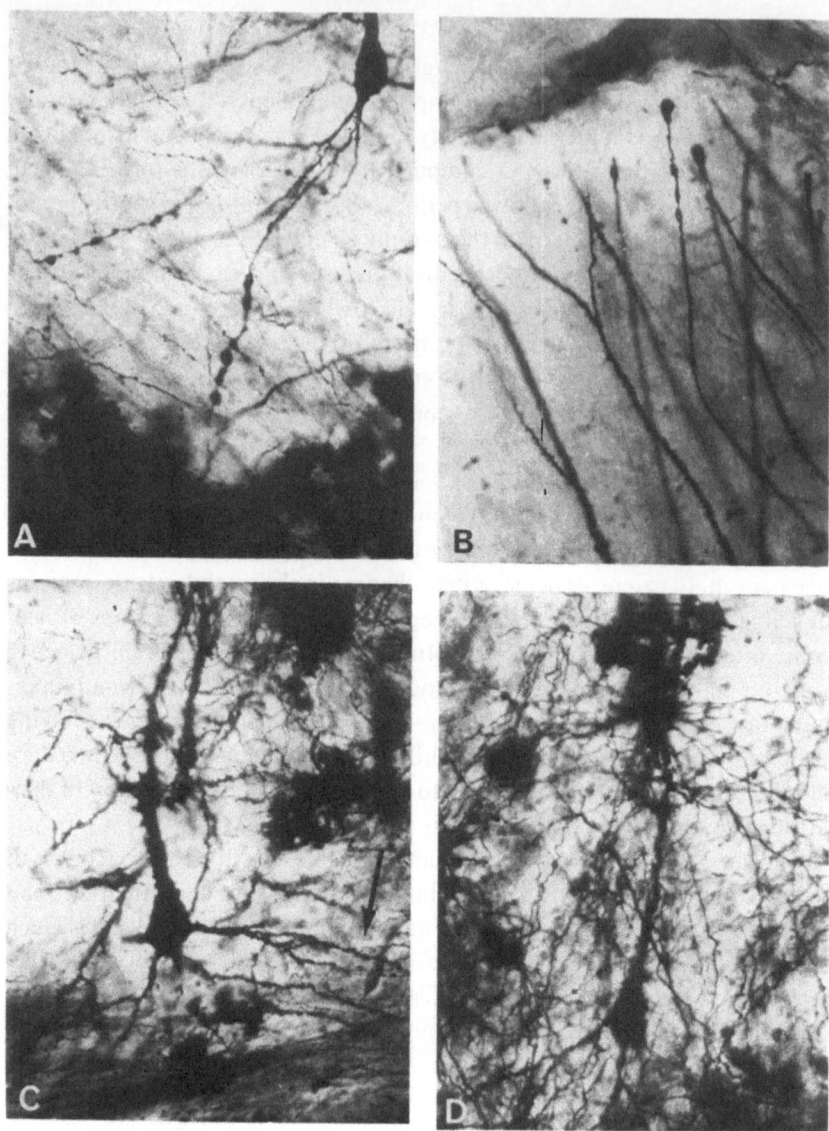

Fig. 1. Early changes of a presumed pathologic nature in human hippocampal pyramids. (*A*) Basilar dendrites of CA$_1$ pyramid showing nodular changes and loss of spines. (*B*) Loss of spines and development of terminal nodules on apical shafts of CA$_1$ pyramid. (*C*) Large CA$_1$ pyramid showing somewhat more generalized changes with loss of spines along basilar dendrites, small nodules, and coarsening of apical shaft. (*D*) Single CÅ$_1$ pyramid sitting in a nest of fibrillary astrocytes. This probably represents a moderately scarred cortical focus. Rapid Golgi variants. Magnification; *A, C,* and *D* 250x; *B*, 450x. (From Scheibel et al. 8.)

of short-axoned (local circuit) cells of the cortex never seem to develop spine arrays. (2) Somewhat similarly, most large and medium-sized neurons of the brain stem reticular formation lose their dendrite spines in the postnatal period, resulting in quite smooth spineless silhouettes for the mature dendrite system. (9) On the other hand, abundant experimental data support the observation that dendrite spines disappear following deafferentation. (3, 5, 16) In the young adult material we have studied, although all changes are, of course, frozen in time, and we are as yet unable to follow sequential changes in one cell, we nevertheless believe that spine loss is a harbinger of more profound and possibly irreversible dendritic changes.

Nodulation along both dendrites and axons is also found in non-pathologic states. Short-axoned cells of the cortex in particular may show periodic swellings along their otherwise smooth dendrite systems, as may the axons of many immature neurons. Under the electron microscope, these enlargements have been found crowded with enlarged cisternae of the endoplasmic reticulum, (7) suggesting that they may be the sites of enhanced protein-synthesis. If so, the "string-of-beads" deformity may assume broader significance than simply that of an index of pathology. It may, instead, represent a cell-dendrite system whose protein-synthesizing capacity is under challenge, whether for physiologic or pathologic reasons.

Figures 2, 3, and 4 show examples of more extensive dendritic pathology. In addition to increased spine loss and nodulation, there is distortion and loss of parts of the dendritic tree, along with the appearance of irregular enlargements distributed without obvious pattern along the shafts (Figs. 2, 3 and 4). These curious structures are pleomorphic and may appear as leafy excrescences, large bulbs, or horn-shaped structures as much as 15-20 mμ across. They are reminiscent of, although not particularly similar to, the multifoliate spines or microdendrites which we have seen in one seizure-ridden Down's syndrome child (see below).

Distortions of the dendritic domain range from moderately mild changes in overall domain pattern to bizarre curling and shrinkage of basilar systems and apical systems culminating in fragmentation of the stalks and total disappearance of the neuron. Domain changes have been prominent in the dentate fascia where granule cells often show a tendency to react en masse. Groups of them sometimes appear bent as if in a high wind (fig. 6). We have suggested that this windswept deformity may be due to traction from an adjacent gliotic scar. (12) A second type of aberrant domain pattern involves a tight cylindrical grouping of the dendrites instead of the inverted cone which constitutes the usual granule cell dendritic domain. The cause of this "closed-parasol

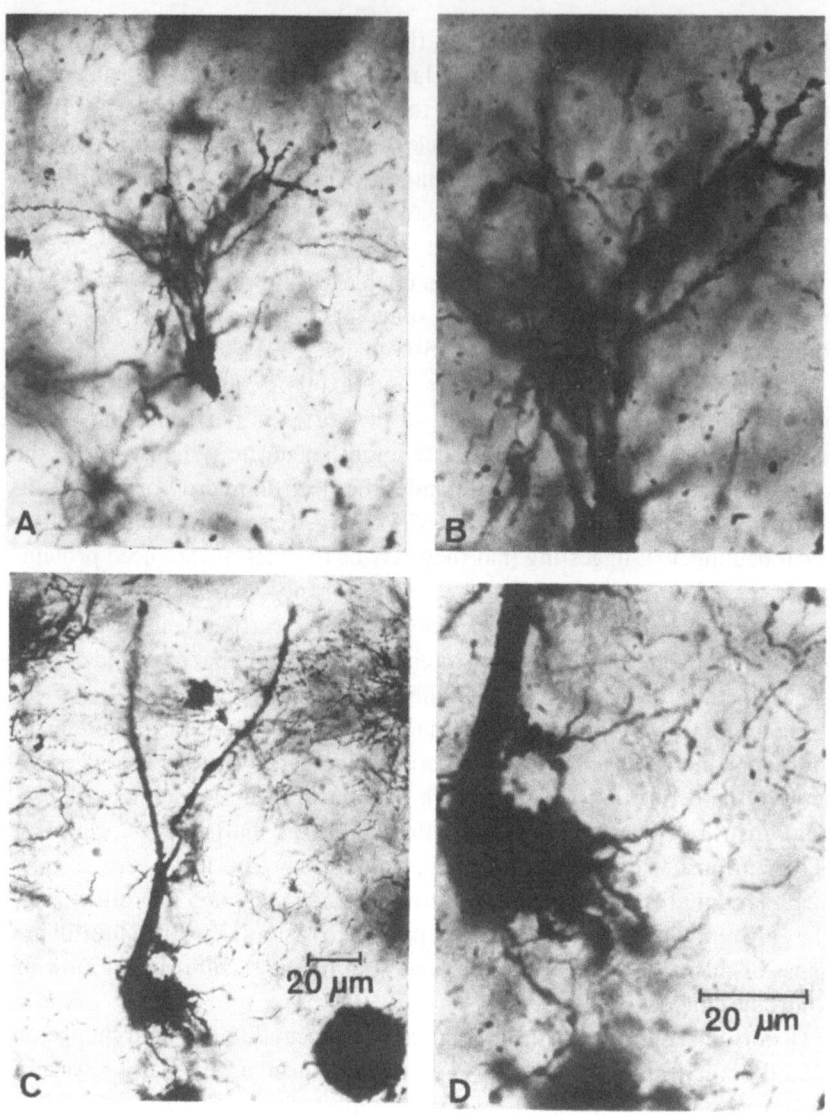

Fig. 2. Advanced changes in 2 pyramids from human hippocampus. A and B, small Ca$_2$ cell showing marked shrinkage and distortion of apical dendrite system with loss of spines and development of nodules. C and D, degenerating cell from CA$_1$ which has lost most of its dendritic demain, and with extensive shriveling of basilar dendrites. Rapid Golgi variants. Magnification A and C, 200 x; B and D, 450x. (From Scheibel et al. 8.)

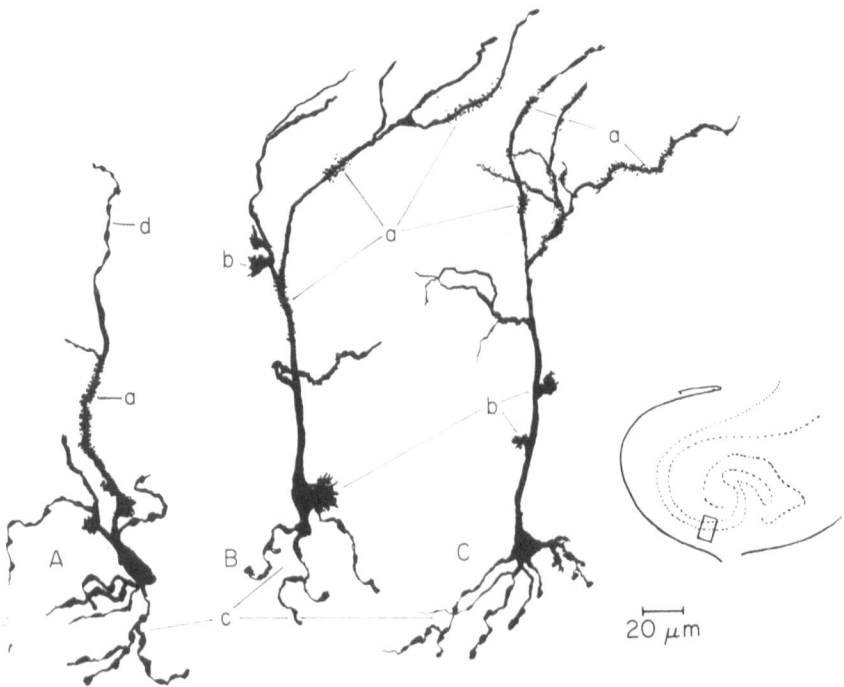

Fig. 3. Drawings of three human hippocampal pyramids from CA$_3$ showing advanced stage of pathologic change. Each dendrite system contains patchy remaining areas of dendrite spines, *a*, and scattered leafy protrusions, *b*. The apical shaft of neuron *a*, also shows marked terminal nodulation, *d*. Basilar dendrites are shrivelled and nodulated, *c*, while apical shafts are shortened and distorted. Rapid Golgi variant. (From Scheibel and Scheibel 12.)

deformity" (fig. 5d) is not known, nor is it completely clear that its presence inevitably indicates pathology. More advanced dentate pathology invariably includes loss of spines and nodulation with a characteristic thinning in diameter of the dendrite stalks. The dentate granules themselves seldom degenerate completely and disappear as hippocampal pyramids do, so the eventual result is a palisade of tightly packed somata emitting their nodulated fibers similar to fibrillary astrocytes.

The various aspects of dendrite pathology already described are based largely on examination of temporal lobe tissue blocks resected from adults. Somewhat different data have been obtained from formalin fixed tissue obtained from autopsy material on two Down's syndrome children, aged nine and twelve, with various related congenital defects. Exemplifying both cases,

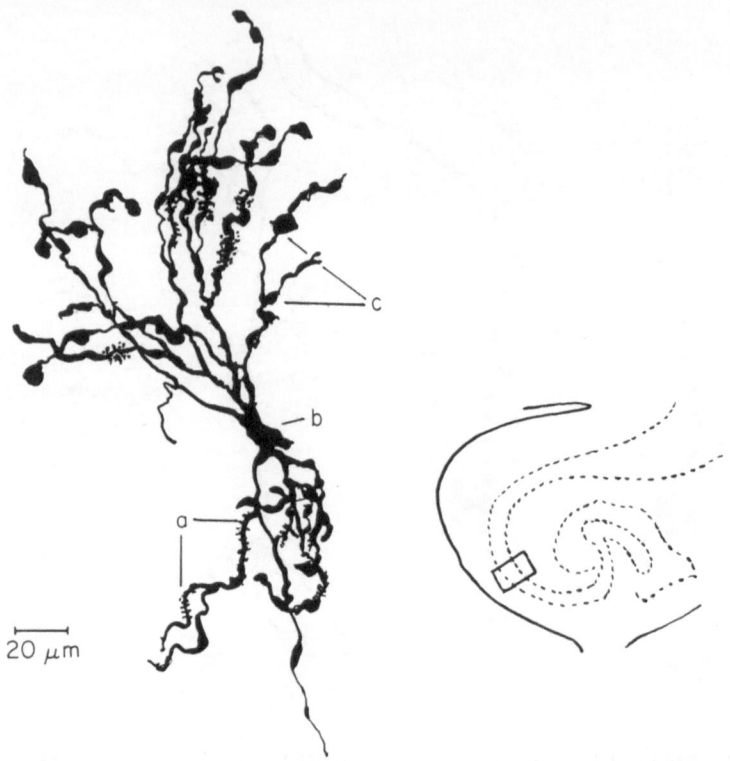

Fig. 4. Drawing of human hippocampal pyramid from CA$_2$ showing advanced changes in the entire dendrite system. There are remaining patches of dendrite spines, *a*; profuse nodulation, *b*; and an eccentrically shaped, probably pyknotic cell body. Rapid Golgi variant. (From Scheibel and Scheibel 12.)

the 12-year-old suffered from a progressive seizure disorder which had initially become clinically obvious three years before his death. Examination of the hippocampus revealed many unusual filiform extension from the surfaces of both pyramidal cell bodies and dendrites (fig. 8). These structures averaged 5-20 mµ in length and usually possessed a stem or pedicle and a broadened, multiple branched terminal structure. It was difficult to decide whether they represented some type of ectopic microdendrites, or a late variant of the heteromorphic protospine which we have described along the surfaces of many cortical pyramids in the late prenatal period. (11) In any case, 1-4 of these structures have been identified on individual hippocampal pyramidal cell bodies, and larger numbers of them along dendrite surfaces. We can only speculate whether they represent archaic structures held over long beyond

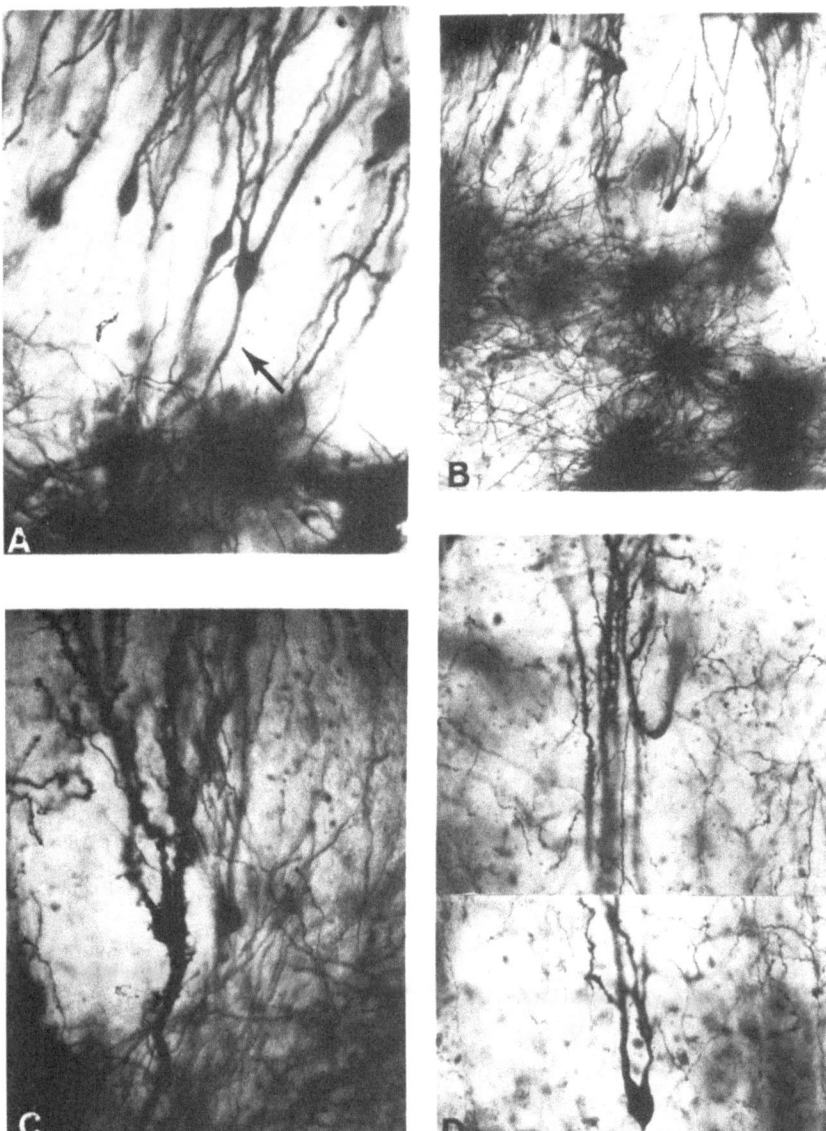

Fig. 5. Dentate granules from several young adults with temporal lobe epilepsy. *A*, group of granules showing some decrease in spines and an unusual dendrite (arrow) protruding from the lower pole of a cell body; *B*, dense fibrillary astrocytosis deep to granule layer; *C*, dentate granule with large lower pole dendrite and many excrescences along dendrite shafts; *D*, dentate granule with marked side-to-side compression of dendrite domain forming 'closed parasol' deformity. Scattered nodulation is also present. Rapid Golgi variants. Magnification; *A*, *C*, and *D*, 250x; *B*, 100x. (From Scheibel and Scheibel 12.)

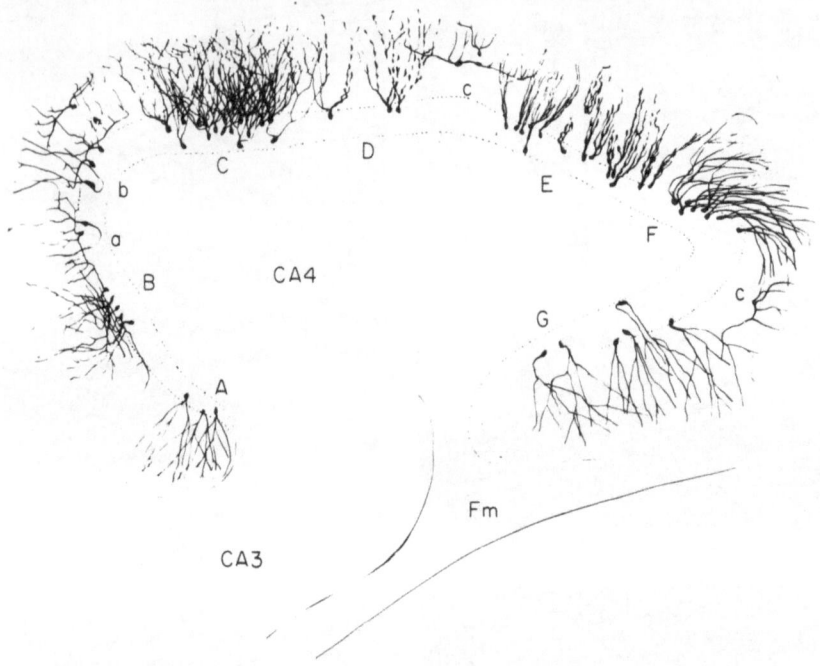

Fig. 6. Composite drawing of the human dentate fascia showing a variety of patterns found in tissue removed surgically from temporal lobe epilepsy patients. *A*, usual appearance; *B*, granules with extensive horizontal ramifications; cells at *Ba* and *Bb* have atypical dendrites springing from opposite end of somata; *C*, mass of apparently normal granules showing density of dendrite plexus; *D*, granules with nodulated dendrites; *E*, cells with 'closed parasol' deformity; *F*, granules with 'windblown' look; *G*, presumably normal granules with pleomorphic patterning of dendrites. Cells marked *c*, are probably displaced granules with horizontal arbors. Rapid Golgi variants. (From Scheibel and Scheibel 12.)

their appropriate period, or are subsequent structural additions to enhance the neuronal receptive surface.

Neuroglial changes, usually of a proliferative nature, are classic concomitants of neural pathology, particularly in those syndromes which include clinical seizures. Our own studies bear this out, and dense fields of fibrillary astroglia are frequently seen (fig. 7d), especially in those areas where there has been a marked reduction in total numbers of neurons. The stalks of the individual gliocytes are frequently studded with nodules of various sizes, some appearing half as large as the glial cell body.

We have stressed the variation in intensity of pathology in the neurons of many patients, some of whom have suffered clinical seizures for ten years or

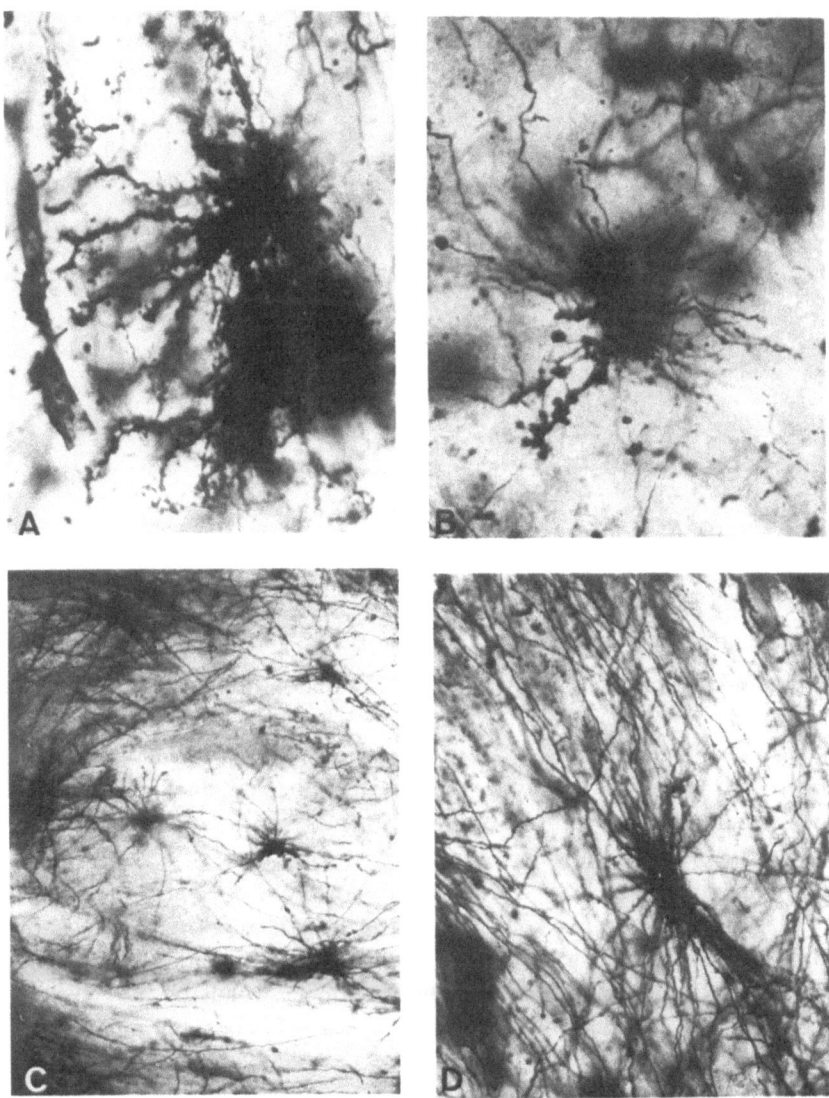

Fig. 7. Glial patterns in human hippocampus. (*A*) normal protoplasmic astrocyte; (*B*) small fibrous astrocyte with nodulated processes; (*C*) field of fibrillary astrocytes in CA$_4$; (*D*) hippocampal scarred area made up of many fibrillary astrocytes. Rapid Golgi variants. Magnification, *A*, *B*, and *D*, 250x; *C*, 100x. (From Scheibel and Scheibel 12.)

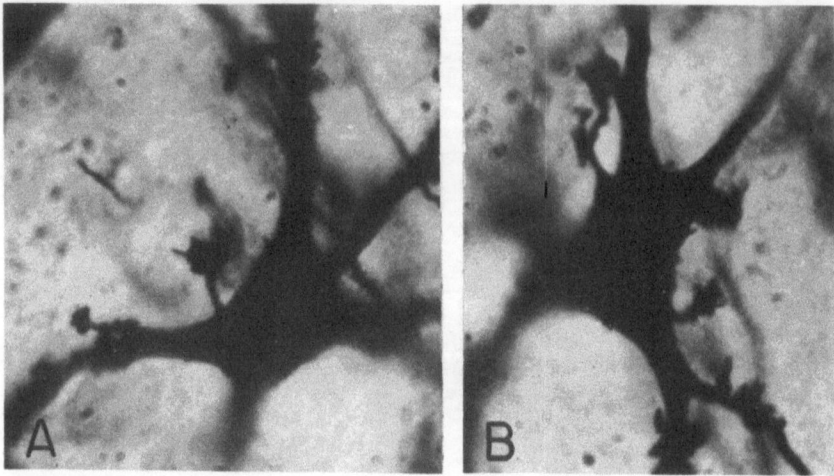

Fig. 8. Hippocampal pyramid cell bodies from a Down's syndrome child with seizures. Unusual excrescences grow out of the somata and initial portions of the major dendrites. See text. Rapid Golgi variant. 450x.

more. While many cells may have disappeared entirely, or show marked pathology, some neurons show changes so localized that it is difficult to escape the inference that they are in the earliest stages of pathologic involvement. Our Golgi studies leave us with the impression of an ongoing process whose pathology proceeds slowly but inexorably, with progressive involvement of previously uninvolved neurons.

In a parallel study using electron microscopy, our colleague, W.J. Brown (1) has found numerous examples of degenerative changes in what appear to be the smallest dendrite branchlets of hippocampal pyramids and in fine, possibly preterminal, myelinated axons. He too has wondered about the possibility of an ongoing process to account for these minimal morphologic changes.

Since this material is presented within the framework of a textbook on brain damage studies in infancy and childhood, we present some of the longer range pathologic concomitants of the infantile ictal state. The material is from patients with long term partial epilepsy – i.e. psychomotor or equivalent states – who became candidates for surgery following the failure of all attempts at medical control. (17) Temporal-lobe epilepsy is, perhaps more than any other ictal syndrome, believed causally related to events of the perinatal period, whether traumatic or febrile-infectious. Yet, it must be pointed out that approximately 40% of the cases we studied had no recorded history of birth difficulties or early seizures, febrile or otherwise. Furthermore, of the 32 specimens we examined, 14 (44%) showed no pathologic changes of which we could be certain. Such figures are not too different from those of other workers, including Falconer and his associates, (9) and emphasize the problems which still exist in wedding pathology to causation.

Brown (1) has reminded us that there is still no agreement on morphologic criteria which set apart the epileptogenic from the normal neuron. The changes which we have described here may also be seen under certain circumstances in cortices made seizure-sensitive with alumina gel (18) and, in somewhat different form, in various stages of senescence. (10) The dendrite is a sensitive indication of state changes in its immediate environment, but its range of gross alterations is nonetheless limited. Surveys such as this, at the level of the light microscope, are useful in providing an overview, but levels of resolutions which allow more searching study of the membrane and the molecule must undoubtedly be looked to before a more definitive picture of the consequences of disorders such as postnatal seizures, can be drawn.

In the meantime, with regard to those of our temporal lobe patients who did show dendritic pathology, we would stress the multifaceted nature of the changes, combining burnt-out, glial-replaced fields; distorted, shrunken dendritic domains; and minimal changes in spine distribution and density on individual dendritic branches. Certainly one thinkable interpretation of such a picture is that of an ongoing pathologic process continuing to recruit neurons year after year. The concept of the epileptic syndrome as a progressive process is, of course, not new (13-15) but has been related invariably to ischemic processes developing during recurrent ictal attacks. However, the nature and distribution of the early changes we have described appear too discrete to fit this interpretation. It is, at the moment, entirely speculative to relate these apparent evidences of progression to some slow infectious agent, or to a hypothetical genetic-enzymic fault. However, we feel that the data do, at the least, cast serious question on presently fashionable notions of epileptic cortices as monuments to a discrete episode in the early history of the individual.

SUMMARY

The dendritic apparatus of the nerve cell is a sensitive index of neuronal pathology and may be studied readily with the light microscope after impregnation with variations of the Golgi methods. We have examined temporal lobe tissue resected from patients with several patterns of long term partial epilepsy, particularly psychomotor equivalent states. In approximately one half of the cases, there was a history of difficult or prolonged labor or of febrile convulsions during infancy. A variety of changes was noted in the hippocampal-dentate complex ranging from minimal localized variations in the surface morphology of dendrites and neuroglia to profound distortions of soma-dendritic dendritic architecture, extensive loss of neurons and overgrowth of astroglia. The most subtle changes included local areas of dendritic spine loss in conjunction with dendritic nodulation, which might appear on only a few neurons of a large ensemble, or in an even more restricted sense, on one or two dendrite shafts of an individual neuron. More advanced pathologic changes included extensive or total spine loss and nodulation, progressive retraction of dendrite shafts, and pyknotic changes in cell bodies accompanied by apparent overgrowth of fibrillary astroglia with nodulation of many glial processes. In addition, conformational changes were often seen in dendritic domains of dentate granules including the apparent inward collapse of dendrite systems upon the axis of the dendrite domain (closed parasol deformity) and marked lateral deformation of fields of dendrites (wind blown look) in possible response to neighboring loci of scar formation.

The relationship of such pathologic changes to specific postnatal insults is for the present, at best putative. However, in a smaller series of autopsy specimens from infants and children whose premortem clinical histories included repeated convulsions, widespread changes were found in several, characterized by hippocampal somata and dendrites bearing highly abnormal spines with multiple secondary branches radiating from individual spine pedicles. This apparent overgrowth of receptive postsynaptic membrane may represent an immediate morphologic substrate of the seizured state.

The relationship of the infantile ictus to later pathology is far from precise. In addition, most of the pathologic hippocampi that we examined gave evidence of including minimal, early dendritic changes, along with massive burnt-out zones containing only glia. The possibility is raised that the lesions we have described represent ongoing disease processes which, whatever their etiology, continue to recruit intact neurons to the pathologic process for many years.

REFERENCES

1. Brown, W.J.: Structural substitutes of seizure foci in the human temporal lobe. In Brazier, Epilepsy. Its Phenomena in Man, pp. 341-374. Academic Press, Inc., N.Y., 1973.
2. Cajal. S.R.Y.: Histologie du Systeme Nerveux de l'Homme et des Vertébrés. Vol. I. Maloine. Paris, 1909.
3. Colonnier, M.: Experimental degeneration in the cerebral cortex. J. Anat. 98: 47-53.
4. Falconer, M.A., E.A. Serafetinides and J.A.N. Corsellis.: Etiology and pathogenesis of temporal lobe epilepsy. Arch. Neurol. 10: 233-248, 1964.
5. Globus, A. and A.B. Scheibel.: Synaptic loci on visual cortical neurons of the rabbit. The specific afferent radiation. Exp. Neurol. 18: 116-131, 1967.
6. Meyer, A., Beck, E. and Shepherd, M.: Unusually severe lesions in the brain following status epilepticus. J. Neurol. 18: 24, 1955.
7. Peters, A.: Stellate cells of the rat parietal cortex. J. Comp. Neurol. 141: 354-373, 1971.
8. Scheibel, M.E., Crandell, P.H. and Scheibel, A.B.: The hippocampal dentate complex in temporal lobe epilepsy. Epilepsia 15: 55-80, 1974.
9. Scheibel, M.E., Davies, T.L. and Scheibel, A.B.: Maturation of reticular dendrites, loss of spines and development of bundles. Exp. Neurol. 38: 301-310, 1973.
10. Scheibel, M.E., Lindsay, R.D., Tomiyasu, U. and Scheibel, A.B.: Progressive dendritic changes in the aging human limbic system. Exp. Neurol. 53: 520-530, 1976.
11. Scheibel, M.E. and Scheibel, A.B.: Selected structural-functional correlations in postnatal brain. In Sterman, McGinty and Adinolfi (eds), Brain Development and Behavior, pp. 1-21, Academic Press, Inc., N.Y., 1971.
12. Scheibel, M.E. and Scheibel, A.B.: Hypocampal pathology in temporal lobe epilepsy. A Golgi survey. In Brazier, Epilepsy, Its Phenomena in Man, pp. 315-340. Academic Press, Inc., N.Y., 1973.
13. Scholz, W.: Les lesions cerebrales recontrées chez les epileptiques; precisions sur la sclerose de la corne d'Ammon. Acta Neurol. Psychiat. Belg. 56(1): 43-60, 1956.
14. Scholz, W. Selective neuronal necrosis and its topistic patterns in hypoxemia and oligemia. J. Neuropathol. Exp. Neurol. 12: 249-261, 1953.
15. Spielmeyer, W. The anatomic substratum of the convulsive state. Arch. Neurol. and Psychiat. 23: 869-875, 1930.
16. Walberg, F. Role of normal dendrites in removal of degenerating terminal boutons. Exp. Neurol. 120: 1-17, 1963.
17. Walter, Richard D. Tactical considerations leading to surgical treatment of limbic epilepsy. In Brazier, Epilepsy, Its Phenomena in Man, pp. 99-118, Academic Press, Inc., N.Y., 1973.
18. Westrum, L.E., White, L.E. and Ward, A.A. Jr.: Morphology of the experimental epileptic focus. J. Neurosurg. 21: 1033-1046, 1964.

CHAPTER TWENTY-TWO

METABOLIC ASPECTS OF CEREBRAL ANOXIA IN THE FETUS AND NEWBORN*

Thomas E. Duffy, Ph.D. and Robert C. Vannucci, M.D.

Developing animals tolerate anoxia and asphyxia better than adults, and fetuses are more resistant than newborns (Glass et al., 1944; Dawes et al., 1959; Duffy et al., 1975). The greater resistance of the immature cardiovascular system partly accounts for this decreased vulnerability, but the resistance of the developing nervous system seems to be the major determinant of anoxic morbidity and mortality. The greater susceptibility of the nervous system, compared to the heart, is emphasized by the fact that when newborn dogs or rats are exposed to an atmosphere of nitrogen, the heart continues to beat long after respiratory movements cease (Fazekas et al., 1941; Swann et al., 1954; Adolph, 1974), implying loss of central respiratory control prior to circulatory failure.

Several factors have been suggested to account for the enhanced tolerance of the immature brain to anoxic insults. The developing brain, presumably owing to its poor synaptic organization (Aghajanian & Bloom, 1967), has a lower metabolic requirement than that of the adult (Thurston & McDougal, 1969; Duffy et al., 1975); it may also contain larger endogenous energy stores (Mayman & Tijerina, 1971). Young animals are relatively poikilothermic and lower body temperatures during anoxia would be expected to further reduce overall energy demands. Differences in regional cerebral blood flow may also play some role.

The survival times of fetal rats at term and of postnatal rats during exposure to nitrogen at 37°C are depicted in figure 1. The fetuses, aged 12-18 hours prior to expected delivery, were obtained following decapitation of the dam and rapid hysterotomy. Comparing the duration of anoxia required to kill 85% of the animals, fetuses lived 50 times longer than young adult rats, 5 times longer than 7-day-olds, and twice as long as 1-day-old neonates.

* These investigations were supported in part by U.S. Public Health Service Grant NS-03346 from the National Institutes of Health. Dr. Duffy is an Established Investigator of the American Heart Association.

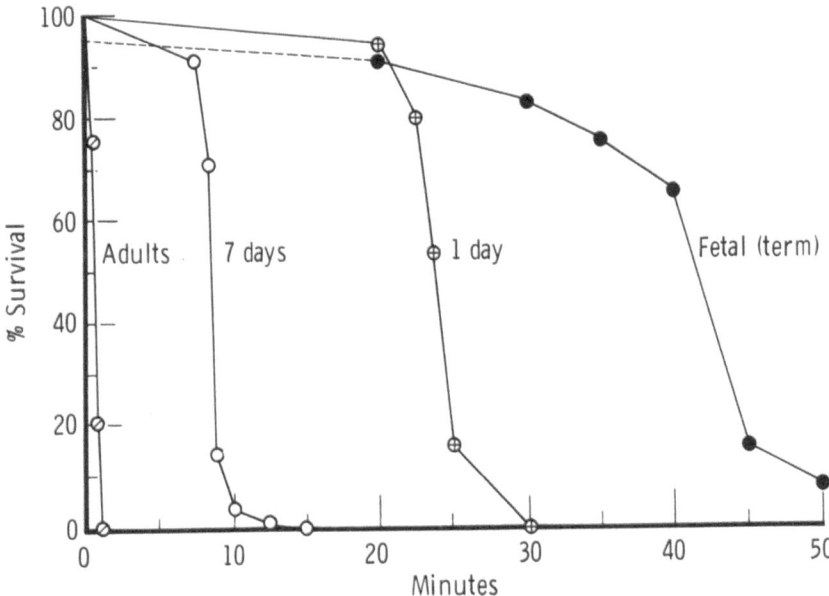

Fig. 1. Survival of rats in nitrogen as a function of age. Animals were placed in a humidified chamber at 37° C that was continuously flushed with nitrogen gas. Survival was assessed on the basis of the animal's ability to recover spontaneously when returned to room air. With every litter there is an expected infant mortality of 5%; therefore, fetal survival without anoxia is extrapolated to 95% initially. Data of Duffy, Kohle & Vannucci (1975).

Since anoxia blocks oxidative energy production by the brain, animals with greater endogenous cerebral energy stores should be more resistant to oxygen lack. Compared in table 1 are the concentrations in whole brain (rat) and cerebral cortex (dog) of the major energy precursors measured in animals of different ages. In rats, glycogen was highest in fetuses and declined with age, whereas phosphocreatine was lowest in fetuses and highest in adults. ATP levels were found to be similar in all age groups. Compared to rats, lower concentrations of ATP were observed in dog brain, but the values for ATP were similar in newborn and adult dogs. Despite some variations in concentrations of individual substrates, however, total preformed and potential energy reserves were similar in newborn and adult rats, and only slightly higher in fetuses. It seems likely that total energy stores in the newborn and adult dog cerebral cortex are also comparable, although we can't conclude this with certainty because values for glycogen in the newborn dog are not available.

318   THOMAS E. DUFFY, ROBERT C. VANNUCCI

Table 1. Cerebral metabolite levels in rat and dog at different ages.

Metabolite	Rat Fetus	Newborn	Adult*	Dog Newborn	Adult**
	(mmol/kg)				
Glycogen	5.77	3.99	3.20	–	(5.16)
Glucose	1.39	1.89	1.39***	2.55	2.19
P-creatine	1.74	3.16	4.90	2.38	2.70
ATP	2.66	2.63	2.76	2.14	2.12
ADP	0.43	0.25	0.38	0.20	0.32
2Total ~ P 24.0	27.0	24.0	22.9	12.0	11.6

Values for fetal and newborn rats represent concentrations in whole forebrain; values for adult rat and for newborn and adult dogs represent concentrations in cerebral cortex. Total potential energy reserves (~P) were calculated as the sum of 2 × ATP + ADP + Phosphocreatine + 2 × Glucose + 2.9 × Glycogen. Data for fetal and newborn rats are from Vannucci & Duffy (1974); data for newborn dogs are from Vannucci & Duffy (1977). Abbreviation: P-creatine, phosphocreatine.
   * Data of Duffy, Howse and Plum (1975).
  ** Data of Drewes & Gilboe (1973) and of Minsker et al. (1970).
 *** Data of Miller et al. (1973).

Energy utilization in brain can be estimated from the cerebral oxygen consumption in vivo, or, as suggested by Lowry et al. (1964), by measurement of the rate of change in the pool of the major endogenous energy precursors when the brain is converted to a "closed system" by decapitation. When blood flow to the brain is cut off, functional activity in the isolated brain must be maintained at the expense of the available high-energy phosphates (ATP, phosphocreatine) and by the energy derivable from the anaerobic breakdown of glucose and glycogen to lactic acid. Thus from measurements of the changes in concentrations of these components during ischemia, the initial rate of cerebral energy use (metabolic rate) can be calculated.

Illustrated in table 2 are the cerebral energy use rates of rats of different ages compared with the animals' tolerance to nitrogen anoxia and to total cerebral ischemia, as reflected by the time to last gasp of the decapitated head. Resistance to anoxia was inversely related to the energy demands of the brain and, except for fetuses, was nearly quantitatively so. The greater tolerance of fetuses to anoxia compared to newborns, despite similar resting cerebral energy use rates in these age groups, results in part from the fact that fetuses develop extreme hypercapnia during exposure to nitrogen (PaCO$_2$ 150-200 mm Hg) which depresses their cerebral metabolism (Vannucci & Duffy, 1976).

In the early 1940's, Himwich and colleagues (Fazekas et al., 1941; see also Himwich, 1951) showed that as resistance of developing animals to anoxia declined with maturity, the oxidative metabolism of their brains progressively

META BOLIC ASPECTS 319

Table 2. Relationship between tolerance to anoxia-ischemia and cerebral energy consumption in rats of different ages.

Age	Survival Time (LD$_{85}$) LD$_{85}$ in Nitrogen (min)	Last Gasp of the Decapitated Head (min)	~P Use Rate (mmol kg min)
Fetus at Term	45	26.4	1.6
1 Day Old	25	19.5	1.3
7 Days Old	8.8	6.9	2.6
Adult	0.8	0.3	27*

The data are from Duffy, Kohle and Vannucci (1975). Cerebral energy consumption (~P use rate) was calculated according to the procedure of Lowry et al. (1964).
* Data of Swaab & Boer (1972).

increased. From this and other evidence, the concept evolved that the newborn is a "glycolytic" animal, the implication being that a greater capacity to generate energy via anaerobic glycolysis underlies the immature brain's extraordinary tolerance to anoxic insults. However, much data obtained both in vitro and in vivo indicate that aerobic and anaerobic glycolysis are greater in adult brain. Glycolytic flux, expressed either as the rate of glucose consumption or of lactate production, was about twice as fast in slices of adult dog (O'Neill & Duffy, 1966) and adult rat (Itoh & Quastel, 1970) cerebral cortex, incubated in an oxygen atmosphere, compared with similar tissues prepared from 1- to 3-day-old animals (table 3). Moreover, minces of adult dog cerebral cortex produced 5 times as much lactic acid when incubated anaerobically than did comparable tissue from newborn puppies (Chesler & Himwich, 1944).

Table 3. Glycolysis of newborn and adult brain-cortex in vitro.

Animal	Conditions (mmol/kg wet wgt/hr)	Glucose Consumption	Lactate Production
Rat (newborn)*	Aerobic	15.1	15.0
(adult)	Aerobic	35.5	31.8
Dog (newborn)**	Aerobic	9.7	9.7
(adult)	Aerobic	17.9	23.6
Rat (newborn)***	Anaerobic	—	20.4
(adult)	Anaerobic	—	99.8

* Data of Itoh & Quastel (1970).
** Data of O'Neill & Duffy (1966).
*** Data of Chesler & Himwich (1944).

Table 4. Glycolytic capacity, energy ($\sim$ P) use, and the percentage of energy demands supplied anaerobically at maximal glycolytic flux in developing rat brain.

Age	Glycolytic Capacity* (nmol lactate/kg/min)	$\sim$ P Use Rate** (nmol $\sim$ P kg min)	$\dfrac{\text{P via Glycolysis}}{\text{P use rate}}$
Fetus at Term	0.85	1.6	53%
1 Day Old	0.69	1.3	53%
7 Days Old	1.40	2.6	54%
Adult***	12.1	27	45%

* Lactate accumulation in brain following decapitation, calculated from 0-5 min (fetuses), 0-5 min (1-day-olds), 0-2 min (7-day-olds), and 0.8-8 sec (adults). The data are from Duffy, Kohle and Vannucci (1975) and Swaab & Boer (1972).
** Energy use rates from Table 2.
*** The values are calculated from the data of Swaab & Boer (1972).

The maximal glycolytic capacity of the brain in vivo can be estimated directly by measurement of the rate of accumulation of lactic acid in brain over a short interval following decapitation (Lowry et al., 1964). Assuming that for each mole of lactate formed, one mole of ATP has been synthesized via glycolytic reactions, the energy derivable from anaerobic glycolysis can be estimated. The data in table 4 show the relationship between the maximal rate of energy production via anaerobic glycolysis in brain and the cerebral energy requirements of rats of different ages. Note the much greater glycolytic capacity of the adult rat brain, and that the percentage of the cerebral energy demands that could be supplied by anaerobic glycolysis are nearly the same for all ages.

Compared with rodents, newborn dogs offer advantages for studies of cerebral metabolism in anoxia because i) physiologic monitoring is possible, ii) brain maturation at birth is nearer to that of human infants, and iii) the larger brains permit regional comparisons of metabolism and blood flow. Puppies (1 to 2 days old), that were paralyzed and artificially ventilated with 30% oxygen-70% nitrous oxide, tolerated 15 minutes, but not 20 minutes, of total respiratory arrest with apparently normal recovery. Asphyxia was always accompanied by bradycardia, hypotension and a progressive fall in arterial blood pH (figure 2). The electroencephalogram invariably became isoelectric within 2 minutes of respiratory arrest.

Because of the marked hypotension, it seemed likely that blood flow to the puppy brain would be compromised during the asphyxial insult. To qualitatively assess cerebral blood flow, finely-divided carbon particles were injected intravenously into puppies at intervals during the period of respiratory arrest; after 3 minutes of circulation time, the animals were decapitated and

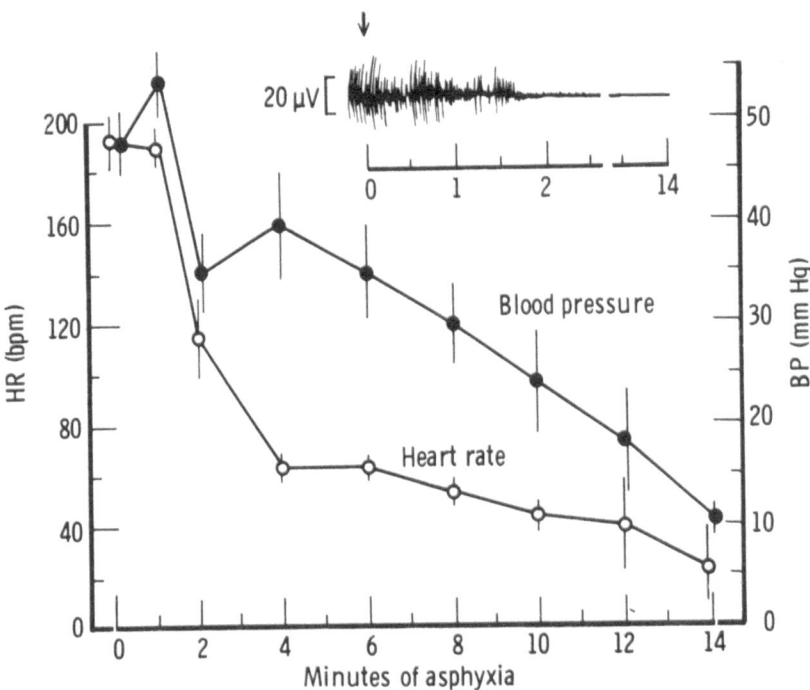

Fig. 2. Heart rate (HR) and mean arterial blood pressure (BP) in newborn dogs during asphyxia. Puppies, aged 1 to 2 days, were paralyzed and passively ventilated to a physiologic steady state, and were then subjected to respiratory arrest for 15 minutes. Points denote means of at least 6 measurements ± S.E.M. Values for heart rate are expressed as beats per minute (bpm). A representative EEG showing the effect of asphyxia on the puppy is superimposed at the top of the figure; the arrow denotes the start of respiratory arrest. Data of Vannucci & Duffy (1977).

the brains fixed in formaldehyde: acetic acid: methanol (1:1:8). After 2.5 minutes of asphyxia, the brains appeared uniformly darkened, though less intensely so compared to control animals. After 5 minutes of asphyxia, there were pale areas throughout the cerebral cortex, and by 15 minutes of asphyxia the brains appeared completely pale. When coronal sections were prepared from animals that had been asphyxiated for 10 minutes (fig. 3), the carbon particles were found to be distributed non-uniformly with minimal staining of the cerebral cortex and preferential staining of the diencephalon and lower brain stem. Selective perfusion of the brain stem during asphyxia would be expected to provide temporary protection of the medullary respiratory and vasomotor centers, and may thereby prolong the newborn animal's survival.

In summary, immature animals exhibit extraordinary tolerance to anoxia and asphyxia compared to adults, largely reflecting the enhanced resistance of the developing nervous system. The explanation for this increased resistance seems to reside, not in any unique abiiity of the infant brain to generate energy anaerobically, but in its very low metabolic requirements. Differences in regional cerebral blood flow and/or metabolism during anoxic insults may also be contributory.

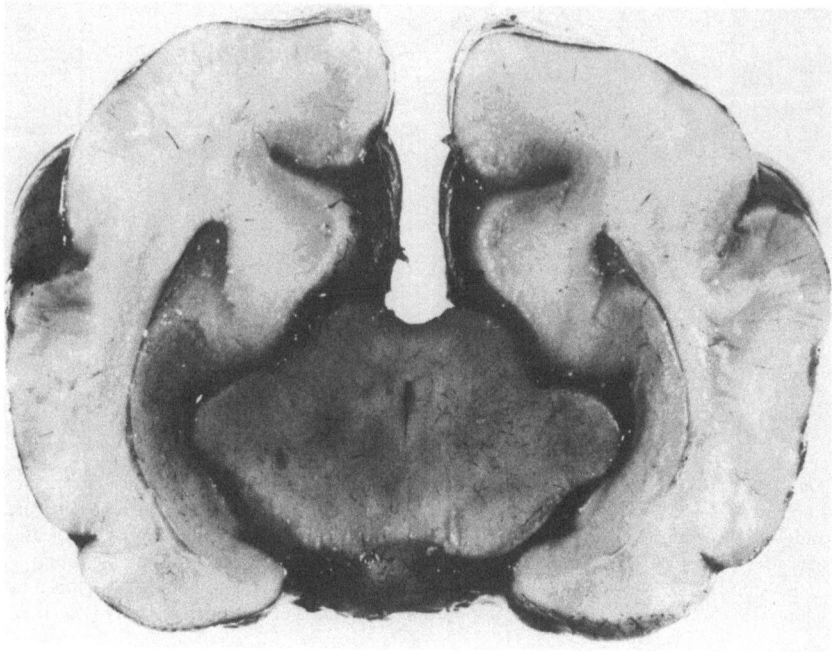

Fig. 3. Coronal section at the level of the superior colliculus from a newborn dog that was injected with carbon black after 10 minutes of respiratory arrest. Note the cortical and subcortical pallor relative to the brain stem (Magnification: 5 times). From Vannucci & Duffy (1977).

REFERENCES

Adolph, E.F.: Kinetics of adaptation to hypoxia in infant rats. Am. J. Physiol. 227: 1030-1032 (1974).
Aghajanian, G.K. & Bloom, F.E.: The formation of synaptic junctions in developing rat brain: A quantitative electron microscopic study. Brain Res. 6: 716-727 (1967).
Chesler, A. & Himwich, H.E.: Comparative studies of the rates of oxidation and glycolysis in the cerebral cortex and brain stem of the rat. Am. J. Physiol. 141: 513-517 (1944).
Dawes, G.S., Mott, J.C. & Shelley, H.J.: The importance of cardiac glycogen for the maintenance of life in foetal lambs and newborn animals during anoxia. J. Physiol. 146: 516-538 (1959).

Drewes, L.R. & Gilboe, D.D.: Glycolysis and the permeation of glucose and lactate in the isolated, perfused dog brain during anoxia and postanoxic recovery. J. Biol. Chem. 248: 2489-2496 (1973).

Duffy, T.E., Howse, D.C. & Plum, F.: Cerebral energy metabolism during experimental status epilepticus. J. Neurochem. 24: 925-934 (1975).

Duffy, T.E., Kohle, S.J. & Vannucci, R.C.: Carbohydrate and energy metabolism in perinatal rat brain: Relation to survival in anoxia. J. Neurochem. 24: 271-276 (1975).

Fazekas, J.F., Alexander, F.A.D. & Himwich, H.E.: Tolerance of the newborn to anoxia. Am. J. Physiol. 134: 281-287 (1941).

Glass, H.G., Snyder, F.F. & Webster, E.: The rate of decline in resistance to anoxia of rabbits, dogs, and guinea pigs from the onset of viability to adult life. Am. J. Physiol. 140: 609-615 (1944).

Himwich, H.E.: Brain Metabolism and Cerebral Disorders, (Williams & Wilkins, Baltimore 1951).

Itoh, T. & Quastel, J.H.: Acetoacetate metabolism in infant and adult rat brain in vitro. Biochem. J. 116: 641-655 (1970).

Lowry, O.H., Passonneau, J.V., Hasselberger, F.X. & Schulz, D.W.: Effect of ischemia on known substrates and cofactors of the glycolytic pathway in brain. J. Biol. Chem. 239: 18-30 (1964).

Mayman, C.I. & Tijerina, M.L.: The effect of hypoglycemia on energy reserves in adult and newborn brain. In Brain Hypoxia (Brierley, J.B. & Meldrum, B.S., Eds.), pp. 242-249. Lippincott, Philadelphia (1971).

Miller, A.L., Hawkins, R.A. & Veech, R.L.: Phenylkentonuria: Phenylalanine inhibits brain pyruvate kinase in vivo. Science 179: 904-906 (1973).

Minsker, D.H., Gilboe, D.D. & Stone, W.E.: Effects of shock and anoxia on nucleotides and creatine phosphate in the isolated brain of the dog. J. Neurochem. 17: 253-259 (1970).

O'Neill, J.J. & Duffy, T.E.: Alternate metabolic pathways in newborn brain. Life Sciences 5: 1849-1857 (1966).

Swaab, D.F. & Boer, K.: The presence of biologically labile compounds during ischemia and their relationship to the EEG in rat cerebral cortex and hypothalamus. J. Neurochem. 19: 2843-2853 (1972).

Swann, H.G., Christian, J.J. & Hamilton, C.: The process of anoxic death in newborn pups. Surg. Gynecol. Obstet. 99: 5-8 (1954).

Thurston, J.H. & McDougal, D.B., Jr.: Effect of ischemia on metabolism of the brain of the newborn mouse. Am. J. Physiol. 216: 348-352 (1969).

Vannucci, R.C. & Duffy, T.E.: Influence of birth on carbohydrate and energy metabolism in rat brain. Am. J. Physiol. 226: 933-940 (1974).

Vannucci, R.C. & Duffy, T.E.: Carbohydrate metabolism in fetal and neonatal rat brain during anoxia and recovery. Am. J. Physiol. 230: 1269-1275 (1976).

Vannucci, R.C. & Duffy, T.E.: Cerebral metabolism in newborn dogs during reversible asphyxia. Ann. Neurol., 1: 528-534 (1977).

CHAPTER TWENTY-THREE

POSTNATAL MATURATION OF THE LOCAL CEREBRAL CIRCULATION

LOUIS SOKOLOFF

In most mammals, especially primates, the brain is highly underdeveloped and relatively undifferentiated at birth, and a major portion of its anatomical, chemical, and functional differentiation is achieved postnatally early in its extrauterine life. It is to be expected that changes in the cerebral circulation accompany and reflect the maturational and developmental processes. Since periods of rapid change are generally quite vulnerable to all sorts of noxious influences, it can further be assumed that there are pathologic conditions which are associated with and, perhaps, even consequences of aberrations in the pattern of development of the cerebral circulation. In order to establish the normal pattern of postnatal development of the cerebral circulation, local blood flow of the individual structures of the brain was studied as a function of postnatal age.

Local cerebral blood flow was measured by means of the [^{14}c]antipyrine modification (Reivich et al., 1969) of the autoradiographic method of Kety and his associates (Landau et al., 1955; Freygang and Sokoloff, 1958; Kety, 1960). Additional modifications adapted it for use with animals of small blood volume (Kennedy et al., 1972). The studies were carried out in beagle puppies of certified pedigree. The choice of animal was made on the basis of the appropriateness of the time span of the process of cerebral maturation, convenience of neonatal size, viability of newborn pups under conditions of hand raising, and relative uniformity of genetic strain. Femoral arterial and venous catheters to be used for sampling of arterial blood and intravenous infusion of the tracer solution were inserted under halothane anesthesia. The procedure for the measurement of blood flow was executed 3-4 hours after recovery from the general anesthesia. Studies were carried out in 35 animals varying in age from one day to one year. Blood flow was determined in a total of 35 structures in the brain.

Figure 1 represents a considerable reduction in the vast amount of data accumulated, but it illustrates some of the essential features of the patterns of change in local cerebral blood flow during postnatal development and ma-

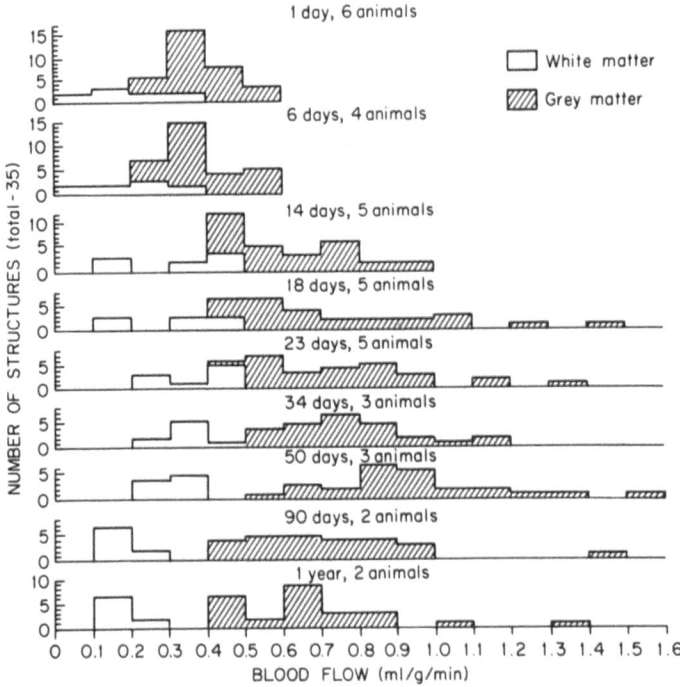

Fig. 1. Changes in distributions of the rates of local cerebral blood flow in the dog during postnatal maturation. From Kennedy et al., 1972.

turation in the dog. Blood flow is relatively uniform throughout the brain at birth. Although average blood flow in gray matter is about twice that of white matter, there is considerable overlap, and both are quite low. In parallel with increasing morphologic and chemical differentiation, the distribution of the values for blood flow in the various parts of the brain also exhibits a progressively increasing heterogeneity with postnatal age. Blood flow rates increase in both gray and white matter; the changes in gray matter are greater so that the degree of overlap becomes less and eventually disappears altogether between 4 and 7 weeks of age. The development of heterogeneity in perfusion rates is particularly prominent in the gray matter during this period, and structures with exceptionally high blood flow, such as the inferior colliculus and superior olive, separate and become clearly distinguishable from the others. By one year of age the distribution of blood flow values throughout the brain has achieved its multimodal distribution, previously observed in the conscious adult cat (Sokoloff, 1961), with non-overlapping ranges for gray

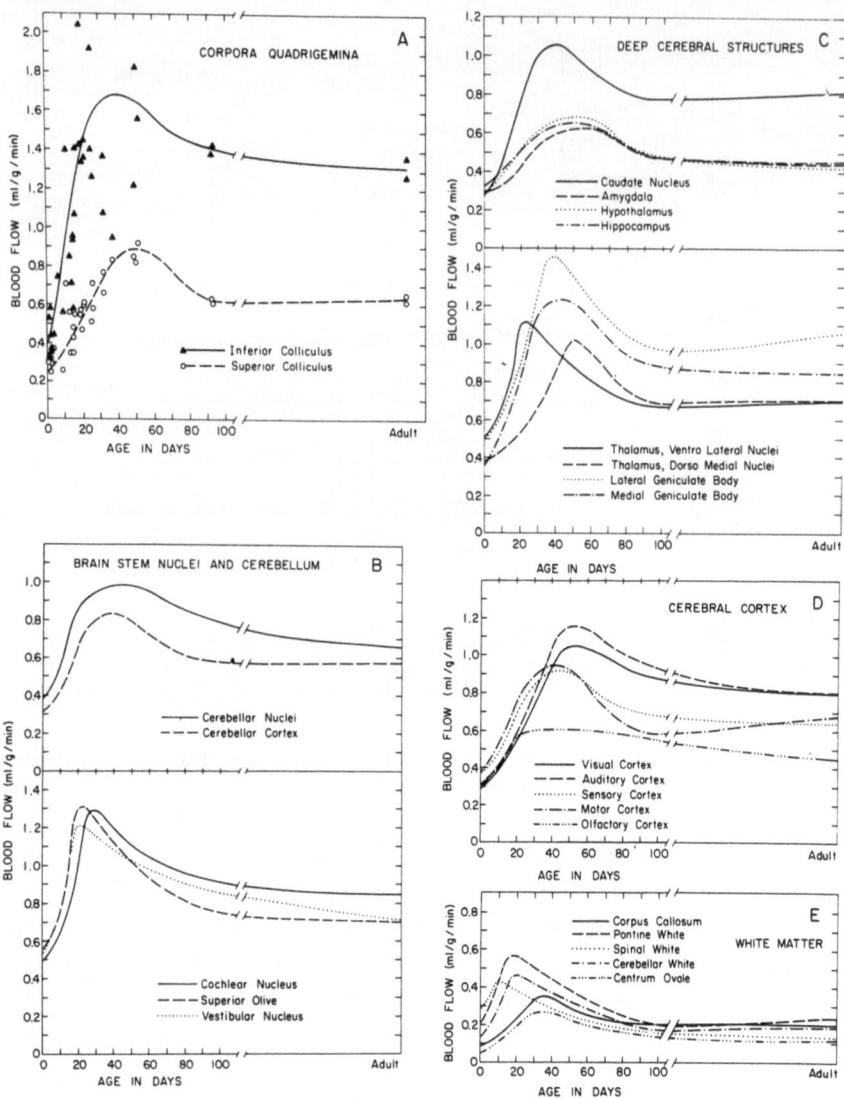

Fig. 2. Time courses of the postnatal developmental changes in blood flow in a variety of gray and white structures of the dog. From Kennedy et al., 1972.

matter and white matter and the distribution in gray matter itself bimodal with the higher values in the primary sensory areas of the cerebral cortex and associated structures subserving sensory functions.

Together with the progressively increasing heterogeneity, there is a pattern of change in blood flow common to all structures (fig. 2). Blood flow is low in all the structures at birth, rises to a peak at some time in the first few weeks of life, and then declines again to the level characteristic of the mature brain. In white matter the decline from the peak is close to the amount of rise so that in the mature brain the blood flows in white matter have fallen to about the same low levels as at birth. The declines from the peaks in gray structures, though definite and considerable, are far less than those in white matter so blood flow in gray matter at maturity is at least twice that present at birth. The peak is achieved at different times in the various structures; in general, the peaks appear earlier in white matter than in gray matter, and occur progressively earlier the lower the level in the neuraxis (fig. 2).

The present studies confirm the finding made earlier in human children that there is a period during growth and development of the brain when cerebral blood flow and metabolic rate are considerably higher than their levels at maturity (Kennedy and Sokoloff, 1957). The patterns of change in individual cerebral structures suggest further that the declines in blood flow from the peak levels during maturation reflect predominantly the changes in white matter and to a small extent those in gray matter (Kennedy et al., 1970).

The rises in blood flow from their low levels at birth to peak levels reached at different times in the first few weeks followed by declines to the levels of the mature brain, as well as the progressive development of heterogeneity in the values among the various structures, parallel comparable changes observed in the oxygen consumption of some of these structures of the dog brain studied in vitro (Himwich and Fazekas, 1941). They also parallel morphologic evidence of maturational changes in these structures. For example, the peak levels of blood flow occur earlier in the white structures lower in the neuraxis which undergo myelinization earlier in postnatal life. There were no systemic changes in blood composition which could fully account for the changes in blood flow. Arterial pCO_2 did not change throughout the life span studied. Arterial pO_2 rose during the period of rising cerebral blood flow. There were changes in hemoglobin concentration, and presumably also blood viscosity, which could account at least partly for the changes in blood flow, but they would be expected to have similar effects throughout the brain and could not explain the heterogeneity of the changes in the various structures of the brain. It is presumed, therefore, that the patterns of change of local blood flow are the consequences of the metabolic changes occurring in the tissues and reflect the extra energy demands of the tissues for biosynthetic processes during the period of growth, myelinization, and maturation. Recent studies of local cerebral glucose utilization in infant monkeys by means of the recently developed [^{14}c]deoxyglucose technique demonstrate that the differences in the rates of blood flow in the component structures of the brain do, in fact, correlate with comparable differences in local cerebral energy metabolism (C. Kennedy, M.H. Des Rosiers and L. Sokoloff, unpublished observations).

SUMMARY

In addition to the usual factors that regulate cerebral blood flow, local cerebral circulation in the postnatal period is also regulated by the maturation of the local tissue that it perfuses. The various structural components of the brain do not appear to mature at the same rate. Blood flow was measured quantitatively in 35 structures of the brains of dogs of various ages from birth to maturity. In general, the values were low at birth and rose to maximal levels between 3 and 7 weeks of postnatal age; declines from the peak levels then followed until values characteristic of maturity were attained by 13 weeks of postnatal age. From relatively uniform perfusion rates throughout the brain at birth there gradually emerged a marked heterogeneity, in parallel with the structural and functional maturation and differentiation known to occur in the brain during this period of life. Our observations may reflect the summation of the changes in energy demands associated on the one hand with biosynthetic processes essential for growth and development and with the support for progressively increasing functional acitivities on the other. These results suggest that the circulatory requirements of the different structural components of the brain and their relative vulnerabilities to circulatory insufficiency vary with the degree of their maturation and, therefore, also postnatal age.

REFERENCES

Freygang, W.H. and Sokoloff, L.: Quantitative measurements of regional circulation in the central nervous system by the use of radioactive inert gas. Adv. biol. med. Physics 6: 263-279 (1958).

Himwich, H.E. and Fazekas, J.F.: Comparative studies of the metabolism of the brain of infant and adult dogs. Am. J. Physiol. 132: 454-459 (1941).

Kennedy, C., Grave, G.D., Jehle, J.W. and Sokoloff, L.: Blood flow to white matter during maturation of the brain. Neurology 20: 613-618 (1970).

Kennedy, C., Grave, G.D., Jehle, J.W. and Sokoloff, L.: Changes in blood flow in the component structures of the dog brain during postnatal maturation. J. Neurochem. 19: 2423-2433 (1972).

Kennedy, C. and Sokoloff, L.: An adaptation of the nitrous oxide method to the study of the cerebral circulation in children; normal values for cerebral blood flow and cerebral metabolic rate in childhood. J. clin. Invest. 36: 1130-1137 (1957).

Kety, S.S.: Measurement of local blood flow by the exchange of an inert, diffusible substance; in Bruner, Methods in Medical Research, Vol. VIII, pp. 228-236 (Year Book Publishers, Chicago 1960).

Landau, W.M., Freygang, W.H., Rowland, L.P., Sokoloff, L. and Kety, S.S.: The local circulation of the living brain; values in the unanesthetized and anesthetized cat. Trans. Am. neurol. Ass. 80: 125-129 (1955).

Reivich, M., Jehle, J.W., Sokoloff, L. and Kety, S.S.: Measurement of regional cerebral blood flow with ^{14}C-antipyrine in awake cats. J. appl. Physiol. 27: 296-300 (1969).

Sokoloff, L.: Local cerebral circulation at rest and during altered cerebral activity induced by anesthesia or visual stimulation; in Kety and Elkes, The Regional Chemistry, Physiology and Pharmacology of the Nervous System, pp. 107-117 (Pergamon Press, Oxford 1961).

CHAPTER TWENTY-FOUR

METHODS FOR PROMOTING FUNCTIONAL RECOVERY FOLLOWING BRAIN DAMAGE

B. E. WILL

The treatment generally used for pathologic conditions following CNS lesions in children or adults pre-supposes, and seems to prove the effectiveness of more or less specific training (29, 50). It should be added, however, that this idea is based on a small number of controlled studies (24), whose conclusions don't always agree and whose interpretation is still seing debated as far as both humans and animals are concerned.

There does exist a general consensus concerning the importance of "experience" pre or inter-lesional, despite several exceptions (88). Most of the studies show that experiences rich in sensory or pharmacologic stimulation and/or more or less specific training produce beneficial effects on the speed and the amount of functional recovery (13, 18, 19, 22, 31, 37, 39, 56, 62, 63, 85).

If one considers the role played by post-lesional experience one is struck by the small number of controlled studies which, until recent years, have tried to determine the factors which aid in recovery from cerebral impairment, particularly after delayed manifestations. One can readily note that the role, specific or not, of this post lesional experience is debatable.

We are interested primarily in the factors which promote improvement in behavior during recovery.

This approach seems justified by the fact that it is often easier to judge the beneficial aspects of behavioral recovery than other aspects of recuperation. For example, neuroanatomically or neurophysiologically it is difficult to know whether axonal growth or collateral germination result in the establishment of functioning synapses; if so, there is generally nothing to prove that they play a role which is beneficial to the functioning of the organism.

Since a cerebral lesion often effects several functions and since improvement in one function may be at the expense of others (57, 84), it seems important to estimate the overall degree of behavioral recovery. To do this we are interested especially in the recovery of training capacities as good training capacities require balanced sensori-motor functioning whether of memory, motivation or attention. A malfunction of even one of these necessarily entails a fall in performance in re-training. We shall say that a factor facilitates functional recovery when it acts in an opposite direction; that is to say when it corrects (speed or magnitude of recovery) the behavioral deficit noted; deficit being defined as a significant deterioration.

Several methods have been considered for stimulating behavior recovery after a cerebral accident, particularly that of learning capacities; essentially this is a question of environmental control and the use of pharmacologic products.

CONTROL OF THE ENVIRONMENT

Habituation to the post-lesional Environment: To facilitate the habituation of lesioned subjects to their new perception of the environment, certain authors have placed their patients in surroundings which constitute a progressive transition between the pre and post lesional milieux. It is possible that this habituation reduces the degree of the postlesional perceptual change and that of the stress which may result therefrom. It was used with some success to rehabilitate dogs and monkeys who had previously suffered from social privation (28, 72, 78, 79, 80). Behavioral deficit due to prolonged isolation may be considered, according to Suoni et al. (80), as a deficit in learning capacity; numerous studies (15) have shown that these behavior modifications are accompanied by serious disturbances, in cerebral functioning. The cerebral deficiency, long considered irreversible, may be corrected, however, if the isolated subjects are confronted with their congeners (members of their own species) only very gradually, according to a sequence such as that used by Novak and Harlow (58):
1. visual interaction between isolated monkeys,
2. visual interaction between an isolated monkey, and a young "therapist" monkey
3. physical interactions between isolated monkeys and finally,
4. physical interactions between an isolated monkey and a "therapist" monkey.

It should be noted that in this procedure the quality of stimulation is at least as important as the amount. Other studies have been done to habituate the subjects to certain special aspects of the post-operative environment as, for example, the reinforcement used in the test situation. Thus, Dicava (17) showed that post-operative habituation to milk accelerated the recovery for taking milk among those rats which had been lesioned at the level of the lateral hypothalamus. As to the effects of habituation to the experimenter by postoperative handling, they appear (superficially) contradictory: excessive handling immediately after the operation prolongs the hyperexcitability and the rage behavior of the rats, subjected to septal lesions, whereas progressive handling facilitates recovery (34).

In fact, excessive handling probably doesn't furnish the gradual transition needed to permit habituation to the experimenter. Habituation to the test situation by pre-training has been studied by several authors. Their experiments dealt almost exclusively with rats which had been subjected to hippocampal lesions, which produce, among other results (54), a drop in inhibitory capacities (20, 21, 42) and hence an increase in perseveration. It is not surprising, under these conditions, to note, that postoperative pretraining in a certain situation (operative conditioning with a continuous reinforcement program or CRF) has adverse effects on the performance of the subjects in a situation with almost the opposite demands (operant conditioning with a schedule of differential reinforcement of low rates or DRL).

This injurious effect of pretraining with a CRF program has been observed each time the hippocampal lesions were large (12, 25, 70). But, does this pretraining truly constitute habituation to the test situation of the DRL program. In short, habituation to a specific environment does not appear to improve performance unless it is evaluated in the same environment. This result is by no means therapeutically negligible, but remains nevertheless of limited interest.

Stimulation by the Environment: A second approach, also based on control of the environment follows from a completely opposite hypothesis. Instead of treating the patient by graduated transitional stages and by progressive habituation to one or several aspects of a fixed environment, this second method consists on the contrary in submitting an individual who is suffering from a cerebral impairment to mildly stressful environments or to frequent physical and/or social environmental changes which may also be stressful.

The aim, more or less clearly explained by its authors, seems to be to stimulate the subjects, and thus, to provoke central activation which allows them to achieve multiple learning. We should note that certain habituation methods previously described may equally produce stimulation, but the underlying hypothesis of these studies is different from the experiments which we shall now describe. Let us add that the habituation experiments remain limited to a very precise type of stimulation, which perhaps explains why the effects of this kind of stimulation on behavior recovery remain confined to very specific aspects.

The oldest experiments which attempted to stimulate cerebrally injured subjects are probably these of Smith (73) and Schwartz (71). Smith's work, having been too inadequately described, Schwartz's is the experiment frequently cited. Schwartz operated on one day old rats, producing bilateral occipital lesions by section. The lesioned rats and those which had pseudo-operations were then raised from 5 days of age, and for 90 days, in milieux either deprived of or very rich in physical and social stimulation. These environments, which resemble the extreme conditions described by Rosenzweig et al. (68) are generally and respectively described as "impoverished" and "enriched". When they were tested subsequently in the maze of Hebb and Williams (cf. 64), significant differences appeared on the one hand between the lesioned and the control groups, and on the other hand between the "enriched" and "impoverished." Schwartz pointed out, furthermore, was significant and that this was due to a great reduction (in absolute values) in the number of errors among the lesioned subjects as compared to the controls.

Rearing them in an enriched milieu had compensated so strongly for the operative effects that the lesioned rats in the "enriched" milieu made less errors than the intact rats raised in an "impoverished" milieu.

Since Schwartz's work on the role of the environment in functional recovery, after neonatal cerebral lesioning, several other experiments have shown that stimulation by the complexity of environment, physical and/or social helps to facilitate recovery following neonatal lesions (83), after early malnutrition (27, 47, 81, 89) and after perinatal hypothyroidism (14). Let us recall on the one hand that many published works have shown that early malnutrition (53), as well as prenatal hypothyroidism (2, 23) results in cerebral functional, as well as structural modifications, and on the other hand, that the observed "facilitating" effects apply only to the recoveries after very early cerebral impairments, usually neonatal.

As this limiting factor is probably not due to chance, one may suggest that the therapeutic effectiveness of the stimulation studied diminishes rapidly with the decrease in "genetic" plasticity (60) of the central nervous system, probably disappearing after its maturation. This supposition, as we shall see, seems at least partly in error: stimulation by an enriched environment facilitates recovery – perhaps more definitely among elderly patients, though this remains to be shown – but its effectiveness certainly doesn't disappear after maturation of the central nervous system.

We have obtained results similar to those of Schwartz with neonatal lesions (95) also with lesions performed after weaning (96) or even among adult rats (94). The effects of stimulation by a complex and changing environment are similar in each sex and in the two rat strains studied. This "therapy" by the environment remains effective even if the duration of exposure to the enriched environment is reduced to 2 hours per day for one month and even if the enrichment period doesn't begin until 25 days after the operations. These results were obtained in the Hebb-Williams maze test, but we have recently obtained comparable results (92) with operant conditioning involving auditive discrimination ("mult. CRF-Ext.") and described in detail elsewhere (91, 93). This experiment showed that rats stimulated both by the physical and social environments exhibited the highest rate of responses, whereas those most deprived of social and physical stimulation showed the lowest degree of responses; the behavior of the subjects reared in environments enriched only in one factor, social or physical, falls between the 2 extremes. This last point needs to be verified, since Rosenzweig has noted, in 3 different experiments, that the performances of rats lesioned and then raised for 60 days in a standard milieu, enriched only socially (3 rats in an ordinary laboratory cage), were scarcely better than subjects lesioned and then raised in an impoverished milieu (personal communication).

If rearing in an enriched milieu facilitates the recovery of learning capacities among individuals with visual cortical lesions, it must nevertheless be remembered that this progress is evaluated by tests in which the animal may use one or several senses other than vision. This improvement could then also be due to the compensatory mechanisms of other functions as well as to a restoration of primitive visual functions. The work of Bland and Cooper (9, 10) as well as Tees (82) seem to support only the first of these hypotheses.

The experiment of Stricker and Zigmond (77) would lead one to believe that it is the stressing aspect of the enriching situation which is the basis for the observed results. These authors noted that behavioral recovery after intraventricular injections of hydroxydopamine is more complete in animals placed in mildly stressing environments than among those kept in more conventional environments. However, one should recall that the experiments of Riege and Morimoto (65) showed that the effects produced by rearing in a stressed milieu and those of enriched environment are different. Riege and Morimoto have shown that enrichment of the environment didn't lead to the modifications observed after stress, in the adrenals and in the cerebral amines.

Other interpretations have been suggested (66). The one which now seems most plausible attributes the effectiveness of "therapy" by enrichment of the environment to an increase in the number of learning experiences, correlated, probably, with an optimal increase in the level of activation. This is the way Teuber (84) interprets the notable observation of Wepman (90), who because of insufficient personnel found it impossible to treat all the military aphasics at the end of the second world war. About half his patients had to wait 6 to 12 months for treatment, whereas the other half were able to receive immediate therapy. Wepman noted that the delayed treatment group showed a lesser degree of recovery than the other.

The training experience offered the aphasics was quite non specific and consisted more of stimulation and activation than of selective re-learning. It is possible, consequently, to interpret these observations in terms of enrichment of the environment. Besides, this type of interpretation could equally be applied to the experiment of Chow and Stewart (11) which showed that in cats the forced use of an eye, which had been previously sutured for the first 20 (± 4) postnatal months facilitates recovery, structural as well as functional. After 6 to 10 months of forced training, especially in discriminating configurations, their behavior strongly resembles that of normal subjects and it seems that treatment had even stimulated the growth of the cells of the external geniculate bodies. It seems likely that the forced sensori-motor training is at the origin of effective enrichment, at least from the point of view of the subject.

In this regard, one can observe that a simple sensory stimulation without feedback by afferent proprioceptives, whether or not it is forced, may not suffice to produce the effects observed when subjects are in sensori-motor contact with the stimulating objects (22, 26). In this case, the degree of activation perhaps remains insufficient due to an absence of retroaction by the proprioceptive afferents, or also because of faulty motivation. A rat which may perceive "enriched" congeners through a grid, but who can't himself move around in an enriched environment, is probably not motivated in the same fashion as they to work out the multiple associations which, for him, remain without any effect.

Cerebral activation seems, in any case, to be the key to the problem, if one refers to the unique and remarkable experiment of Harwell et al. (36). The workers have described an intriguing effect of experience on the recovery of subjects having lateral hypothalamic lesions. They found that one hour per day of stimulation of the zone adjoining the lesioned area decreased the recovery period for food intake. After these results one might think that an appropriate post lesional experience would involve simply the activation of the affected area.

In view of these encouraging results, several authors have conjectured whether it might not be possible to obtain similar effects, indeed superior ones, by pharmacologic means, or even to potentiate in this way the action of the preceding method.

USE OF PHARMACEUTICALS

The effects of drugs and the manifestations of cerebral malfunctioning may be studied simultaneously in the same experimental context for reasons quite different (32). The approach which interests us here is that which aims at facilitating recovery by the use of drugs and which seeks to explain by this method the very mechanisms involved in recovery.

These types of research may be classed as: 1) those in which the subject is tested while he is under the action of the product being studied, and 2) those in which the substance is used well in advance of the test, but during the recuperative period following the cerebral damage.

Short term pharmacologic effects: In the first kind of experiment, the substance is given just prior to the behavior test (the time necessary for it to work) and one is interested in the action which it may exercise on function. This type of experiment shows that the effects of some cerebral lesions may be attenuated, or abolished, by administering a stimulating drug, or neurotransmitter; but this improvement is only transitory (5, 32, 38, 51, 55). Thus one may consider that this kind of treatment is only symptomatic and that it does not act on the recovery process itself, which, moreover, doesn't detract from an interest in this kind of study: these experiments make it possible to show that better function remains possible and thereby demonstrate that nerve pathways necessary for this expression are always present. They permit, likewise, a better understanding of the possible causes of the deficit provoked by the cerebral lesion.

Long term pharmacologic effects: The second kind of experiment, perhaps more ambitious, seeks to facilitate long term recovery, whether by direct or indirect action on cerebral functioning, or by direct or indirect action on nerve regeneration.

An older publication, by Ward and Kennard (87) summarizes a number of clinical observations and indicates that cholinergic agents, as strychnine, contribute to the improvement of motor functions in monkeys subjected to unilateral necortical motor ablation. Later, Cole et al. (13) showed that the continuing administration of d-amphetamine to rats raised in darkness between two cortical ablations 12 days apart induced an improvement in performance comparable to that noted among rats raised in light. The performance was measured by an avoidance task with a light warning stimulus. The interpretation of the results remain discutable, but it seems possible to explain them by an increase in the activation level, similar to that obtained by stimulation in an enriched environment. This type of stimulation, remaining frequently at a sub-optimal level (44, 46), it seemed reasonable to try to increase the stimulating effects of the environment by using drugs. In fact it was shown that stimulants, as amphetamine and strychnine, were able to increase the cerebral effects (4, 67) and behavior effects (45, 61) of an enriched environment. Based on these results, we have tried to potentiate the effects of "therapy" by the environment, by combining its stimulating effects and those of metamphetamine at the time of rearing these subjects under different conditions. In 3 successive experiments there were no effects of the drug on rat performance in a Hebb-Williams maze. If in addition one considers that the probability of the publication of negative results is generally more unlikely than that of positive results, it must be admitted that the effectiveness of this method remains disputable, to say the least.

Some other substances have been studied, such as α-methyltyrosine, trypsin, hyaluronidase, nerve growth factors (NGF). These substances are supposed to facilitate nerve regeneration, directly or indirectly.

According to Osterholme and Mathews (59), the administration of α-methyltyrosine would prevent or minimize the paraplegia caused by trauma resulting in hemorrhagic necrosis of the spinal cord.

According to Glick and Greenstrein (33) α-methylparatyrosine would speed recovery of food taking by rats subjected to lateral hypothalamic lesioning, by increasing the hypersensitivity of denervation of the remaining catecholaminergic neurones. Unfortunately, the beneficial effects of this substance now seem to be doubtful (86).

Soviet workers have reported that injecting trypsine or a hyaluronidase into the site of the transverse section of the spinal cord would speed its regeneration and functional recovery among rats (52). These proteolytic or mucoproteolytic enzymes reduce the density of the scar thus facilitating the growth of axones across this zone. Stimulation caudal to the section site results in evoked potentials in the cerebral cortex, which shows that the regenerated pathways are functional. A significant proportion of the animals recovered a remarkable degree of locomotor capacity. These original results are promising but should be regarded cautiously until confirmed by others.

As to nerve growth factors, they were used recently by Berger, Wise and Stein (6), who showed that their intracerebral injection at the time of operation facilitates behavior recovery among rats rendered anorexic by lateral hypothalamic lesions. The NGF does not have the same mode of action as trypsine or hyaluronidase. Initially isolated from mouse salivary glands by Levi-Montalcini and her colleagues, the NGF has metabolic and trophic effects on the nervous system, particularly on the autonomic nervous system in vitro as well as in vivo (46). Since the brain levels of these proteins are very low, one may doubt their effectiveness at this site. Recent experiments (2, 24) have shown that these factors are likewise capable of facilitating the growth of central monoaminoergic neurones.

Certains drugs, especially those which facilitate nerve regeneration (proteolytic or NGF) seem, consequently, to contribute, to the long term functional improvement of organisms affected by CNS lesions. However, it should be noted that the studies done thus far are too few and that the conditions giving successful results by this method seem, at this time, to require that the substance be administered close to the site of the lesion, almost immediately after the operation. This would undoubtedly limit its possible use.

Before concluding, it should be emphasized that the effectiveness of some of the treatments under consideration may be modified by various conditioning factors, as the degree of motivation of the subjects (1, 2, 9), the length of treatment or the lapse of time following the cerebral damage (1, 30, 84), the age of the subjects at the time of the lesion onset (41, 98), or the type or localisation of the lesion (49). In certain special cases, other methods than those analyzed above may be considered. Thus, effects related to hypophyseal ablation may be temporarily compensated for by hormone administration (16). Recently nerve tissue transplants appear to have been used with some success (8, 48, 75).

The methods presented in detail and which encourage a relatively lasting recovery may be classed in 2 categories: those which act after a delay following cerebral lesions and those which are probably ineffective unless applied immediately after damage.

The first are essentially methods of stimulation and optimal activation by enrichment of the environment. It is not yet certain that their mode of action differs basically from the methods of habituation to the post-lesional environment, the progressive aspect of the stimulation perhaps permitting it to be maintained at an optimal level. Nor is it certain that the mode of action is characteristic among the healthy or among lesioned subjects. No regeneration having been observed at the lesioned site after this kind of treatment (94, 95, 96, 91) one may suppose that the recuperative process involved is essentially a process of substitution which at present, remains difficult to differentiate from processes observed in normal subjects (3, 96).

The long acting pharmacologic methods seem to act more directly on recovery by facilitating 2 different ways of nerve regeneration. These methods are of interest since they offer new possibilities, but they run into the difficulty that they have not yet been sufficiently verified by replication experiments. In many cases, these 2 types of methods could be combined and one may suppose that overall recovery, in this case, would be even more distinct than by using one method or the other separately.

SUMMARY

Postlesional experience is important for the functional rehabilitation of subjects who are suffering central nervous system damage. Several methods have been proposed in order to facilitate recovery, but they differ with respect to duration and generality of the effects, and can probably be attributed to different underlying processes: These methods are essentially founded either upon environmental control or the use of drugs.

The experiments aimed at controlling the environment try to progressively habituate the subjects to various aspects of the postlesional environment or, conversely, to stimulate them by a complex, changing and even mildly stressful environment. Both methods show beneficial effects but, in the case of the former, the effects remain more specific and limited. It can be assumed that the facilitative effects caused by environmental stimulation can be explained by structural and functional modifications of undamaged cerebral areas. It seems that this substitution can even be fostered some time after operation or trauma.

The pharmacologic approach has, up until now, been concentrated upon the use of two types of drugs: First, those which promote only a transient and symptomatic recovery and, second, those which promote more persistent recovery, either by reproducing or enhancing the activation caused by a stimulant environment, or by facilitating the regenerative growth of the damaged neurons. Drugs which may promote long term recovery are rare and have been little studied. The results obtained with stimulating drugs, such as methamphetamine, are disappointing and, at present, drugs which facilitate nervous regeneration seem more promising. These drugs are, 1: the nerve growth factors, whose efficiency appears restricted to monoaminergic neurons, and, 2: enzymes such as trypsin or hyaluronidase.

The significance of other conditioning factors and some research possibilities are analysed.

REFERENCES

1. Bach-Y-Rita, P.: Plasticity of the nervous system; in Zulch, Creutzfeld and Galbraith, Cerebral localization, pp. 313-327 (Springer Verlag, New-York, 1975).
2. Balasz, R.: Biochemical effects of thyroid hormones in the developing brain. UCLA Forum in Medical Sciences 14: 273-320, 1971.
3. Bennett, E.L., Diamond, M.C., Krech, D. and Rosenzweig, M.R.: Chemical and Anatomical plasticity of brain. Science 146: 610-619, 1964.
4. Bennett, E.L., Rosenzweig, M.R. and WU, S.-Y.C.: Excitant and depressant drugs modulate effects of environment on brain weight and cholinesterases. Psychopharmacologia (Berlin) 33: 309-328, 1973.
5. Berger, B.D., Wise, C.D., Stein, L.: Norepinephrine: Reversal of anorexia in rats with lateral hypothalamic damage. Science 172: 281-284, 1971.
6. Berger, B.D., Wise, C.D. and Stein, L.: Nerve growth factor: Enhanced recovery of feeding after hypothalamic damage. Science 180: 506-508, 1973.
7. Bjorklund, A. and Stenevi, U.: Nerve growth factor: stimulation of regenerative growth of central noradrenergic neurons. Science 175: 1251-1253, 1972.
8. Bjorklund, A., Stenevi, U. and Svendgaard, N.A.: Growth of transplanted monoaminergic neurones into the adult hippocampus along the perforant path. Nature 262 (5571): 787-790, 1976.
9. Bland, B.H. and Cooper, R.M.: Posterior neodecortication in the rat: Age at operation and experience. J. Comp. Physiol. Psychol., 69: 345-354, 1969.
10. Bland, B.H. and Cooper, R.M.: Experience and vision of the posterior neocorticate rat. Physiol. Behav. 5: 211-214, 1970.
11. Chow, K.L. and Stewart, D.L.: Reversal of structural and functional effects of long-term visual deprivation in cats. Exp. Neurol., 34: 409-433, 1972.
12. Clark, C.V.H. and Isaacson, R.L..: Effect of bilateral hippocampal ablation on DRL performance. J. Comp. Physiol. Psychol., 59: 137-140, 1965.
13. Cole, D.D., Sullins, W.R., Jr., and Isaac, W.: Pharmacological modification of the effects of spaced occipital ablations. Psychopharmacologia 11: 311-316, 1967.
14. Davenport, J.W., Gonzales, L.M., Carey, J.C., Bishop, S.B. and Hagquist, W.W., Environmental stimulation reduces learning deficits in experimental cretinism. Science 191: 578-579, 1976.
15. Defeudis, F.V.: La biologie de la solitude: La Recherche 55: 344-356, 1975.
16. De Wied, D.: Pituitary-adrenal system hormones and behavior; in Smitt and Worden, The Neurosciences, Third study program, pp. 653-666 (Rockefeller Univ. Press, New-York 1974).
17. Dica, L.V.: Role of postoperative feeding experience in recovery from lateral hypothalamic damage. J. Comp. Physiol. Psychol. 72: 60-65, 1970.
18. Donovick, P.J., Burright, R.G., Fuller, J.L. and Branson, P.R.: Septal lesions and behavior: effects of presurgical rearing and strain of mouse. J. Comp. Physiol. Psychol. 89: 859-867, 1975.
19. Donovick, P.J., Burricht, R.G. and Swidler, M.A.: Presurgical rearing environment alters exploration, fluid consumption, and learning of septal lesioned and control rats. Physiol. Behav. 11: 543-553, 1973.

20. Douglas, R.J.: The hippocampus and behavior. Psychol. Bull. 67: 416-442, 1967.
21. Douglas, R.J.: Pavlovian conditioning and the brain; in Boakes and Halliday, Inhibition and learning, pp. 327-361 (Academic Press, New York 1972).
22. Dru, D., Walker, J.P. and Walker, J.B.: Self-produced locomotion restores visual capacity after striate lesions. Science 187: 265-266, 1975.
23. Eayrs, J.T.: Thyroid and developing brain: anatomical and behavioral effects; in Hamburgh and Barrington, Hormones in development, pp. 345-355 (Appleton-Century-crofts, New-York 1971).
24. Eidelberg, E. and Stein, D.G.: Functional recovery after lesions of the nervous system. Neurosciences Research Program Bull. 12: 190-303, 1974.
25. Ellen, P., Aitken, W.C., Jr. and Walker, R.: Pretraining effects on performance of rats with hippocampal lesions. J. Comp. Physiol. Psychol. 84: 622-628, 1973.
26. Ferchmin, P.A., Bennett, E.L. and Rosenzweig, M.R.: Direct contact with enriched environment is required to alter cerebral weights in rats. J. Comp. Physiol. Psychol. 88: 360-367, 1975.
27. Frankova, S.: Influence of early social environment on behaviour of the protein-calorie malnourished rats. Act. Nerv. sup. (Prague) 16 (2): 98-99, 1974.
28. Fuller, J.L.: Experimental deprivation and later behavior. Science 158: 1645-1652, 1967.
29. Gazzaniga, M.S.: Determinants of cerebral recovery; in Stein, Rosen and Butters, Plasticity and recovery of function in the central nervous system, pp. 203-215 (Academic Press, New York 1974).
30. Geschwind, N.: Late changes in the nervous system: an overview; in Stein, Rosen and Butters, Plasticity and recovery of function in the central nervous system (Academic Press, New York, 1974).
31. Glendenning, R.L.: Effects of training between two unilateral lesions of visual cortex upon ultimate retention of black-white discrimination habits in rats. J. Comp. Physiol. Psychol. 80: 216-229, 1972.
32. Glick, S.D.: Changes in drug sensitivity and mechanisms of functional recovery following brain damage; in Stein, Rosen and Butters, Plasticity and recovery of function in the central nervous system, pp. 339-372 (Academic Press, New York 1974).
33. Glick, S.D. and Greenstein, S.: Facilitation of lateral hypothalamic recovery by postoperative administration of alpha-methyl-p-tyrosine. Brain Res. 73: 180-183, 1974.
34. Gotsick, J.E. and Marshall, R.C.: Time course of the septal rage syndrome. Physiol. Behav. 9: 685-687, 1972.
35. Greenough, W.T., Fass, B. and Devoogd, T.J.: The influence of experience on recovery following brain damage in rodents: hypotheses based on development research; in Walsh and Greenough, Environments as therapy for brain dysfunction, pp. 10-50 (Plenum Press, New York 1976).
36. Harrell, L.E., Raubeson, R. and Balagura, S.: Acceleration of functional recovery following lateral hypothalamic damage by means of electrical stimulation in the lesioned areas. Physiol. Behav. 12: 897-899, 1974.
37. Hebb, D.O.: Organization of behavior (John Wiley and Sons, New York 1949).
38. Herndon, J.G. and Neill, D.B.: Amphetamine reversal of sexual impairment following anterior hypothalamic lesions in female rats. Pharmac. Biochem. Behav. 1: 285-288, 1973.
39. Isaac, W.: Role of stimulation and time in the effects of spaced occipital ablations. Psychol. Rep. 14: 151-154, 1964.
40. Isaacson, R.L.: The myth of recovery from early brain damage; in Ellis, Aberrant development in infancy (Lawrence Erlbaum Associates, Potomac 1975).
41. Kennard, M.A.: Age and other factors in motor recovery from precentral lesions in monkeys. Amer. J. Physiol. 115: 138-146, 1936.
42. Kimble, D.P.: Possible inhibitory functions of the hippocampus. Neuropsychologia 7: 235-244, 1969.

43. Kircher, K.A., Braun, J.J., Meyer, D.R. and Meyer, P.M.: Equivalence of simultaneous and successive neocortical ablations in production of impairments of retention of black-white habits in rats. J. Comp. Physiol. Psychol. 71: 420-425, 1970.

44. Kuenzle C.C. and Knusel, A.: Mass training of rats in a superenriched environment. Physiol. Behav. 13: 205-210, 1974.

45. Leboeuf, B.J. and Peeke. H.V.S.: The effect of strychinine administration during development on adult maze learning in the rat. Psychopharmacologia (Berl.) 16: 49-53, 1969.

46. Levi-Montalcini, R. and Angeletti, P.U.: Nerve growth factor. Physiol. Rev. 48: 534-569, 1968.

47. Levitsky, D.A. and Barnes, R.H.: Nutritional and environmental interactions: in the behavioral development of the Rat: Long-term effects. Science 176: 68-70, 1972.

48. Lund, R.D. and Hauschka, S.D.: Transplanted neural tissue develops connections with host rat brain. Science 193: 582-584, 1976.

49. Luria, A.R.: Restoration of function after brain injury (Macmillan, New York 1963).

50. Luria, A.R., Naydin, V.L., Tsvetkova, L.S. and Vinarskaya, E.N.: Restoration of higher cortical function following local brain damage: in Winken and Bruyn. Handbook of Clinical Neurology, vol. 3 (North-Holland, Amsterdam, 1969).

51. Macht, M.B.: Effects of d-amphetamine on hemidecorticate, decorticate, and decerebrate cats. Fed. Proc. 63: 731-732, 1950.

52. Matinian, L.A. and Andreasian, A.S.: Enzyme therapy in organic lesions of the spinal cord (en Russe; résumé Anglais); in Akademia Nauk Armenian SSR. pp. 1-94, 1973.

53. McKhann, G.M., Coyle, P.K. and Benjamins, J.A.: Nutrition and brain development: in Nurnberger. Biological and environmental determinants of early development. pp. 10-22 (The Williams and Wilkins Company, Baltimore 1973).

54. Meissner, W.W.: Hippocampal functions in learning. J. Psychiat. Res. 4: 235-304, 1966.

55. Meyer, P.M., Horel, J.A. and Meyer, D.R.: Effects of dl-amphetamine upon placing responses in neodecorticate cats. J. Comp. Physiol. Psychol. 56: 402-404, 1963.

56. Meyer, D.R., Isaac, W. and Maher, B.: The role of stimulation in spontaneous reorganisation of visual habits. J. Comp. Physiol. Psychol. 51: 546-548, 1958.

57. Milner, B.: Hemispheric specialization: Scope and limits; in Schmitt and Worden. The neurosciences: Third study program, pp. 75-89 (M.I.T. Press, Cambridge Mass. 1973).

58. Novack, M.A. and Harlow, H.F.: Social recovery of monkeys isolated for the first year of life. I. Rehabilitation and therapy (En préparation, cité dans 80).

59. Osterholme, J.L. and Mathews, G.L.: Altered norepinephrine metabolism following experimental spinal cord injury. II. Protection against traumatic spinal cord hemorrhagic necrosis by norepinephrine synthesis blockade with alpha-methyltyrosine. J. Neurosurg. 36: 395-401, 1972.

60. Paillard, J.: Réflexions sur l'usage du concept de plasticité en neurobiologie. J. Psychologie 1: 33-47, 1976.

61. Peeke, H.V.S., Leboeuf, B.J. and Herz, M.J.: The effect of strychnine administration during development on adult maze learning in the rat. II: Drug administration from day 51-70. Psychopharmacologia (Berl.) 19: 262-265, 1971.

62. Petrinovich, L. and Bliss, D.: Retention of a learned brightness discrimination following ablations of the occipital cortex in the rat. J. Comp. Physiol. Psychol. 61: 136-138, 1966.

63. Petrinovich, L. and Carew, T.J.: Interaction of neocortical lesion size and interoperative experience in retention of a learned brightness discrimination. J. Comp. Physiol. Psychol. 68: 451-454, 1969.

64. Rabinovitch, M.S. and Rosvold, H.E.: A closed-field intelligence test for rats. Can J. Psychol. 5: 122-128, 1951.

65. Riege, W.H. and Morimoto, H.: Effects of chronic stress and differential environments upon brain weights and biogenic amine levels in rats. J. Comp. Physiol. Psychol. 70: 396-404, 1970.

66. Rosenzweig, M.R.: Effects of environment on development of brain and of behavior: in

Tobach, Aronson and Shaw, The biopsychology of development, pp. 303-342 (Academic Press, New York 1971).
67. Rosenzweig, M.R. and Bennett, E.L.: Cerebral changes in rats exposed individually to an enriched environment. J. Comp. Physiol. Psychol. 80: 304-313, 1972.
68. Rosenzweig, M.R., Bennett, E.L. and Diamond, M.C.: Brain changes in response to experience. Scientific American, 22-30, 1972.
69. Sarno, M.T., Silverman, M. and Sands, E.: Speech therapy and language recovery in severe aphasia. J. Speech Hear. Res. 13: 607-623, 1970.
70. Schmaltz, L.W. and Isaacson, R.L.: The effects of preliminary training conditions upon DRL performance in the hippocampal destruction. Physiol. Behav. 1: 175-182, 1966.
71. Schwartz, S.: Effect of neonatal cortical lesions and early environmental factors on adult rat behavior. J. Comp. Physiol. Psychol. 57: 72-77, 1964.
72. Scott, J.P.: The process of primary socialization in the dog: in Newton and Levine, Early Experience and behavior (C.C. Thomas, Springfield, Ill. 1968).
73. Smith, C.J.: Mass action and early environment in the rat. J. Comp. Physiol. Psychol. 52: 154-156, 1959.
74. Stenevi, U., Bjerre, B., Björklund, A., Mobley, W.: Effects of localized intracerebral injections of nerve growth factor on the regenerative growth of lesioned central noradrenergic neurons. Brain Res., 69: 217-234, 1974.
75. Steveni, U., Björklund, A. and Svendgaard, N.A.: Transplantation of central and peripheral monoamine neurons to the adult rat brain: techniques and conditions for survival. Brain Research, 114: 1-20, 1976.
76. Stern, P., McDowell, F., Miller, J.M. and Robinson, M.: Factors influencing stroke rehabilitation. Stroke 2: 213-218, 1971.
77. Stricker, E.M. and Zigmond, M.J.: Recovery of function following damage to central catecholamine-containing neurons: A neurochemical model for the lateral hypothalamic syndrome: in Sprague and Epstein, Progress in Psychobiology and Physiological Psychology, vol. 6 (Academic Press, New York – in press).
78. Suomi, S.J. and Harlow, H.F.: Social rehabilitation of isolate-reared monkeys. Develop. psychol., 6: 487-496, 1972.
79. Suomi, S.J., Harlow, H.F. and McKinney, W.T., Jr.: Monkey psychiatrists. Amer. J. Psychiat., 128: 927-932, 1972.
80. Suomi, S.J., Harlow, H.F. and Novak, M.A.: Reversal of social deficits produced by isolation rearing in monkeys. J. Hum. evol., 3: 527-534, 1974.
81. Tanabe, G.: Remediating maze deficiencies by the use of environmental enrichment. Develop. Psychol. 7,: 224, 1972.
82. Tees, R.C.: The effects of neonatal striate lesions and visual experience on form discrimination in the rat. Canad. J. Psychol. 29: 66-85, 1975.
83. Tees, R.C.: Depth perception after infant and adult visual neocortical lesions in light- and dark-reared rats. Dev. Psychobiol. 9: 223-235, 1976.
84. Teuber, H.L.: Recovery of function after lesions of the central nervous system: History and prospects; in Eidelberg and Stein, Functional recovery after lesions of the nervous system. Neurosciences Research Program Bulletin 12: 190-303, 1974.
85. Thompson, R.: Retention of a brightness discrimination following neocortical damage in the rat. J. Comp. Physiol. Psychol. 53: 212-215, 1960.
86. De La Torre, J.C., Johnson, C.M., Harris, L.H., Kajihara, K. and Mullan, S.: Monoamine changes in experimental head and spinal cord trauma: failure to confirm previous observations. Surg. Neurol. 2: 5-11, 1974.
87. Ward, A.A., Jr. and Kennard, M.A.: Effect of cholinergic drugs of the central nervous system in monkeys. Yale J. Biol. Med. 15: 189-229, 1942.
88. Weinstein, S. and Teuber, H.L.: The role of preinjury education and intelligence level in intellectual loss after brain injury. J. Comp. Physiol. Psychol. 50: 535-539, 1957.

89. Wells, A.M., Geist, C.R. and Zimmerman, R.R.: Influence of environmental and nutritional factors on problem solving in the rat. Percep. Mot. Skills 35: 235-244, 1972.
90. Wepman, J.M.: Recovery from aphasia (Ronald Press. New York. 1951).
91. Will, B.: Les différences individuelles dans les capacités d'apprentissage: contribution à leur étude comportementale et biochimique chez le rat. Thèse de doctorat non publiée. Université Louis Pasteur. Strasbourg, 1974.
92. Will, B.E.: Effets de l'environnement physique et social sur la récupération comportementale après lésions cérébrales chez le rat (en préparation).
93. Will, B. and Checchinato. D.: Effects of d-amphetamine on the operant behavior in the rat. Psychopharmacologia 29: 141-149. 1973.
94. Will, B.E. and Rosenzweig. M.R.: Effets de l'environnement sur la récupération fonctionnelle après lésions cérébrales chez des rats adultes. Biol. Behav. (Paris) 1: 5-16. 1976.
95. Will, B.E., Rosenzweig. M.R. and Bennett. E.L.: Effects of differential environments on recovery from neonatal brain lesions, measured by problem-solving scores and brain dimensions. Physiol. Behav. 16: 603-611. 1976.
96. Will, B.E., Rosenzweig. M.R., Bennett. E.L., Hebert. M. and Morimoto. H.: Relatively brief environmental enrichment aids recovery of learning capacity and alters brain measures after postweaning brain lesions in rats. J. Comp. Physiol. Psychol. 91: 33-59 1977.
97. Will, B.E., Sutter, A.R. and Offerlin. M.R.: Effets de la méthamphétamine et d'un environnement complexe sur la récupération comportementale après atteinte cérébrale. Psychopharmacologia (Berl.). 51: 273-277. 1977.
98. Zilbert. D.E. and Riesen. A.H.: A comparison of the effects of infant and adult retinal lesions upon visual acuity in the rabbit. Exp. Neurol. 33: 445-458. 1971.

INDEX